Analytical Chemistry of Macrocyclic and Supramolecular Compounds

Analytical Chemistry of Macrocyclic and Supramolecular Compounds

S.M. Khopkar

Springer-Verlag

Narosa Publishing House

S.M. Khopkar
Professor Emeritus
Department of Chemistry
Indian Institute of Technology, Bombay
Mumbai-400 076, India

Copyright © 2002 Narosa Publishing House

All rights reserved. No part of this publication may be reproduced, stored in a retrieval system, or transmitted in any form or by any means, electronic, mechanical, photocopying, recording or otherwise, without the prior written permission of the publishers.

Exclusive distribution in North America (including Canada and Mexico), Europe and Japan by Springer-Verlag Berlin Heidelberg New York.

All export rights for this book vest exclusively with Narosa Publishing House. Unauthorised export is a violation of Copyright Law and is subject to legal action.

ISBN 3-540-64695-7 Springer-Verlag Berlin Heidelberg New York
ISBN 0-387-64695-7 Springer-Verlag New York Berlin Heidelberg
ISBN 81-7319-236-7 Narosa Publishing House, New Delhi

Printed in India.

This research monograph is dedicated to the band of my brilliant research scholars, who had actively collaborated with me in the research projects pertaining to crown ethers, cryptands and calixarenes

Preface

Analytical chemistry has made a spectacular progress mainly due to two reasons. The first was due to advances in design and development of sophisticated analytical instruments permitting analysis at microgram concentrations. While second reason was that the synthesis of novel organic ligands permitting the quantitative analysis by complexation of element with these ligands at trace concentration. Although large number of organic reagents were synthesised and characterised, unfortunately very few of them were used for the quantitative analysis of either anions or alkali and alkaline earths. The analytical chemistry of s-block elements as well as that of anions was totally neglected. The developments in the coordination chemistry favoured formulation of newer methods for analysis of transition and main group elements. Only after the discovery of crown ethers there was surge of activity in the chemistry of s-block elements. The anions were best analysed by means of ion chromatography. In 1967 Pedersen discovered crown ethers while in 1987 Cram synthesised new macrobicycle compounds. Simultaneously with the brisk activity in the laboratory, Lehn prepared cryptands. Preciously at this time attendation of large number of chemist was focused to the coordination chemistry of alkali metal such as potassium complexing with dibenzo 18 crown-6. This in-turn led to the award of Noble Prize in 1987 to Pedersen, Cram and Lehn. Then there was spur of activity in the field of synthetic chemistry for design of new ligands for molecular recognisation. In 1981 Gutsche synthesised new cyclic compounds analogues to spherands and called them as calix(n)arenes. These compounds constitute the back bone of supramolecular chemistry. Much activity was initiated in the field of host guest chemistry, with the discovery of concept that crown ethers formed stable complexes with alkali metals provided if cavity size in crown ether or cryptand matched with ionic size of the metal. With the generation of large size cyclic compounds analogues to porphyrine or fullerene a new field was opened in the chemistry of the supramolecular compounds. The compounds synthesised by Gutsche and his coworkers as calixarene or calixresorcinarene were excellent supramolecular compounds for complexation of metals.

A large number of review articles and specialised papers are available in the field of synthesis, characterisation and complexion of crown ethers, cryptands or calixarenes. However book on the analytical applications of these crown compounds is not available. There are excellent review articles for the use of these compounds in solvent extraction, ion selective electrodes or in membrane chemistry. When I made first attempt to write a review article followed by a concise monograph on applications of these compounds in liquid-liquid extraction, I strongly felt the need of having a research monograph covering entire analytical chemistry of macrocyclic and supramolecular compounds. There was need of comprehensive treaties covering analytical applications of these compounds in separation science, analytical absorption or emission spectroscopy or electroanalytical chemistry and allied techniques. The concept of host guest chemistry of these compounds was well understood with the knowledge of the analytical aspects of these compounds. Though there are excellent monographs describing coordination chemistry of macrocyclic compounds or synthesis and structure elucidation of supramolecular compounds or those dealing with thermodynamic aspects such as stability of complexes and cation binding capacity of crown compounds, a comprehensive monograph encompassing analytical chemistry of macrocyclic and supramolecular compounds is totally lacking. Therefore I have made this modest endeavour to fill this gap in analytical chemistry.

With this view in mind efforts are made to present the entire work on analytical applications in ten independent chapters. The chapter on Introduction covers historical development, classification, nomenclature, important properties and toxicological manifestations of crown compounds. The second chapter on synthesis cover the preparation of crown ethers, cryptands, aza and thia crown ethers, proton ionisable and diester crown ethers and calixarenes. The understanding on synthetic aspects certainly gives better view of the applications of these compounds in analytical chemistry. The third chapter pertains to characterisation and metal complexation studies involving absorption spectroscopy methods as well as NMR and the Mass spectral technique. The fourth chapter deals with elucidation of structure of complexes by classical, X-ray crystallographic, thermal and radiometric methods.

The real major breakthrough in applications of macrocyclic and supramolecular compounds was accomplished in solvent extraction separations. The presentation is simplified throughout this entire monograph by description of methods inform of Table covering characterisation, extraction conditions of these crown ethers, cryptands calixarenes, and aza and thia crown ethers. The various parameters influencing such extraction interms of extraction equibra are well covered in this chapter. The next chapter covers use of these chromatographic methods such as columnar methods, HPLC, GC, extraction chromatography, ion exchange, ion chromatography and capillary electrophoresis.

The next two chapters deal with solvent extraction of elements with crown compounds followed by their determinations by extractive photometry or by extractive emission spectroscopic methods such as fluoroscence or atomic absorption spectroscopy. Some information on applications of chromoionophores or fluoroionophores and photoresponsive switching crown ethers in analytical chemistry is also covered. In comparison to spectral methods of analysis electroanalytical techniques were less commonly utilised. The most popular methods were potentiometry, polarography, conductometry and voltammetry in chemical analysis. The real impetus to the electroanalytical methods was given by the discovery of use of these crown compounds in ion selective electrodes for ion recognisation. Therefore the final chapter pertains to characterisation of electrode and use of crown ethers as neutral carriers to examine the Nernestian response for particular cation. The calixarenes show a promise for analysis of metals, organic compounds and anions. That this area is expected to grow in leaps and bound in next century. An impressive investigations were made in biommics and large scale separations which were carried out with membrane transport involving the use of crown ethers, cryptands or calixarene as liquid membranes. Such transport of ions through membrane was possible with the use of selective crown compounds. Large number of inorganic separations involving heavy metals were accomplished with the crown compounds. Infact there appears to be real challenge to chemist in next two decades for use of these compounds in molecular recognisation in biological fluids.

Every effort is made to incorporate pertinent references till early part of 1999. It is quite possible that inadvertently some important references might have been left out but not intentionally. The monograph is made up to date by incorporating recent advances from modern field of chromoionophores, fluoroionophores, switchon crown ethers, and large number of representative supramolecular compounds in various analytical techniques including ionchromatography, capillary electrophoresis, plasma spectroscopy and membrance transport. The monograph should prove useful to practising chemist, research workers and students advancing in the field of macrocyclic, macrobicyclic and supramolecular chemistry.

I take this opportunity to thank Mr. Hemant Kubal, a project assistant and Prof. A. Chatt Dalhousi, University Halifax, Canada for collating the major references. My thanks are due to staff members of Indian Institute of Technology, Bombay, and that of Department of Chemistry and Central Library for providing me all facilities for the completion of this monograph. I am grateful to Department of Science and Technology, Govt. of India for providing me all financial assistance in the form of DST USERS project.

Finally, I take this opportunity to thank family of my son Samir-Madhari-Rohan and that of my daughter Dr. Supriya-Ketan and, of course, my wife Dr. Sucheta for their cooperation, patience and keeping conducive atmosphere necessary for completion of this monograph.

S.M. KHOPKAR

Acknowledgement

This book research monograph was produced with the Financial Assistance from Department of Science and Technology, Science and Engineering Research Council, Government of India in Department of Science and Technology, New Delhi on utilisation of scientific expertise of retired scientist (DST-USERS) in Project No. HR/UR/18 of 1998–1999.

Symbols and Abbreviations

Org	= organic phase		W_L	= base of elution curve
aq	= aqueous phase		N	= number of theoretical plates
K_d	= distribution coefficient		σ	= standard deviation
D	= distribution ratio (overall)		R	= resolution factor
K_{ex}	= distribution ratio (overall)		nm	= nanometer = 10^{-6} Å
K_a	= acid dissociation constant		µg	= microgram = 10^{-6} g
K_{DR}	= partition coefficient of ligand		pg	= picogram 10^{-11} g
K_{DX}	= partition coefficient of complex		fg	= femagram 10^{-15} g
			F	= Faradays constant
α	= separation factor		i_d	= diffusion current
γ	= activity coefficient of species		t_d	= time between successive drops
β	= stabilization factor in synergic extraction		m	= rate of flow of mercury
			DME	= dropping mercury electrode
[ ]	= active mass of species		D	= diffusion coefficient (Ficks)
n	= valency of metal			
M	= metal		$T_{1/2}$	= half time in chronopotentiometry
HR	= ligand or extractant			
λ	= wavelength for measurement		K_B^A	= selectivity coefficient for A and B
λ_{max}	= wavelength maxima		K_M^{Pot}	= potential of cation 'M'
ε	= molar absorptivity		a.c, AC	= alternating current
s	= Sandell's sensitivity		Å	= angstorm unit
E_M°	= standard electrode potential of 'M'		concn	= concentration
			diam	= diameter
$E_{1/2}$	= halfwave potential		dil	= dilute
ln	= natural logarithm		e.m.f.	= electromotive force
T°	= absolute temperature		ev	= electron volts
R	= universal gas constant		g	= gram
SCE	= saturated calomel electrode		15C5	= 15 crown 5
			18C6	= 18 crown 6
K_{sp}	= solubility product		24C8	= 24 crown 8
LD_{50}	= lethal dose at 50% consumption		30C10	= 30 crown 10
			DB 18C6	= dibenzo 18 crown 6
V_{max}	= peak elution volume		DC 18C6	= dicyclo 18 crown 6
V_t	= total volume of eluant		A15C5	= aza 15 crown 5
atto	= 10^{-18} g		DA 18C6	= diaza 18 crown 6
Δ	= height of theoretical plate		Thia 15C5	= thia 15 crown 5
W_e	= width of elution curve			

Symbols and Abbreviations

Cryptand 221	= Cy-221		TOPO	= trioctyl phosphine oxide
Cryptand 222	= Cy-222		TEP	= tris(ethyl)hexyl phosphoric acid
Cryptand-222B	= benzocryptand 222			
Calix(n) arene	= n = 2, 4, 6, 8		TOA	= trioctyl amine
Calix(n)	= calixarene		TIOA	= triiso octyl amine
THF	= tetrahydrofurane		EDTA	= ethylene diamino teracetic acid
TPB	= tetraphenylborate			
DMF	= dimethyl formamide		PAN	= 1(2 pyridylazo) napthol
GLC	= gas liquid chromatography		PAR	= 4(2 pyridylazo) resorcinol
HETP	= height equivalent of theoretical plate		TAR	= 4-thiloyiazo resorcinol
MS	= mass spectrometry		TAN	= tholylal azo naphthol
NAA	= neutron activation analysis		PMBP	= 3-methyl-4 benzyl pyrozolone
NMR	= nuclear magnetic resonance		ONPOE	= orthonitrophenyl octylether
PIXE	= protoninduced X-ray emission		PTFE	= poly (tetrafluoro ethylene)
psi	= pounds per square inch		Ox	= 8-hydroxyquinoline
SIMS	= secondary ion mass spectrometry		O-phen	= 1–10 phenanthroline
			MIBK	= methylisobutyl ketone (Hexanone)
SFC	= supercritical fluid chromatography		DBSO	= dibutyl sulphoxide
TLC	= thin layer chromatography		TLA	= tri lauryl amine
			HDz	= dithizone
XRF	= X-ray fluorescence		CHEMFELT	= chemical field effect transitors
UV	= ultraviolet			
soln	= solution		AAS	= atomic absorption spectroscopy
HAA	= acetylacetone			
HTTA	= 2-thenoyltrifluoro acetone		AES	= atomic emission spectroscopy
HDBM	= dibenzylmethane		DSC	= differential scanning colorimetry
HDEHP	= bis(2-ethylhexyl)-phosphoric acid		ICP-AES	= inductively coupled plasma atomic emission spectroscopy
DNNS	= dinonly sulphonate			
TBP	= tributyl phosphate			
TBPO	= tributyl phosphine oxide		ODS	= octadecyl silanised silica

Contents

Preface   *vii*
Acknowledgement   *xi*
Symbols and Abbreviations   *xiii*

1. Introduction   1
    1.1 Introduction *1*
    1.2 History *2*
    1.3 Classification of compounds *5*
    1.4 Nomenclature *5*
    1.5 Characteristics of crown compounds *7*
    1.6 Solubility in solvents *7*
    1.7 Significant properties of crown compounds *7*
    1.8 Recent developments in macrocyclic and supramolecular compounds *9*
    1.9 Toxicological manifestations *11*
    1.10 Physiological properties of calixarenes *14*
    1.11 Toxicity of cryptands and crown compounds *14*
    1.12 Handling precautions *14*
    Conclusion *14*
    References *15*

2. Synthesis of Crown Compounds   18
    2.1 Introduction *18*
    2.2 Synthesis of crown ethers *19*
    2.3 The template effect *20*
    2.4 Recent developments in the synthesis of crown ethers *22*
    2.5 Synthesis of cryptands *24*
    2.6 New developments in the synthesis of cryptands *27*
    2.7 Synthesis of aza crown ethers *28*
    2.8 Synthesis of thia crown ethers and related compounds *30*
    2.9 Chiral crown ethers *31*
    2.10 Proton ionisable crown ethers *31*
    2.11 Diester crown ethers *31*
    2.12 New developments in thia and aza crown ether synthesis *32*
    2.13 Synthesis of calix (n) arenes *33*
    2.14 Recent developments in synthesis of calixarenes *41*
    Conclusion *44*
    References *45*

3. Characterisation and Metal Complexation — 51
   3.1 Introduction *51*
   3.2 Metal complexes with crown ethers *52*
   3.3 Metal complexes with cryptands *57*
   3.4 Characterisation of complexes by absorption spectroscopy *58*
   3.5 Characterisation of crown complexes by NMR spectroscopy *59*
   3.6 Metal complexes with calixarenes *65*
   3.7 Characterisation of calix(n)arene complexes by NMR spectroscopy *69*
   3.8 Characterisation of calixarene complexes by mass spectrometry *73*
   Conclusion *74*
   References *76*

4. Metal Complexes and Their Structure — 82
   4.1 Introduction *82*
   4.2 Structure in the solution *87*
   4.3 Complexation and structure elucidation by classical methods *90*
   4.4 Metal complexes structure studies by X-ray methods *94*
   4.5 Structure of complex by thermal methods *101*
   4.6 Structure of the complex by radiochemical methods *103*
   Conclusion *103*
   References *105*

5. Solvent Extraction Separations — 113
   5.1 Introduction *113*
   5.2 Extraction equilibria with crown ethers[5] *114*
   5.3 Interpretation of equilibria *114*
   5.4 Factors influencing extraction *117*
   5.5 Solvent extraction separations with crown ethers *125*
   5.6 Solvent extractions with cryptands *127*
   5.7 Solvent extractions with aza and thia crown ethers *135*
   5.8 Solvent extractions with calixarenes *137*
   Conclusion *144*
   References *144*

6. Chromatographic Separations — 155
   6.1 Introduction *155*
   6.2 High performance liquid chromatography *159*
   6.3 Gas chromatographic separations *162*
   6.4 Reversed phase extraction chromatography *167*
   6.5 Ion exchange chromatography *169*
   6.6 Ion chromatography separation *171*
   6.7 Capillary electrophoresis with crown compounds *175*
   Conclusion *178*
   References *179*

7. Extractive Spectrophotometry 186
  7.1 Introduction *186*
  7.2 Classification of chromoionophores *188*
  7.3 Neutral chromoionophores *188*
  7.4 Monoprotonic crown ether dyes *189*
  7.5 Diprotonic crown ether chromoionophores *192*
  7.6 Other protonic chromoionophores *193*
  7.7 Analytical applications of crown ethers in extraction photometry *194*
  7.8 Application of cryptands in extractive photometry *197*
  7.9 Application of thia and aza crown ethers in extraction photometry *204*
  7.10 Applications of calixarene in extractive photometry *205*
  Conclusion *209*
  References *209*

8. Extractive Emission Spectroscopy with Crown Cryptands and Calixarenes 215
  8.1 Introduction *215*
  8.2 Fluoroionophores *217*
  8.3 Ring substituted fluorogenic crown ethers *218*
  8.4 Luminescence characterisation of lanthanide by crown ethers *219*
  8.5 Lanthanide complexes as supramolecular photochemical devices *221*
  8.6 Analytical application in fluorescence spectroscopy *223*
  8.7 Atomic emission spectroscopic analysis with crown compounds *224*
  8.8 Applications of emission spectroscopy for analysis *225*
  8.9 Atomic absorption spectroscopy with crown compounds *228*
  8.10 Analytical application of AAS with crown ethers and cryptand extractions *228*
  8.11 Recent advances in switched on crown ethers *231*
  8.12 Photoresponsive switching crown ethers *233*
  Conclusion *235*
  References *235*

9. Electroanalytical Methods with Crown Compounds 240
  9.1 Introduction *240*
  9.2 Potentiometry *241*
  9.3 Analytical applications of potentiometry *243*
  9.4 Conductometric studies of complexes with crown ethers *245*
  9.5 Analytical applications of conductometry with crown compounds *247*
  9.6 Polarography *248*

- 9.7 Cyclic voltammetry *249*
- 9.8 Analytical applications of voltammetry with crown compounds *251*
- 9.9 Coulometry and its analytical applications *254*
- 9.10 Electrochemistry of supramolecular compounds *255*
- Conclusion *256*
- References *256*

10. Ion Selective Electrodes and Membrane Transport with Crown Compounds **261**

- 10.1 Introduction *261*
- 10.2 Characteristics of the electrode *262*
- 10.3 Crown ethers as neutral carriers *265*
- 10.4 Bis (crown ethers) in ion selective electrodes *268*
- 10.5 Analytical applications of crown compounds in ion selective electrodes *270*
- 10.6 Membrane transport *273*
- 10.7 Separations with liquid membranes *278*
- 10.8 Experimental set up for ion transport work *279*
- 10.9 Applications of membrane transport *280*
- 10.10 Novel role of crown and pondant as ionophores *281*
- 10.11 Transportation of ions across membrane *281*
- 10.12 Analytical applications of membrane transport with crown compounds *282*
- Conclusion *285*
- References *287*

INDEX **293**

Analytical Chemistry of
Macrocyclic and Supramolecular Compounds

1
Introduction

1.1 Introduction

Analytical chemistry made spectacular progress in the last two to three decades not only due to the advancement in the field of the instrumentation but also due to successful synthesis of novel organic ligands enabling chemist to form useful compounds by complexation with metals. The spectral methods of analysis specially spectrophotometry or fluorimetry became most popular with chemist on account of the availability of the selective, specific and sensitive organic reagents. Even classical chemical methods of analysis like gravimetry or precipitation titrations depended heavily upon the use of novel complexing organic ligands. Several separation techniques including chromatography and specially solvent extraction methods could sustain development due to easy availability of organic ligands. Therefore, the entire credit for advancement of analytical science must go to synthetic chemists for their contribution.

However, in spite of availability of special reagents the analytical chemistry of certain elements were neglected, e.g s-block or alkali and alkaline earths. Similarly, to some extent, the chemistry of anions was also ignored from development point of view the methods for the quantitative analysis. Even plethora of reagents like oxines, oximes, naphthol, azonaphols, β-diketones or thio-substituded ligands like dithizone or dithol were not useful for the analysis of s-block metals. They provided excellent methods for the quantitative analysis of transition and the main group elements. Such methods included gravimetry spectrophotometry polarography techniques of analysis. The separations could be carried out by ion exchange, extraction chromatography and liquid-liquid extraction. It was only with the discovery of crown ethers, cryptands as well as calix (n) arene, that a new era for the separation and analysis of the alkali and alkaline earth elements at tracer concentrations was dawned.

Crown ether also put the study of mechanism of the complexation of elements on sound theoretical footing, or the first time due weightage was given to the considerations of interrelations between atomic, or ionic size of the metal cavity, or hole on the crown ether, calix(n) arenes or cryptands. Secondly, first time a due consideration and importance was given to the study of nature of the counter anion used during complexation and ion pair formation. Further, during extraction studies of such complexes, great attendation was paid to the dielectric constant of the solvent used for attaining equilibria in the quantitative extraction. Thus, 1968 was epoch making year in chemistry with the discovery of macrocyclic compounds by Pedersen. Lehn and Gutsche contributed with upsurge of activity in chemistry of macrobicyclic and supramolecular componds. The cone shaped conformation of the calix(n) arenes provided three dimensional structures favouring the separation of metals belonging to same group or same period of classification of elements.

Large number of books [1-12] that appeared in last three decades on the chemistry of crown ethers and cryptands and few on chemistry of calixarenes have not only covered coordination chemistry of complexes, but also provided for their applications in chemistry for separation and determination. Several review articles also appeared [13-34] in last two decades.

Recent review articles [50-80] encompass several aspects like use of these compounds in analytical chemistry with special reference to their applications in solvent extraction, ion selective electrods, electroanalytical techniques and spectral methods of analysis. The highlight of some of these is taken into consideration at the end of the sections of this chapter [50-80]. Before considering these aspects it is worthwhile to consider historical development in the discovery of the macrocyclic and supramolecular compounds.

1.2 History

The discovery of crown ether (1967) by Pedersen [2, 5, 34] was great revolution in chemistry of the cyclic compounds.

Catechol   Dichlorodiethyl ether   Bis phenol   DB18C6 (I)

Thus it was accidentally, discovered during synthesis of bisphenol from catechol and dichlorodiethylether [21]. The attendation which was considerably drawn towards the new compound, was named by Pedersen as Dibenzo 18 crown 6(I). It had considerable solubility in water in presence of sodium hydroxide. In fact this observation gave him due recongnition of his discovery of complexation of ligand with sodium.

In fact Luttringhaus [2, 35] had great fascination for the ring compounds. He was interested in preparing large rings molecules. Therefore, he reacted nucleophile with substitutional diol derivative to get a cyclic structure (II). Unfortunately this compound had no donor group to exhibit the binding properties. That is where Pedersen scored to induct other species.

(II)

However, it was only after the 1987 Nobel Prize in chemistry to Pedersen, Cram and Lehn, there was upsurge of activities in the area of macrocyclic chemistry and synthesis of novel compounds. The presence of sodium hydroxide enhanced the rate of formation of cyclic compound such as DB 18C6. This effect is termed as the template effect [1].

The discovery of crown ethers was followed by cryptands by Lehn [11]. He recognised that the presence of donor group within three dimensional network not only favoured complexation of cation but also promoted its encapsulation in the cavity of the crown compound.

The third donor group strand to the crown ethers made the molecule an analogue of first solvation shell of cation. Such compound having the properties of bicyclic amines as well as crown ethers are shown in (III) to (V).

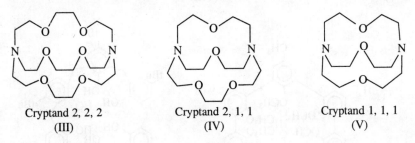

Cryptand 2, 2, 2 (III)   Cryptand 2, 1, 1 (IV)   Cryptand 1, 1, 1 (V)

They were called cryptands [36] because *cryptos* means 'hidden'. Thus, during complexations of cations which were bound were actually hidden from bulk of the solvent which had previously stabilized it. Like Luttringhaus who missed the bus for getting the credit for the synthesis of first novel macrocyclic compounds, Simmons also narrowly missed the credit for the discovery of cryptands. In fact Simmons was busy in preparing in-out bicyclic amines. Cycloproponation was discovered by Simon and Smith. Simon was also interested in synthesis of ring structures with three dimension and pointed out that in such compounds electron pairs of nitrogen were pointed inward, the electron pairs of both nitrogen atoms were pointed outward and the conformation was one which had both possibility for N-atoms electrons. The structure he discovered were in-in (VI) an in-out (V) compounds (VI-VII).

(VI) in-in   (VII) out-out

These compounds appeared like cryptand but were not symmetrical. However, in 1977 Lehn declared the discovery of cryptands. These compounds had added advantage over crown ethers as they formed three dimensional rigid structures with strong bond between metal and ligand. In the synthesis of crown ether (or macrocyclic polyether) with donor atom as oxygen and cryptands were termed

4 Analytical Chemistry of Macrocyclic and Supramolecular Compounds

as macrobicyclic polyethers with donor atoms being both oxygen and nitrogen. There was great surge of activity for the preparation of compounds with large ring size apart from 18C6, 24C8 and 30C10 [37]. Attempts were made to synthesise large compounds such as spherands. The large compound obtained is shown in (VIII) and (IX).

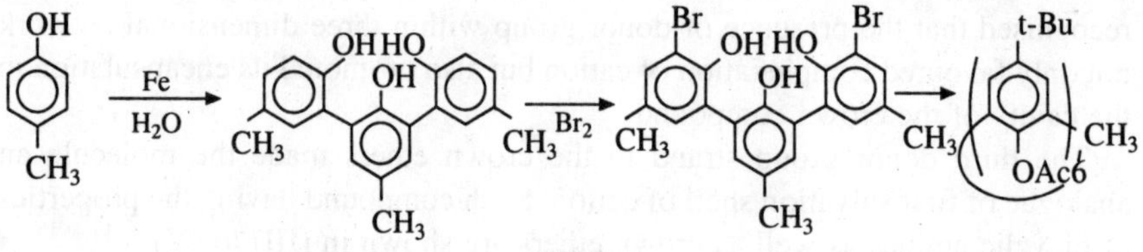

(VIII)

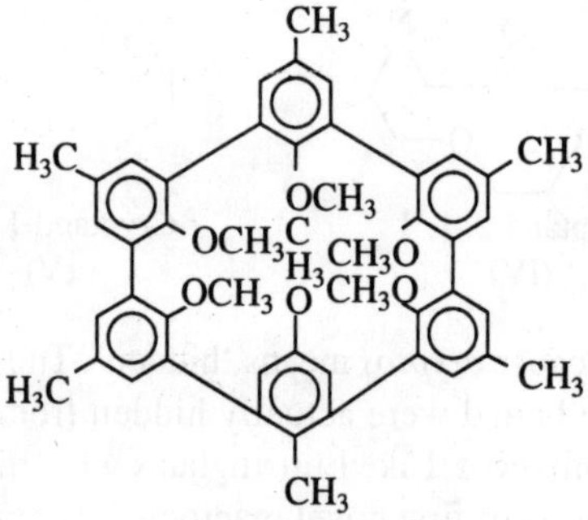

Spherands (IX)

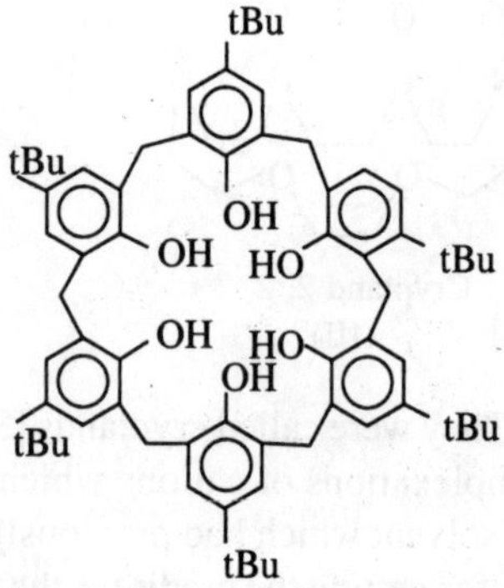

p-tertiary butyl calix (6) arene (X)

This compound being cyclic in structure was called spherands (IX). In 1988, a German chemist Gutsche demonstrated that the base catalysed reaction between t-butylphenol and formaldehyde gave cyclic compound called calix (n) arene as (X) in spite of the discovery of spherands by Cram and his group these compounds did not find many applications in coordination chemistry due to the absence of donor group, Gutsche synthesised substituted derivatives wherein complexing group was acetate. For example, a compound hexaacetato calix (6) arene which could easily form complexes with several transition metals [12]. This firm foundation led to chemistry of crown ethers, cryptands and calixarenes.

After the discovery of such compounds the focus shifted from covalent to non-covalent interactions. The cyclio oligomerisations of ethylene oxide led to the discovery of array of interesting compounds. Subsequent efforts were made for the preparation of chiral macrocycles which could distinguish between optically active primary ammonium salts.

Crown ethers are macrocyclic polyether compounds containing oxygen atom but could also have nitrogen, sulphur as heteroatom separated by two carbon atoms. In cyclic structure oxygen acts as the donor atom. The replacement of oxygen by sulphur, nitrogen and heteroatoms led to the discovery of thia crown or aza crown ethers [37] as shown in (XI) to (XIV).

Thia crown ether (XI) Aza crown ether (XII) Aza crown ether (XIII) Thia aza crown ether (XIV)

The cryptands represent three dimensional cation binder. Many such compounds had more carbon chains and donor groups. Three dimensional enveloping ligands led to compounds like spherand, cavitands and lariat ethers. Such compounds which can bind and transport cations were called 'ionophores'.

1.3 Classification of Compounds

Compounds were classified [38] in the following manner depending upon the kind of ring or chain.

(1) Podands: Chain compounds.
(2) Coronands: Cyclic compounds like crown ethers.
(3) Cryptands: Compound containing bridge like cryptands.
(4) Supramoleculor entity: Three dimensional structures like calix(n)arenes.

Thus, 12C4, 18C6, DB18C6, DC18C6 or 24C8, DB24C18 all belonged to category of coronands containing oxygen as the donor atom. While cryptand 221, cryptand 222, cryptand 222B or cryptand 1, 1, 1 were termed as closed ring bridge spherands commonly called "cryptands" [26, 27], compounds like calix (n) arenes where n = 4, 5, 6 or calix (n) resorcinarene (n = 4, 5, 6) were giant molecules with symmetrical cone shaped structure called "supramolecular compounds". The common thing in all these compounds was that they had cavity (or hole) in their structure. Further if the hole matched with the ionic size of the cation then it was capable of forming complexes. The chemistry of such complexation was called 'Host guest' chemistry, [39-43]. Here crown ethers, cryptands, (or calixarene) act like host to receive guest like metals or neutral organic molecule to form big, coordination complex or entity called 'Host guest' compounds [44]. The podants were worthless as complexing ligands, as they were open chain compounds incapable of forming cyclic or closed ring complexes. Lariat ethers were exception to this definition as they had tail or wing attached to aromatic entity which in turn helped the complexation phenomena while interacting with the metal.

1.4 Nomenclature

Ever since Pedersen discovered crown ether [45] or when Lehn discovered cryptands or Gutsche discovered calixarenes, all of them had complex problem of naming. Pedersen called them crown compounds because when these compounds encapsulated metal, in their cavity they appeared like crown. Lehn called 'cryptand' as they could hide and encapsulate metal away from the solvent. The name calixarene was used to denote shape of phenol derived cyclic tetramer in conformation in which all four aryl groups were oriented in same direction, to accommodate the other oligomers. They were bracketed with numeral between calix and arene, e.g. calix [4] arene.

Since it was difficult to deal in absence of any of the systematic nomenclature, the IUPAC system [1, 2] was only choice but it was cumbersome and impracticable. For example, DB18C6 would have been named as 2, 3, 11, 12, dibenzo-1, 4, 7, 10, 13, 16 hexaoxacyclooctadeca-2, 11 diene. Similarly, cryptand 2, 2, 1 would have been named as 4, 7, 13, 16, 21, 24-hexaox 1-10 diazabicyclo 8-8, 8 hexacosane. Similarly, calix (6) arene hydroxyderivative as 37, 38, 39, 40, 41, 42-hexahydro calix (6) arene or t-butyl calix (6) arene would be named as 5, 11, 17, 23, 29, 35 tertiar butyl 37, 38, 39, 40, 41, 42 hexahydroxy calix (6) arene [4]. To avoid this, Pedersen evolved a new system for crown ethers. In this system if a compound contained 18 carbon atoms and has 6 donor oxygen atoms as in (XV) it was named 18C6 where 18 carbon and 6-oxygen atoms were present. If there was substituent like benzo group, it was called DB18C6 (XVIII) and if substituent was dicyclohexyl group, then it was DC18C6. Similarly, if there are six sulphur atoms in thia crown it was named as thia 18C6 (XVII). With six nitrogen atoms it was named as aza 18C6 (XVI).

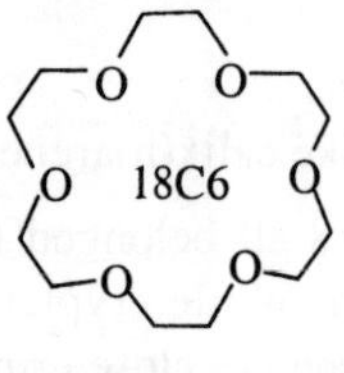

(XV)	(XVI)	(XVII)	(XVIII)
18C6	N Aza crown N	Thia crown	DB18C6

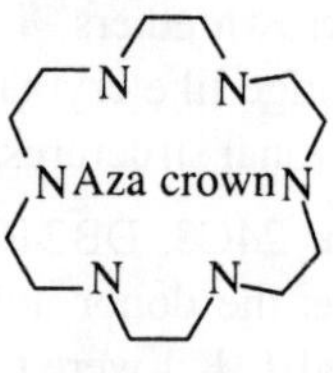

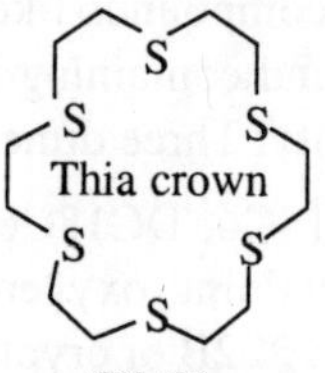

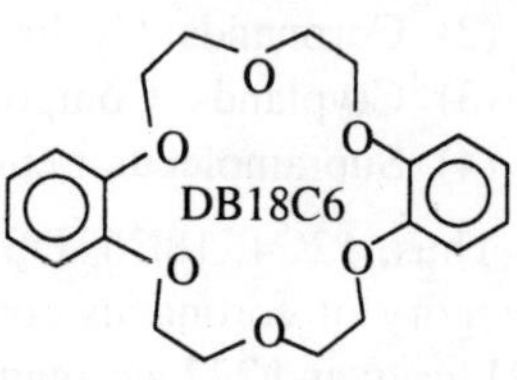

Lehn also accepted similar principle for adopting nomenclature for cryptands [46].

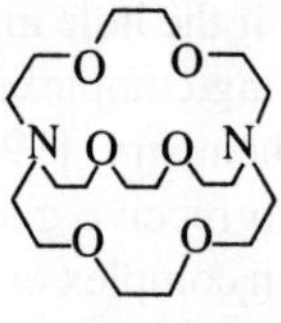

Cryptand 2, 2, 2
(XIX)

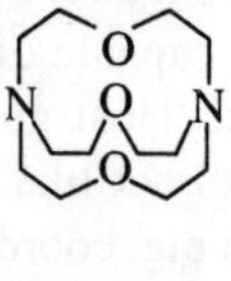

Cryptand 1, 1, 1
(XX)

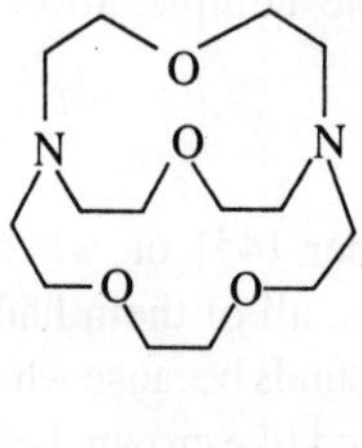

Cryptand 2, 1, 1
(XXI)

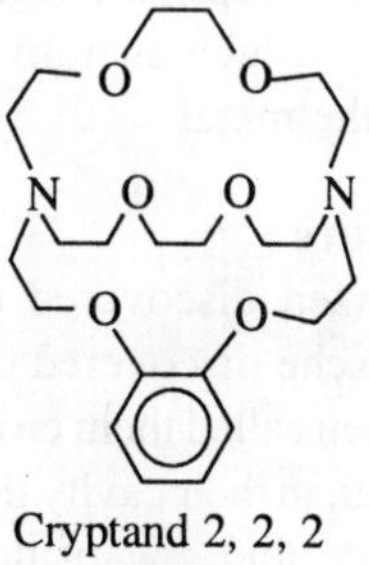

Cryptand 2, 2, 2
(XXII)

Introduction 7

In compounds 2, 2, 2 or 1, 1, 1 as shown in (XIX and XX) note the carbon spacing between two oxygen atoms but the number of oxygen atom in each chain is to be counted. In each chain there were eight atoms - six carbon and two oxygen. In first (XIX) compound we had two, oxygen atoms in each bridge. In second (XX) we had one oxygen atom within each bridge while in third (XXI) we had two oxygen atom in one bridge and one oxygen atom each in two bridges as called cryptand 2, 2, 2 or 1, 1, 1 or 2, 1, 1. In last compound since benzene was attached to one side we name it as benzo 222 B cryptand. It could also be referred as [2, 2, 2B] cryptand [46] (XXII).

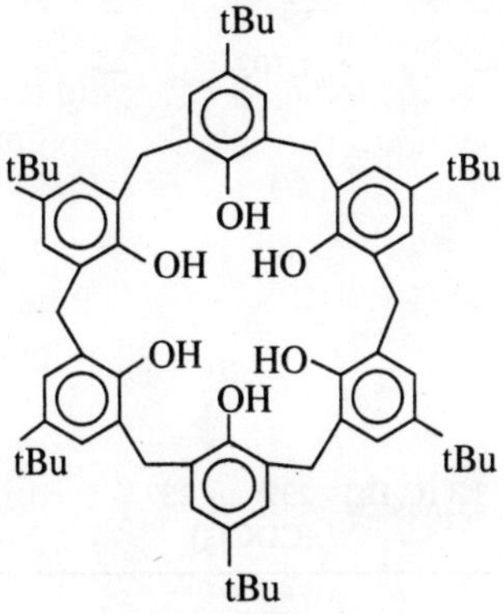

(XXIII) 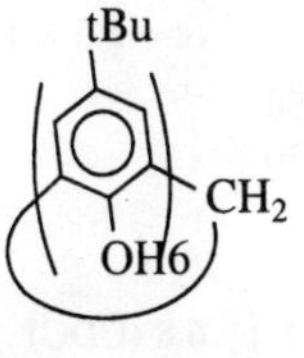 (XXIV) 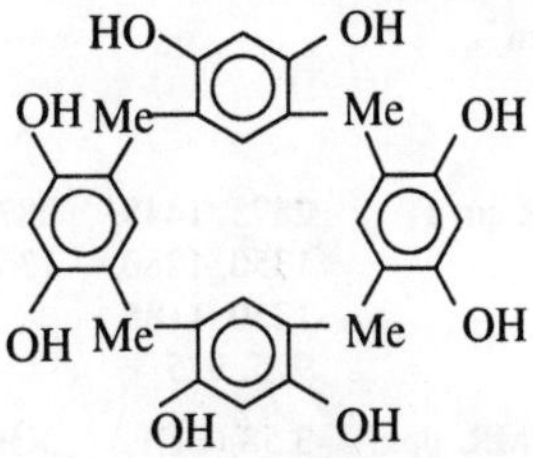 (XXV)

Finally for calix arenes the nomenclature was simplified [4]. The hexamer with six aromatic rings each attached with t-butyl group para in respect of hydroxy group was termed as 'tert butyl' calix (6) arene as in (XXIII) or (XXIV). The term (n) was used for general. Also calix resorcinarene were designed as shown in (XXV) and was called C-methyl calix (4) resorcinarene. The substituent methylene carbon introduced by aldehyde was indicated by prefix C-substitute.

1.5 Characteristics of Crown Compounds
Some important physical properties are summarised in Tables 1.1 and 1.2.

1.6 Solubility in Solvents
It is listed in Table 1.2 [47].

The solubility in the solvents is of prime consideration while selecting a diluent during solvent extraction procedure.

1.7 Significant Properties of Crown Compounds
The most important property of crown ethers was their selective complexing ability. They bound cations in cavity of crown ether due to ion dipole interaction between cation and negatively charged donor atoms in ring structure of the cyclic polyether. Such selectivity depend upon relative size of cavity and diameter of cation, number of donor atoms in crown ring, the relationship between cation donor atoms and charge of cation. The formation of molecular complex obtained by molecular interaction in hostguest, molecule was called molecular recognisation.

Analytical Chemistry of Macrocyclic and Supramolecular Compounds

Table 1.1 Physical characteristics

Property	15C5	18C6	DB18C6	DC18C6	DB24C8	DC24C8
Mol. Formula	$C_{10}H_{20}O_6$	$C_{12}H_{24}O_6$	$C_{20}H_{24}O_6$	$C_{20}H_{36}O_6$	$C_{24}H_{32}O_8$	$C_{24}H_{44}O_8$
Mol Wt.	220.27	264.32	360.41	372.47	448.52	360.61
Physical State	Liquid	Solid	Solid	Solid	Solid	Liquid
mp°C	−32.4	43.42	163	–	103	–
bp°C	102	117	380	–	–	–
Sp gravity	1.113	–	–	–	1.102	–
UV peaks, nm			223 ($\varepsilon = 1.7 \times 10^4$) 275 ($\varepsilon = 5.5 \times 10^4$)			
IR cm-1	2875, 1445, 1350, 1280, 1250, 1185, 975, 925	2875, 1450, 1350, 1120				
NMR, ppm	3.58 (CCl_4)	3.56 (CCl_4)	6.8 ($CDCl_3$)	3.3 (C_6H_6)	232 cps ($CDCl_3$)	237 –

Such molecular recognisation led to host guest chemistry and discovery of supramolecular chemistry. This was the field where synthesis and applications of molecular complexes were formed by recognisation and matching guest with host molecule with specific cavity. The supramolecular chemistry [49, 50] was an extension of hostguest chemistry. It was field of high order molecular aggregates formed by molecular action between two or more molecules. The following tendencies were emerging fast, viz. diversification of guest, rigorous molecular recognisation, tight binding, produces stimulus responsive compounds and formation of molecular aggregate or supramolecules. Such complexes were soluble in various organic nonpolar solvents because these compound possessed hydrophobic groups. In such reactions counter anions were highly active as they existed in the solvent as naked anions which were not dissolved in solvents.

Table 1.2 Solubility (moles/L or g/L) in solvents [47, 48]

Crown ether	$CHCl_3$	Benzene	Water
DB18C6	0.2 mg/L	–	–
DC18C6	16.7 g/100	50 g/100	1 g/100
DB24C8	19.3 g/100	3.6 g/100	0.1 g/100
DC 24C8	5 g/100	5 g/100	0.5 g/100
15 C 5	< 50 g/100	< 50 g/100	50 g/100
18 C 6	50 g/100	50 g/100	50 g/100

The details of various parameters governing the complex formation specially with inorganic materials like metal cations will be discussed in Chapter 3.

1.8 Recent Developments in Macrocyclic and Supramolecular Compounds

In 1923 Leregue and Robsset [51] predicated the possibility of use of cryptand to form cryptates with alkali metals. Subsequently a review appeared in Russian Journal about the utility of crown ethers and cryptands in analytical chemistry [52]. Around same time the Chinese workers indicated the use of macrocyclic polyethers, i.e. crown compounds in analytical chemistry [53]. A comprehensive review appeared in topics in current chemistry [20]. However most extensive review (213 references) was published by Yoshio, Naguchi [13] indicating applications of these compounds with special reference to solvent extraction of metals, organic cations and metallic anionic complexes. Use of thia crown ethers in ion selective electrodes was also demonstrated, subsequently. Zolotov et al [54] showed use of macrocyclic extractants with 10-30 membered rings with donor atoms as nitrogen or sulphur in addition to usual oxygen for photometric determination of elements from sample of water. A Japanese worker [55], showed the general features of crown ethers and their use in the analytical chemistry. In 1984 Weber [56] showed for the first time how crown ether could be used in membrane transport, chromatography, electroanalytical techniques apart from solvent extraction. Takagi [57] demonstrated use of macrocyclic polyamines and their derivatives for the complexation of metals, with evaluation of formation constant of complex and the stability constants of ligand. Although 15C5 was less studied due to small cavity size the study of ultraviolet spectra [58] had showed that definite relationship existed between its absorption spectra and chemical constitution. Blanco [59] made outstanding contribution to the chemistry of cryptands and their application in fluorimetric analysis. He published a review for natural polyethers like cyclodextrin and antibiotics. A systematic investigations [60] on the effect of substitution of 16C5 in 14, 15, 16 position indicated decrease in extractibility of metals due to less access of counter anion and inability of ligand to adjust conformation during extraction, however bridging substituent enhanced extraction due to favourable entropy changes in conformation. Balzani [61] was first to study photochemistry of supramolecular species including covalently linked components in ionpair hostguest system and caged complexes. 'Crown ether and analogues' was brought out by Weber [62] and his group. At the same time Cooper [64] published a manual on 'Crown compounds and their future applications'.

There was upsurge of activity in synthesis of supramolecular compounds after their discovery by Gutsche [4]. On regular basis symposium on supramolecular compounds with special reference to molecular architecture [63] were arranged. He indicated supramolecular chemistry as a ripple of future and truly interdisciplinary field. Calix(n) arene with large molecular weight were discovered with H-bonding receptor and with π-bond stacking interactions [64]. An attempt was made to examine crown ether cation complex in extraction and transport through membrane [65] with special reference to counter anion in high lipophillic organic solvent. Calix(n) arene as molecular receptor underwent readily electrophilic substitution and the Classian rearrangement routes [66] to give variety of compounds as shown in (XXVI) to (XXVII).

[Scheme showing synthesis: (XXVI) with R, OH, CH₂, OHₙ, R₁ reacting with compound containing OH, OH, R₁, R₂ and Br, OH, Br, R to give product R₄–(OH, OH, HO, HO)–R₂ with R₁ and R₃ substituents; condition: R₁ ≠ R₂ R₃ or R₄]

Such compounds could be easily substituted in lower and upper rims (XXVII). Tsukube [67] demonstrated the use of double armed crown ethers and armed macrocycles as metal selective reagent in various analytical techniques such as extraction, membrane transport, ion selective electrodes and in spectroscopic and chromatographic analysis. Very little work had been carried out on calix(n) resorcinarene [68]. It provided a framework for artificial esterase. The compound shown by Lehn [69] gave novel information in the popular article on supramolecular chemistry, it's molecular assembly chemistry and intramolecular bonding. It was chemistry beyond simple molecule involving intramolecular bonding (noncovalent) interaction. Lindoy [70, 71] reviewed a book on molecular receptors in the round containing lectures by Lehn on macrocyclic synthesis and complexes like cryptates. An extensive information on cryptands and cryptates was presented by the author. The story of calixarene as novel complexing ligand was unfolded [72] to show the use of molecular basket for trapping of metal ions in the three-dimensional network. An international symposium on Macrocyclic (ISMC) chemistry was held in Utah (USA) in 1992 [73]. An authoritative review on self assembling supramolecular complexes was published recently [74]. Shinkai [75] threw light on various conformation of calix(n) arene (if n = 4) (XXIX to XXXI).

[Structure of octa-substituted calix(4)resorcinarene with OR groups and CH₃ groups]

Octa (dimethyl) amino propyl calix (4) resorcinarene

$R = CH\ CONH(CH_2)_3\ N(CH_3)_2$

(XXVII)

The stereochemistry of calix(n) arenes was very important [75] in metal complexation (XXVIII to XXXI).

Matsui [76] stated that the factors influencing ion size recognisation depended on the distance between the donor atoms and the rigidity of the chelate ring structure. A paper on the application of cryptands in the solvent extraction separation of the metal pollutants in aquatic environment was published by Khopkar and Gandhi [77]. Another review on solvent extraction separations with crown ethers and cryptands was also published [78] covering various factors like size of atom, cavity size of ligand, kind of diluent, effect of temperature on entropy changes etc. The possible separations of s, p, d and f block metals was enumerated and birds eye view of their applications in photometry, fluorimetry, extraction chromatography, ion selective electrodes was also included. A review in chinese language on applications of calixarene derivative in analytical chemistry [79] covered the uses of calixarene in solvent extraction, membrane separation, ion selective electrode, capillary electrophoresis, gas chromatography as stationary phase and as fluoroionophore for chiral recognisation. A comprehensive review on electrochemistry of supramolecular system [80] covered the area of self-aggregation on solid electrode, analyses of hostguest complexes in aggregates and similar structure. A classic example was the study of complex formation of mixed metal complex of tetracarboxylate and similar compounds using the potentiometric titrations [81].

(XXVIII) (XXIX) Cone-form (XXX) 1-2 alternate Partial cone (XXXI) 1-3 alternate

It would be seen that the chemistry of the crown ethers, cryptands and calixarenes was most facinating with the spectacular progress in the area of synthesis and analytical applications in chemistry involving various modern methods of analysis. It is true as per Lehn [69] that through supramolecular chemistry there is even more room at the top.

1.9 Toxicological Manifestations

These properties must be throughly examined before embarking on any experimental work pertaining to crown compounds. Unfortunately, few reports were available with virtually no information on cryptands or calixarene. Cryptands were expected to be more toxic than crown ethers on account of presence of nitrogen atom on the ring. As a rule, nitrogen containing compounds like benzidine or α-napthylamine were supposed to be carcinogenic.

Some investigations [82, 83] on toxicity of 18C6 were studied. Similarly, toxicological manifestations of other compounds other crown ethers like 12C4 [84], 15C5 [84], 15C5 [84], 18C6[85] and neopharmacological toxicity of [86, 87] had been thoroughly investigated. A data on LD_{50} dose on mouse is listed in Table 1.3.

DB18C6 in comparison to other crown ethers was less toxic. It was even less poisonous than glyme, ethyleneglycol or ethylene glycolmonoether which had

Table 1.3 Toxic effects of crown ether [88-90]

Structure Nos.	Crown ether	Formula	Toxic effects	LD_{50} mouse
XXXII	DB18C6	$C_{20}H_{24}O_6$	Skin irritation eyes affected conjunctives corneal inflammations	> 300 ms/kg
XXXIII	DC18C6	$C_{20}H_{36}O_6$	Skin and eye irritations corneal injury, irratic problems, absorbed in skin impaired coordination sleepiness, tight chest	> 300 ms/kg
XXXIV	12C4	$C_8H_{16}O_4$	Anorexia, loss of weight, asthma, deafness, tremors, convulsion impaired coordination	3.15 kg
XXXV	15C5	$C_{10}H_{20}O_5$	Affects central nervous system, absorbed in skin, redness of skin	1.02 g/kg
XXXVI	18C6	$C_{12}H_{24}O_6$	Skin irritation, salivation, paralysis of extremities weakness of musles, poor reflex action, tremors	0.705 g/kg

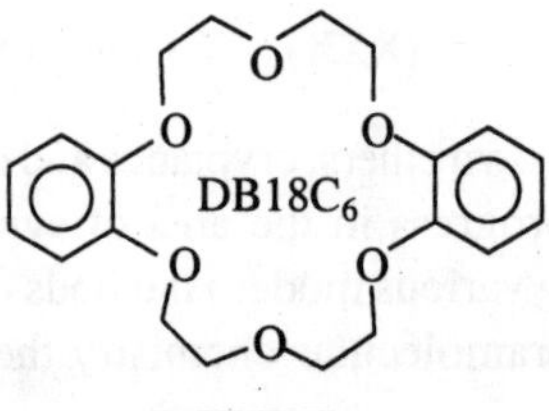

DB18C$_6$ (XXXII)

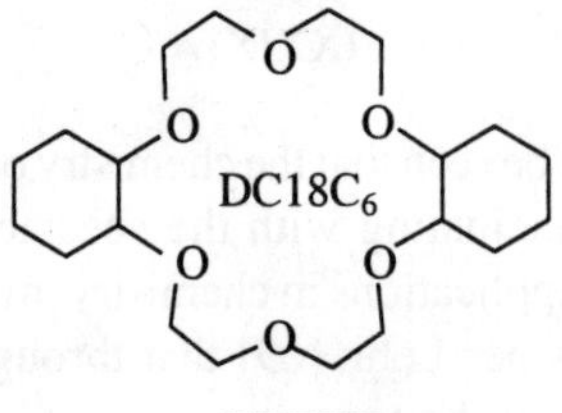

DC18C$_6$ (XXXIII)

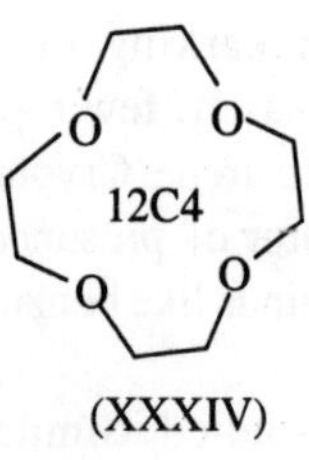

12C4 (XXXIV)

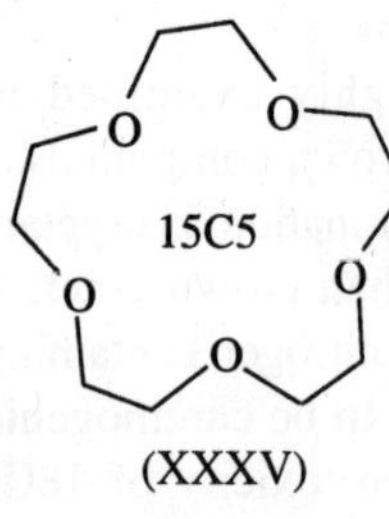

15C5 (XXXV)

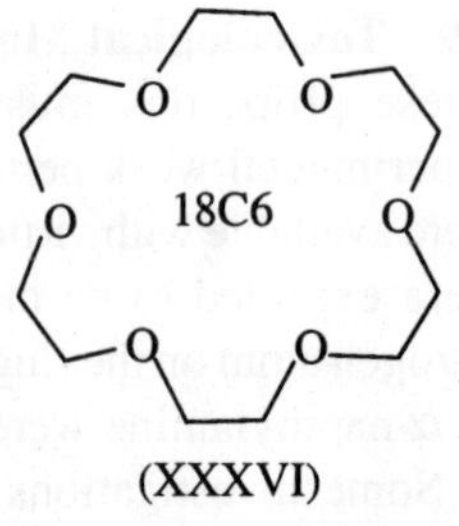

18C6 (XXXVI)

similar structure like other crown ethers. DB18C6 was least soluble [89]. Similarly, polymeric and immobilised crown ethers were relatively less toxic. However, by structural modification the toxicity of the compound could be reduced to some extent.

Hiraoka [91-92] reviewed in Japanese publication systematic investigations on toxicity of crown ethers. Kimura [93] listed biological activity of metal complexes and ionophores [94]. The toxicity of known crown ether would be considered in details.

1.9.1 Toxicity of Dibenzo 18 Crown 6 and Dicyclo 18C6

It was much less for DB18C6 than DC18C6. Lethal dose (Table 1.3) on rat was 11 g/kg. It caused skin irritation, eye effects leading to conjunctival irritation on cornea, while DC18C6 is equally injurious to skin and eyes and not to be inhaled by a person. It produced corneal injury, irratic injury and conjunctivitis in solution form, [89]. Instantaneous washing with water did not help as it caused permanent damage to eye. It was irritating to skin and was readily absorbed in the skin. It caused rapid breathing, tension in chest, impaired coordination and sleepiness affecting mainly central nervous system. In comparison DB24C8 was least toxic as tested on rats [90].

1.9.2 Toxicity of Liquid 12 Crown 4

Leong [95, 96] warned that this compound be handled with extra care. Its inhalation toxicity was high in vapour form as it was liquid at room temperature, with low molecular mass and 0.03 mm vapour pressure. On exposure to 12C4 it caused anorexia, loss of weight, asthma, impaired coordination of organs, listening sensitivity, tremors, convulsions and fatality in large doses exceeding 1.2-64 ppm in the atmosphere. Colour of blood changes to cherry red due to metabolism of carboxyhemoglobin. The central nervous system and testicular atrophy was observed in patients exposed to 12C4 vapours or orally ingested in the body. Hendrixon [84] made comparative study of the toxicity of 12C4, 15C5 and demonstrated that the LD_{50} value was three times higher for 12C4 in relation to that of 15C5 and 18C6. As a rule, toxicity increased with ring size but 12C4 was an exception to this generalisation. It has also a cumulative effect like cadmium toxicity. A constant exposure for prolonged period could cause immunity but it had not been experimentally tested on crown ethers [96].

1.9.3 Toxicity of 15 Crown 5 and 18C6

It affected central nervous system on absorption on skin, it caused redness of the skin. The LD_{50} value was 1.02 g/kg. There was no cumulative effect.

In comparison 18 Crown 6 is equally toxic like 15C5. It also caused skin irritation, with LD_{50} dose more than that of 15C5. The oral toxicity of 18C6 and its complex with sulphomonomethoxime was investigated [85]. The symptoms caused on animal included salivation, paralysis of extremities. The neuropharmacological effects [87] by intravenous injection indicated weakness of muscle, poor reflex action and tremors. The antidote was p-chlorophenylalanine or dibenzyline as they reversability blocked action of 18C6 in tissues.

Amongst noncyclic polyethers, glymes (non-ionic surfactant) did not cause hydropsy of skin [97] injury to lung kidneys and liver [98] but polyethylene glycol structurally somewhat similar to the crown ether were less toxic with no adverse effects on eyes or skin. Therefore, they found extensive practical applications in medicines and cosmetics [99, 100].

1.10 Physiological Properties of Calixarenes

Usually phenolic compounds did possess physiological properties. A dermatic response was noted for p-tertiary-butyl-formaldehyde specially a linear tetramer [101], while p-tert-butyl-calix(4) arene and p-tert-butyl-calix(8) arene gave negative response. Literature on oxyalkyl derivatives of phenols was available. A condensation compound of p-(1, 1, 3, 3-tetramethylbutyl phenol and formaldehyde was studied for tuberculostatic activity. Macrocyclon was tested for the parasitic disease. Jain and Jahagirdar [102] studied effect of phospholipase A_2. They showed that calixarene carrying short polyoxyethylene chain of oxygen inhibited the action of phospholipase A_2, while one with long chain stimulated the action. Physiological action might depend upon size of ring as well as nature of substituent in para position.

1.11 Toxicity of Cryptands and Crown Compounds

The toxicity of 12C4 was thoroughly investigated by Dow Chemical Co., USA DC18C6 was declared bad for eyes, but even in symposia and conference toxicity was never discussed. Luckily none of the crown compounds were carcinogenic in nature. It was believed that a common analgesic compound like aspirin was toxic while toxicity of 15C5 was half the toxicity of aspirin tablets. Although no information is available on cryptands, but precaution should be taken while handling cryptands and their analogues.

1.12 Handling Precautions

Since many crown compounds exhibit oral and skin toxicity, they should be handled with care. A direct contact with skin should be avoided. A protective suit/gloves be worn while dealing with continuous and long exposures to these compounds. If accidentally exposed to skin or eyes, it should be immediately washed with copious stream of water. Generally, crown compounds with ether linkages must not be allowed to come in contact with air at high temperature leading to formation of explosive peroxides. To ensure safety, they should be stored in the inert atmosphere of nitrogen [99, 100].

1.13 Conclusion

We can say discovery of crown ethers had opened a new era in the hostguest chemistry. Although the discovery of DB18C6 was accidental, subsequent planned work led to synthesis of novel compounds like cryptands and calixarenes. They had done justice to coordination chemistry of alkaline earths. The classification was relatively simple but nomenclature (as per IUPAC system) was complicated. Hence simple naming system was adopted. Their physical and chemical characteristics were examined which appear promising for their analytical applications. There is growing development in all the areas of crown compound cryptands and supramolecular compounds. However, one must not forget the toxicological effects of these compounds while handling in the course of experimental work. Fortunately, most of them are not toxic. We expect brisk activity in this area for next two decades.

References

1. G.W. Gokel, Crown ethers and cryptands, Royal Chemical Society, London (1991).
2. M. Hiraoka, Crown compunds: Their characteristics and applications, Elsevier Scientific Pub. Co., Amsterdam (1982).
3. M. Hiraoka, Crown ethers and analogus compounds Elsevier and Co (1992).
4. C. David Gutsche, Calixarenes, Royal Chemical Society, London (1999).
5. F. deJong and D.N. Reinhoudt, Stability and reactivity of crown ether complexes. Academic Press N.Y. (1981).
6. G.W. Gokel, S.H. Korzeniowski, Macrocyclic polyether chemistry, Springer-Verlag, Berlin (1982).
7. Y. Inoue, G.K. Gokel (Ed), Cation binding by macrocycles. Marcel Dekker, New York (1990).
8. R.M. Izatt, J.J. Christensen (Ed), Synthetic multidentate macrocyclic compounds, Academic Press (1978).
9. G.A. Melson (Ed), Coordination chemistry of macrocyclic compounds, Plenum Press, New York (1980).
10. F.Vögtle, E. Weber, Host guest complex chemistry; macrocycles. Springer-Verlag, Heidelberg (1985).
11. J.M. Lehn, Structure and bonding, Volume 16, p-1, Springer-Verlag (1973).
12. S.M. Khopkar and Mayuri Gandhi, Crown ethers and cryptands in solvent extraction (unpublished monograph).
13. M. Yoshio, H.Noguchi, Analytical Letters **15**, A15(1982).
14. Y. Ikeda, Topics in current chemistry **211**, 1 (1984).
15. M. Takagi and K. Ueno, Topics in current chemistry. **121**, 39 (1984).
16. S. Shinkai, Bio. Org. Chem. Frontiers. **1**, 161 (1990).
17. L. Rossa, R. Votle, Topics in Current Chem **113**, 1 (1982).
18. C.J. Pedersen. J. Inc phenomena **6**, 337 (1988).
19. C.J. Pedersen, Organic synthesis. **52**, 66 (1972).
20. E. Blasius, K.P. Janzen, Topics in Current Chem. **98**, 163 (1981).
21. C.J. Pedersen, Aldrchimica **4**, 1 (1971).
22. J.S. Bradshaw, J.Y.K. Hui, J. Hetero Cyclic Chem. **11** 649 (1974).
23. J.J. Christensen D.J. Eatough, R.M. Izatt—Chemical Reviews **74**, 351 (1974).
24. D.J. Cram J.M. Cram, Science **183**, 803 (1974).
25. P.N. Kapoor, R.C. Merhotra, Coordination Chemistry Review **74**, 1 (1974).
26. J.M. Lehn, Pure and applied Chem. **49**, 857 (1977).
27. J.M. Lehn Acct Chem. Res. **11** 49 (1978).
28. N.S. Poonia, A.V. Bajaj, Chemical Revies, **79**, 389 (1979).
29. J.M. Lehn, La Recherche, **127**, 1213 (1981).
30. I.M. Kolthoff, Anal. Chem, **51**, 1R (1979).
31. M.N.H. Irving, Pure and Applied Chem. **50**, 1129 (1978).
32. T. Sekine, Y. Hasegawa KagakuNo Ryoiki, **33**, 464 (1979)..
33. M. Takagi, H. Nakamura, Y. Sanui. K. Ueno, Anal Chim. Acta. **126**, 185 (1981).
34. E. Weber, Crown compounds—Properties and practice, Merck Schuchardt Monograph (1987).
35. A. Luttringhaus, I. Sirchert Modrow, Makromol Chem., 18-19, 511 (1956).
36. B. Dietrich, J.M. Lehn, J.P. Sauvage, Tetrahedron Letters 2885 (1969).
37. Y. Hasegawa, T. Sekine, Bunsekil 55 (1982).
38. F. Vögtle, E. Weber, Anegew Chem. **18**, 753 (1979).
39. F. Vögtle, W. Sieger, W.M. Müller (Ed), Topics in Currents Chem., **98**, 107 (1981).

40. F. Vögtle, (Ed), Topics in Current Chem., 101 (1982).
41. F. Vögtle, E. Weber (Ed), Topics in Current Chem., 121 (1984).
42. F. Vögtle, E. Weber (Ed), Hostguest complex chemsitry. Springer-Verlag, Berlin (1985).
43. F. Vögtle, E. Weber (Ed). Topic in Current Chem., **128**, 132, 136 (1985).
44. F. Vögtle, H. Kagkuto, Kogyo 43, No 8, **12**, 21 (1990), Special number.
45. C.J. Pedersen, H.K. Frensdroff Angew Chem. 11,16 (1972).
46. B. Dietrich, J.M. Lehn, J.P. Sauvage, Chem. uns. Zeit 7, 120 (1973).
47. C.J. Pedersen, J. Am. Chem. Soc. **92**, 386 (1970).
48. E. Kimura, A. Scakonaka, T. Yatsunami, M. Kodama, J. Am. Chem., Soc. **103**, 3041. (1981)
49. J.M. Lehn, Science, **227**, 849 (1985).
50. J.M. Lehn, Angew Chem, **27**, 90 (1988).
51. A. Leregue, R. Rosset, Analysis **2**, 218 (1973).
52. V.P. Antonovich, E.1. Shelkhina-Zhur anal khim **35**, 992 (1980), AA **40**, 2A10 (1981).
53. Hüang Rongji, Zhang Dasheng, Fen Hsi, Hua Hsuch, **9**, 615 (1981), AA **42**, 6A9 (1982).
54. E.A. Krasnushkina, Y.A. Zolotov, Trends Anal. Chem. **2**, 158 (1983).
55. T. Shono Bunseki kagaku 33 E., 449 (1984), AA **47**, 11A5 (1985).
56. E. Weber Kontakle 26 (1984).
57. A. Jyo, M. Takagi, Bunseki 185 (1985) A.A **48** 3A9 (1986).
58. J. Hu, X. Zhang, X. Han, G. Li, Guangpuxue Yu Guangpu Fenxi, **8**, 10 (1988).
59. G.D. Blanco, A.P. Arias, A. Sanz Medel., Quim Anal. 7, 241 (1988), A A **51** 5A5 (1989).
60. Y. Ionoue, K. Wada, Y. Liu, M. Ouchi, A. Tai, T. Hakushi, J. Org Chem. **54**, 5268 (1989).
61. V. Balzani, Pure and Applied Chem. **62** 1457 (1990).
62. E. Weber, J. Taner, I. Goldberg F. Vögtle, D.A. Laidler, J.E. Stoddart, R.A. Bartsch, C.L. Liotta, Crown ether and analogues, Wiley (1990).
63. E. David Gutsche J. Chem. Ed., **67**, 812 (1990).
64. S.R. Cooper, Crown compounds—Towards future applications, VCH Verlag (1992).
65. G.Y. He, M. Kurita, C. Izumi, Ishii, F. Wada., T. Matsuda, J. Memb. Sci., **69**, 61 (1992).
66. A. Mckervey, V. Böhmer, Chem. Brit. **28**, 724 (1992).
67. H. Tsukube, Talanta **40**, 1313 (1993).
68. P. Necmeltin, Z. Flora, W. Andrew, J. Chem. Soc. Perkin Trans. 2, 2561 (1996).
69. J.M. Lehn, Sciences **260**, 1762 (1993).
70. L.F. Lindoy, Nature, **363**, 30 (1993).
71. B. Dietrich, P. Viout, J. M. Lehn, Molecular receptors in the round—Macrocyclic chemistry aspects of organic and inorganic supramolecular chemisrty, VCH publication (1993).
72. S.M. Khopkar, M.N. Gandhi, Bull. Inst Asso. Nucl. Chem. and Allied Science, **10**, 20 (1994).
73. R.M. Izatt, J.J. Christensen-Proceedings of International symposium in Macrocyclic ligands held in Utah, USA in September, Guildfortd Surrey. (1992).
74. D.S. Lawrence, J. Jiang, M. Levett, Chemical Reviews, **95**, 2229 (1995).
75. M. Takeshita, S. Shinkai, Bull. Chem. Soc. Japan, **68**, 1088 (1995).
76. M. Matsui, Bunseki Kagaku, **45**, 209 (1996) AA **58**, 7D1 (1996).
77. M.N. Gandhi and S.M. Khopkar, J. Sci. and Ind. Res., **53**, 630 (1994).
78. M.N. Gandhi and S.M. Khopker, J. Sci. and Ind. Res., **55**, 139 (1996).

79. L. Lin. C.Y. Wa, Fenxi Huaxue **25**, 850, (1997), AA **60** 2D 210 (1998).
80. P.L. Boulas, M. Gomez, Kaifer, L. Echegoyen, Angew. Chemi. **37**, 216 (1998).
81. N. Tarakkol M. Shamsipur, Talanta, **45**, 1219 (1998).
82. C.J. Pedersen, J. Am. Chem. Soc., **89**, 7017 (1967).
83. E. Brown, Toxicity of metals, Butterworth Publication NY (1960).
84. R.R. Hendrixon, M.P. Mack, R.A. Palmer, A. Ottolenghi R.G. Ghivecidelli Toxicological and Pharma applications, **44** 263 (1978).
85. K. Takayama, S. Hasegawa S. Sasagawa, N. Nambu T. Nagari, Chem. Pharm. Bull., **25**, 3125 (1977).
86. F.A. Patty, Industrial Toxicology and hygiene, Inerscience Publication (1958).
87. S.C. Grad, W.J. Conoray, J.A. Mckelvey R. A. Turney, Drug and Chem Toxicology, **1**, 339 (1978).
88. I.M. Kolthoff, P.J. Elving, Treaties in Analytical Chemistry Part 3, Vol. 2, Interscience Publication (1960).
89. E.I. du Pont de nemours, Co-Elastomers, Area Chambers Works Chemical Hazard, Sheet No 148, 1 (1972).
90. Report of the Institute for Life Science Nippen Soda Co, **50**, 60 (1975) (Cited. [3]).
91. M. Hiraoka, Kagaku Keizai, **23**, 61 (1961).
92. M. Hiraoka, Oligomer Handbook, Kagakukogyo Nippo, p. 240 (1977).
93. E. Kimura, Kagaku no Ryoki, **29**, 413 (1975).
94. E. Kimura, Chemistry of crown ethers (Ed). R. Oda, T. Shono, I. Tabushi Kagaku Extra No. 74, Chapter 5, Kagakudojin, cited in [3] (1987).
95. B.K.J. Leong, Chemical and Engineering news p. 5 (1975).
96. B.K.J. Leong, T.O.T.T' so, M.B. Chenoeth, Toxicol and Applied Pharmacol. **27**, 342 (1974).
97. A. Shitamori, Yozai Handbook, (Ed), T. Ashahara Kodansha (1976).
98. M. Hiraoka, Yozai handbook (Ed), T. Ashahara Kodansha (1976).
99. J.S. Bradshaw (Ed) R.M. Izatt, J.J. Christensen in Synthetic multidentae to microcyclic compounds. J.J. Christensen Ed. Acdemic Press (1978), Ref. 3.
100. Y. Inoue, G.W. Gokel, Cation binding by macrocycles, Complexation of Cationic species by crown ethers, Marcel Dekker, New York (1990).
101. H. Schubert, G. Agatha, Dermactosen in Beruf and Umwelt **27**, 49 (1979), cited in No. 4, p. 201.
102. M.K. Jain, D.V. Jahagirdar, Biochem. J., **227**, 789 (1985).

2
Synthesis of Crown Compounds

2.1 Introduction

A spectacular progress was made in the chemistry of macrocyclic, macrobicyclic and supramolecular compounds with the development in the field of synthetic chemistry which gave rise to the newer effects like template or macrocyclic. This led to the discovery of new series of crown compounds like switch on crown ethers, or bis (crown ethers), spherands and calixarenes.

The early efforts were directed towards the development of simple and reliable methods for the synthesis of crown ethers. Influence of ring size, presence of heteroatom on the ring and nature of side arm was of crucial importance in such synthesis which led to the discovery of new synthesis of lariat ethers with side arm. Early developments in the field of crown ethers encompassed study of effect of temperature, inclusion of diethyloxalate and diamines in the structure to promote complexation, synthesis of large members containing 30-60 members, and to produce doubly helical chiral crown thio ethers.

For cryptand synthesis, Lehn (a biochemist) developed large number of macrobicyclic compounds. Such synthetic work also included compounds like aza crown ethers. All attempts were made towards one step synthesis of cyclisation reactions. Early developments included development of simplified method of synthesis involving minimum steps, and preparation of photoactive derivatives.

Earlier workers focussed on the synthesis of aza and thia crown ethers because the complexation through sulphur and nitrogen atoms opened a new field in analytical chemistry. Several ditosylate derivatives of oligoethylene glycols were studied. Apart from these chiral crown ethers originating from pyridino, 18C6 ligand were synthesised in Japan.

The proton ionisable crown ethers have special place in the complexation of crown ethers. On account of the easy mode of replacement of proton by metal cation they form an array of coordination complexes with different metals. In comparison diester crown ethers show less tendency for the formation of complexes. The recent development in this area was confined to the synthesis of macrocyclic compounds and compounds containing pyrazole rings.

However, a new era was ushered in synthetic chemistry with the discovery of calix(n) arenes and calix(n) resorcinarene. This work was further simplified on account of fissibility of the substitution of the functional group on upper or lower rim of basket or caged structured ligand. The base catalysed reactions for calix(n) arene and the acid catalysed reactions for calix(n) resorcinarene yielded new and challenging series of ligands useful in solvent extraction separation as well as in the ion selective electrodes as the ionophores. Chiral calixarenes were also synthesised. The large number of functional groups to be introduced in upper or lower rim with favourable conformation and stereochemistry made these compounds most useful for the analytical chemist. The electrophillic substitution showed many miracles in the synthesis. Large number of calixarene were synthesised. These compounds included chromogenic calixarenes, and rotaxenes, two cyclic molecule containing wheel. Crown imido derivatives proved to be most promising hence attempts were made to prepare the imido analogous of calixarenes. Template effect vigorously followed in synthesis of calixarenes p-iodo calixarene proved to be quite useful. The electrophillic substitution of aromatic ring gave effective tool for the variation of cavity in the supramolecular compounds. The compounds containing trögers base moiety were also synthesised.

Thus the fundamental knowledge of the synthesis of the crown ether, cryptands and calixarenes is essential in order to understand their utility in analytical chemistry. Knowledge of the chemistry of synthesis, conformation will throw much light on the selectivity and stereochemistry of such compounds. This chapter attempts to consider the synthesis of these compounds. Subsequently the new developments in relation to the synthesis have been also incorporated at the end of each section [1-6].

2.2 Synthesis of Crown Ethers

In early sixties two mode of synthesis of crown ethers adopted were simple [7, 8] nucleophillic substitution and the formation of the amid linkages followed by reduction to the saturated system. This was used specially for synthesis of aza crown ethers, using high dilution. The former technique was called template effect' [6] as it involved use of metal cations.

We must examine Pedersen's work [9] at this stage who obtained DB18C6 by interaction of catechol. Thus (1-2 dihydroxybenzene) was treated with base to form dianion. This nucleophile reacted with 2, 2-dichloroethyl ether in quadruple Williamson reaction. Sodium hydroxide was used as base since phenol was acidic in nature. The reactions were carried out in n-butanol as it did not deprotonate and compete in nucleophillic substitution reaction. The reactions proceeded as shown in (I).

(I)

No high dilution was used. The reactant were heated at the reflux temperature of n-butanol. This basic technique was widely used for the preparation of DB18C6 with 40% yield.

Usually ethyleneoxy unit was preferred for synthesis of large ring systems. Thus $-O-CH_2-O-$ was simplest unit. It did not involve $>C=C<$ backbone interactions. Only disadvantage was that this group was acid labile. The other alternative was to use propylenoxy unit. It was $-O-CH_2CH_2\,CH_2-O$ which was not acid catalysed and was suitable for extended arrangement. The unit lowered the oxygen atom density within the ring. The ethylene ethoxy unit was most favoured route for synthesis as it was flexible, strainfree and popular with chemist. This unit was available from percursor ethylene oxide as the latter could be easily oligomerized or polymerised to get repeating unit (CH_2CH_2O) in crown rings system.

2.3 The Template Effect

This effect involved the use of metal ion which in turn directed the transition state leading to the formation of macrocyclic [10] compound. A cyclic material was formed from compound that was nucleophile at one end and electrophile at the other by reacting with itself (e.g. path X) shown in (II).

The nucleophilic, negative end of one molecule might find electrophilic or positive end of another molecule (intermolecular path Y) as shown in (III). The second mode (viz. Y) however led to noncyclic product. To achieve cyclisation high dilution conditions were usually employed. This way cyclisation enhanced over oligomerisation. Template effect influenced base strength rather than complexation. 18C6 was obtained using hexaethylene glycol monochloride and potassium ter-butoxide.

One route using triethyleneglycol and triethylenedioxylate showed sudden increase in yield due to presence of metallic cation. In synthesis of 12C4 presence of lithium ion enhanced the yield. However, such conclusion should be drawn with caution. This phenomena is called 'template effect'.

We now consider some observations made in the early synthesis of crown ethers. As per Pedersen, the process of hydrogenation of DB18C6 with ruthenium tetraoxide catalyst led to the formation of DC18C6 along with other isomers [6]. The compound was more lipophillic [11, 12].

The better complexation properly may be attributed to sp^3 hybridisation. In addition to cyclic oligomerisation, it was possible to obtain 18C6 from triethylene glycol and dichloride triethylene glycol and ditosylate in benzene as shown in (IV). By mere stirring of triethylene glycol, dichloride triethylene glycol and potassium hydroxide in tetrahydrofurane solution, 18C6 is obtained (IV).

(IV)

The three factors worth considering during the synthesis were the ring size, presence of heteroatom apart from oxygen and the nature of side arm. The synthesis of DB18C6 was shotgun reaction as catechol and 2, 2-dichlorodiethyl ether each had two reactive sites. It was possible to get monomer, dimer or trimeric products arising from the reaction of catechol and diethylene glycol dichloride. If the reaction in crown ether synthesis is basic it is termed as "Williamson reaction". Crown ether is a poly(oxygen)macrocycle. The alkoxide were difficult to generate while phenoxides were more nucleophillic. They used weak base sodium or potassium hydroxide. The potassium ion was better template ion for synthesis of many crown ethers. The strong base could be used with phenoxide anion.

The most popular leaving group was tosylate. It was easy to synthesise from primary alcohol and were usually solid and permitted recrystallisation. It could not be distilled. Mesylate CH_3-SO_2- was also useful as alternate to tosylate. They were less crystalline with low molecular weight. Usually tosylate was preferred as leaving group [7]. During synthesis, polar solvent was required. The choice of nucleophile anion decided the nature of solvent to be used. When potassium tertiary butoxide was formed tert butylalcohol was used. Sodium hydride or dimethyl sulphoxide was good combination. Benzene or toluene were preferred for high dilution reactions. Such reaction favoured intramolecular versus intermolecular reaction. The high dilution technique involved use of small reaction volume and needed slow addition of the reactants. For better results one must use sodium hydride as base, tetrahydrafurane as solvent, diol as nucleophile, ditosylate or dimestoylate as the electrophile (i.e. leaving group) with vigorous mixing at reflux temperature and carrying out final purification by chromatography.

The reactions for synthesis can be written as follows:

CH_2-CH_2 $\xrightarrow{Catalyst}$ [CH_2-CH_2-O]
     \ /
      O

R.OH + R'Cl $\xrightarrow{NaOH}$ ROR'

ROH + R'OTs $\xrightarrow{KOH}$ ROR'

ROH_2 + R'-C-Cl → RNH-C-R' $\xrightarrow{H_2}$ $RNHCl_2$R'
          || ||
          O   O

R-SH + R'Cl $\xrightarrow{NaOH}$ RSR'

They either involved: (a) high dilution method, (b) a two step condensation, or (c) template reaction. Reaction (c) was carried out in high diluted solution. While method (b) involved condensation of one of the terminal groups on both bifunctional compounds and separation of reaction product followed by ring closer. While reaction (a) involved use of metal ion whose ionic diameter was just corresponding to the cavity size of the desired crown ring and it thus acted as template. These reactions have been commercialised [13].

Similarly, aromatic crown ethers reaction could be summarised in Williamson condensation reaction using catechol and dichloroether. In such reactions, apart from many advantages cited above, ethylene group [$-(CH_2)_2-$] was obtainable from ethylene oxide. The structure so formed showed less strain and good stability, including complexing property. They were stable and provided suitable ratio of hydrophilic or lipophyllic quotient which actually determined solubility of complex in the organic solvents. In addition to metal cations, guanidium ion [14] was also used as the template in the synthesis of B27C9. Apart from benzene biphenyl [15], 2-3 naphthyl, binaphthyl [16], hydroquinone groups [17] were introduced in aromatic crown ethers. These compounds were soluble in methylene chloride, chloroform, pyridine and acetic acid. In early synthesis, workers were suspicious as to whether crown ethers could be distilled lest it might decompose but later DB18C6 was successfully distilled [18] at 380° C at the atmosphere pressure and such misplaced doubts were thus removed.

Finally, let us consider synthesis of lariat ethers [19]. They were represented by large number of structures. A microring and side arm helped to bind a cation. The structures with side arms were thus lariat compounds, as shown in (V).

$$C_6H_5CH_2NH_2 + \text{[diiodo diether]} \xrightarrow[CH_3CN]{Na_2CO_3} \text{[bis-benzyl diaza crown]}$$

(V)

In (V) the side arm was attached to macroring at N-in lariat ether. When two or more arms were attached to macrocycle the lariat or lasso image was altered. Bibrachial lariat (2 arms) tribrachial lariat (3-arms) were formed [20].

Thus, the synthesis of crown ethers involved all the difficulties usually encountered in organic synthesis. The important factors to be considered were ring size, functional group compatibility and stereochemistry. We would concentrate on specific properties. These highly symmetrical and simple structure present complex synthetic problems. Only fascinating aspect of all labour was to invent novel product which had intended chemical properties as anticipated by theoretical chemistry of the product.

2.4 Recent Developments in the Synthesis of Crown Ethers

In 1974, Pedersen first synthesised 18C6 and DB18C6. It generated a considerable interest in early 1980 on exact mode of the preparation. For instance, hazard of preparing 18C6 was pointed out [21] as during synthesis a violent explosion occurred during thermal decomposition of crown-potassium chloride complex. This was attributed to the ignition of air and 1.4 dioxan mixture in trap. Hence reactions were carried out at low temperature [21]. Later a phenomena of poor distillation was noticed due to explosion. It was suggested that heating should be stopped no sooner the substance was collected in trap and if it was blocked then distillation should be arrested. The atmospheric pressure was regulated by flushing nitrogen [22]. However, this was contradicted by other suggestion to use two independent cold traps in parallel instead of clamping tube between trap and distillation units or flushing nitrogen to restore normal pressure [23]. A method was suggested to determine impurities in crown ethers [24]. Lindoy synthesised a new receptor which was capable of simultaneously binding a cation to its polyether crown ring and an anion to its boron (VI) size.

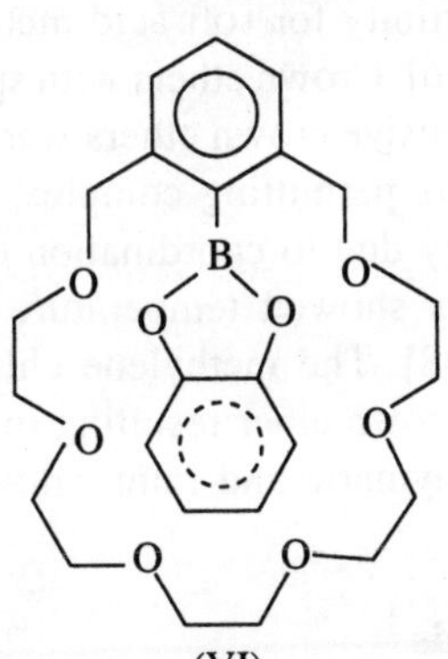

(VI)

They served dual role of trapping cation as well as anion [25]. An attempt was made to synthesise novel crowned coumarins as catalytic fluorescence reagents [26]. Crown ethers had novel application of pretreatment of serine proteases by lyphlisation leading to increased enzyme activity [27].

There was growing trend to synthesis macrocyclic [28] amides from diethyl oxalate and ethereal oxygen containing diamines. They later proved to be promising complexing ligands for the metals. A new linked macrocyclic compound was obtained from selectively protecting $S_2 N_2$ macrocycles [29] shown in (VII) and (VIII).

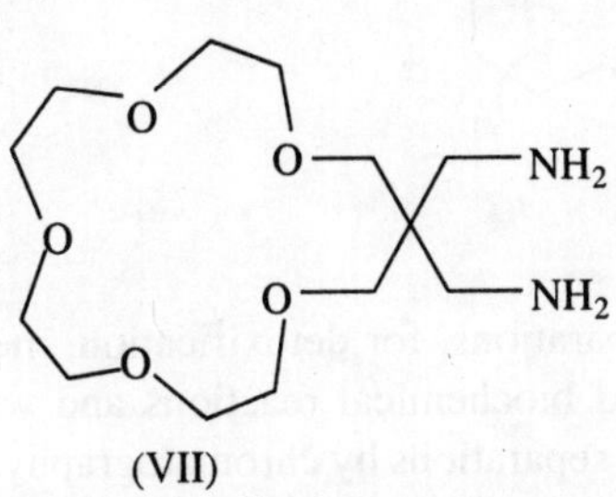

(VII) 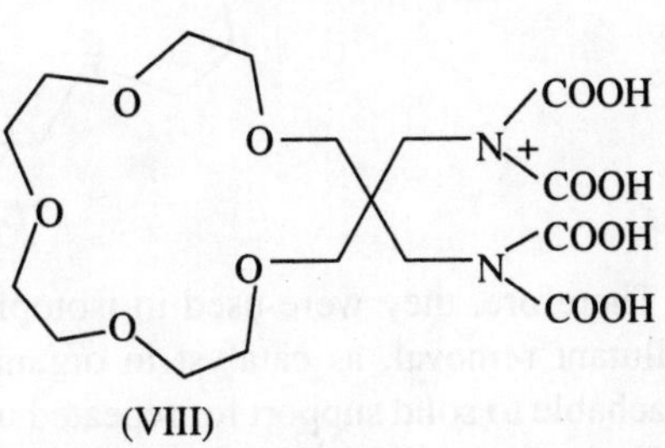   (VIII)

The pendant polyamine crown ethers (VII) and polyaminopoly carboxyllic acid (VIII) were useful reagents for magnetic resonance imaging (MRI) and as

NMR shifts reagents [30]. Template ion had induced cyclisation of oligomeric ethyleneglycol diacrylates which in turn were synthesised to produce pseudo-crown ethers which were useful for binding target cations [31].

There was continuous activity on the synthesis of large crown ethers containing 30-60 members which isolated small and large cyclic polyethers by combining oligo (ethylene glycol) and oligo (ethylene glycol derivative) such as ditosylates. At high temperature, large crown ethers and spherands were formed, viz. 60C20, 48C16, 42C14, 36C12 and 30C10. They decomposed when heated at temperature greater than 200°C [32-34].

There was brisk activity to produce doubly helical chiral crown thio ethers, which were optically active. They readily reacted with ruthenium to give a plannar coordination [35]. Thus to develop ion specific crown ethers, cation complexation and novel organic synthesis was examined. This showed that those containing neutral oxygen donors were best for s-block metals while one containing sulphur atom had highest affinity for soft acid metals like mercury, palladium and to some extent copper [36]. Crown ethers with specialised applications were also synthesised. Photoresponsive crown ethers were prepared which contained intra-annular azo substituents permitting complexation by photoisomerisation indicating high metal affinity due to coordination of one of the azo nitrogens [37]. Phenolic crown ethers showed temperature dependent reversal in the complexation of calcium [38]. The methylene chloride protonated derivative gave adduct with aromatic crown ether resulting in quenching of fluoroscence [39]. For ion separation, polymeric and immobilized crown compounds were synthesised [40].

2.5 Synthesis of Cryptands

Although much information was available on synthesis of crown ethers, aza and thia crown ethers, little information was available on synthesis of cryptands. In fact, the cryptands formed strong complexes with several metals. In (IX) all positions (A, B, C, D) are occupied invariably by oxygen (IX).

(IX)

Therefore, they were used in isotopic separations, for detoxification, metal pollutant removal, as catalyst in organic and biochemical reactions and were attachable to solid support for repeated use in separations by chromatography. In 1968 the first cryptand [41] was prepared by Lehn. Dietrich [42, 43] suggested following steps for the synthesis of cryptand-2, 2, 2 (X).

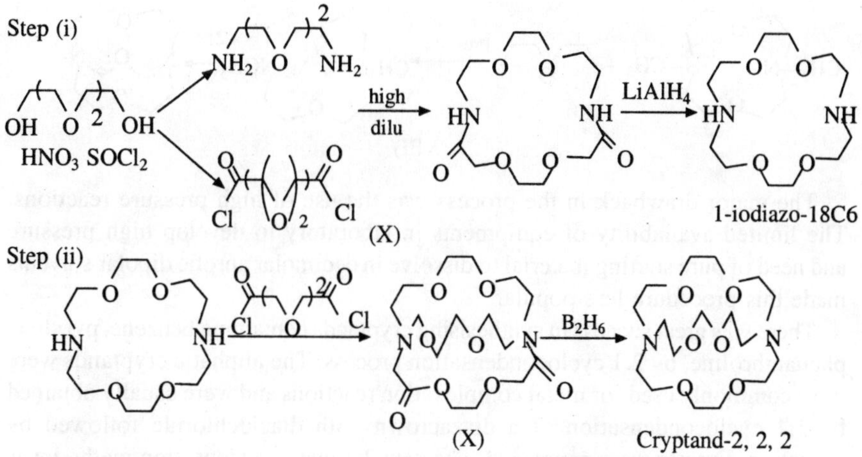

Step (i)

[Scheme showing synthesis: starting from OH-O-OH with HNO₃, SOCl₂ reacting with NH₂-O-NH₂ (×2) under high dilution to form diamide (X), then LiAlH₄ reduction to give 1-iodiazo-18C6]

Step (ii)

[Scheme showing 1-iodiazo-18C6 reacting with acid chloride to form diamide (X), then B₂H₆ reduction to give Cryptand-2,2,2]

The synthesis involved several steps as the reduction of diamide was critical. Hence, number of starting materials were used in ring closer step. The process involved 1:1, 2:1 or 3:1 cyclocondensations. Since starting material like 1-iodiazo-18C6 were readily available, synthesis was simplified. The maintenance of high dilution prevented polymerisation. The simple procedure followed [44, 45] is shown in (XI).

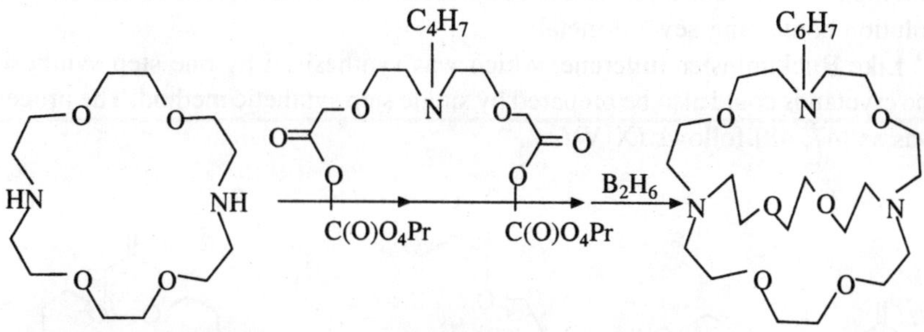

[Scheme XI: 1-iodiazo-18C6 reacting with C₄H₇ and C(O)O₄Pr reagents, then B₂H₆ to give product with C₆H₇ substituent]

Thus, number of cryptands were synthesised by metal template catalysed 1:1 condensation of diaza crown ethers with oligoethylene oxydihalide or disulphonate. The cryptands containing benzene unit formed weak complexes with alkali metals as compared to normal cryptands. They were obtained as [46] follows (XII):

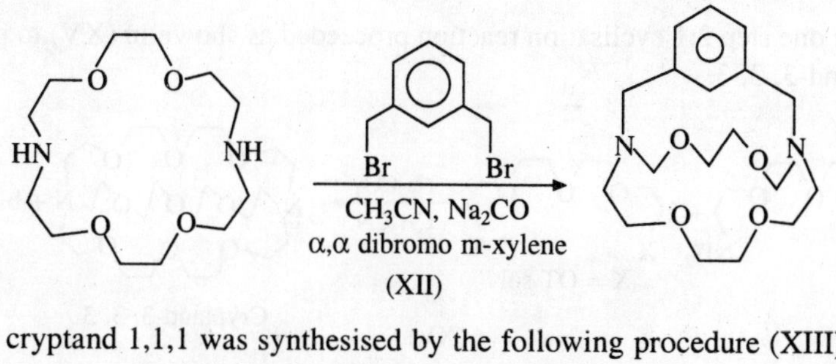

[Scheme XII: diaza crown ether + α,α dibromo m-xylene in CH₃CN, Na₂CO₃ giving benzene-containing cryptand]

The cryptand 1,1,1 was synthesised by the following procedure (XIII):

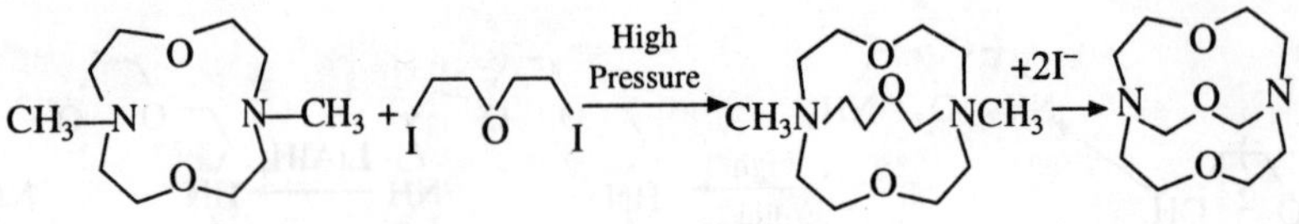

(XIII)

The major drawback in the process was the use of high pressure reactions. The limited availability of equipments in laboratory to develop high pressure and need of pure starting material to dissolve in decimolar aprotic dipolar solvents made this procedure less popular.

There was great success in synthesising cryptands containing benzene, pyridine, phenanthroline, by 2:1 cyclocondensation process. The aliphatic cryptands were most commonly used for metal complexation reactions and were usually obtained by 1:1 cyclocondensation of a diazacrown with diacidchloride followed by reduction. Former were expensive. The new 2:1 cyclocondensation method was useful. It was possible to get cryptand in one step and isolation was simple. In summary, cryptands could be made by variety of methods. The 2:1 or 3:1 cyclocondensation reaction became more popular with chemist, though these reactions gave lower yield [47] during synthesis.

A novel method was proposed for preparation of cryptand 222 and supercryptands. They interacted with many cations and anions with high selectivity and high stability constants and hence were useful in the separation of ions from solution containing several metals.

Like Buckiminster fullerene, which was synthesised by one step synthesis, the cryptands could also be prepared by single step synthetic method. The process was as [47, 48] follows (XIV):

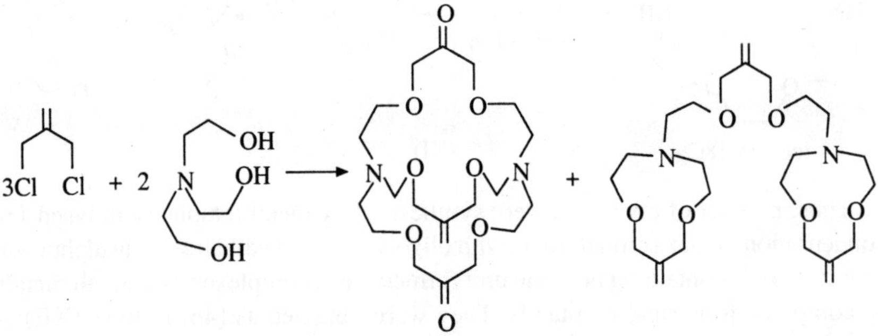

(XIV)

The one step 2:1 cyclisation reaction proceeded as shown in (XV) to procure cryptand-3, 3, 3.

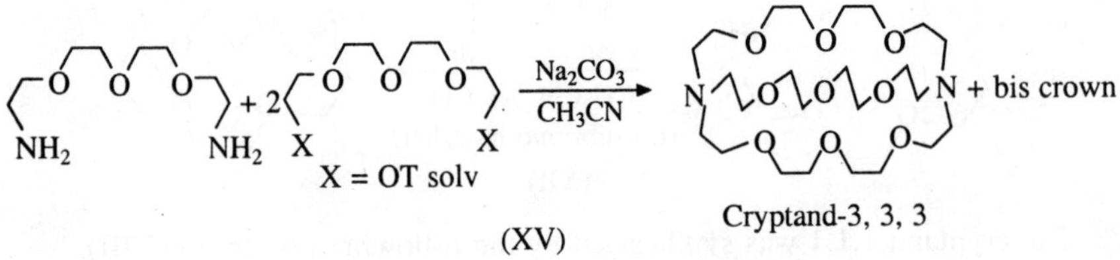

(XV)

The super-cryptand was synthesised as follows {49-52] (XVI):

$$\text{TS-cryptand} \xrightarrow[\text{THF}]{\text{LiAlH}_4} \text{H-cryptand} \xrightarrow[\text{Na}_2\text{CO}_3]{\text{C}_3\text{H}_7\text{CN}} \text{super-cryptand}$$

(XVI)

A large number of cryptands have been prepared by one step bicyclisation of tetra alcohol (prepared from diethanolamine and dichloride, derivative of an oligoethylene glycol with dihalide or dibromide or ditosylate [53]. Thus bi-or tri-cyclic crown compounds with cage structure (whose bridgehead was N-atoms) were formed as cryptands (XVI). They formed complexes by tight binding of metal ions in space lattice. Cryptands containing binaphthyl group were optically active, resulting in more selective complexation with ions and more stable complexes than those obtained with monocyclic crown compounds. The complexes were called 'cryptates' meaning hidden compounds.

2.6 New Developments in the Synthesis of Cryptands

Lehn was first to develop cryptand-2, 2, 2 or kriptofix-2, 2, 2—a new macrobicyclic polyether containing a bridged structure with nitrogen as the donor atom. Several compounds were subsequently developed which showed immense applications in the chemistry of complexes of soft acid metals like mercury, lead and copper. Attempts were made to simplify the method for the synthesis of cryptands. Usually, the simple method consisted of the reaction of a viable oxa-alkylane diamine with dioxide derivative of oligoethylene glycol. The procedure gave cryptands containing two propylene units in bridge. Cryptands like (2, 2, 1), (2, 2, 2) and (3, 2, 2) were synthesised by treating ditosylate derivative of the oligoethylene glycols with the available oligoethylene oxydiamines. The methods became popular [54]. An improved method for the synthesis of cryptand consisted of use of cheap starting material with one or two steps in synthesis [55-57]. A new two nitrogen donor cryptand was synthesised as it provided selective binding

(XVII) + 3 [CH₂ with Cl Cl] → (XVIII) + New compound

for soft cations and isolating anions from surroundings at low pH [58]. A single step method was devised for synthesis of cryptand-222 containing three methylene substituents by reaction of triethanolamine and 3-chloro-2-chloromethyl-1-propane (XVIII). The synthesis was [59] as shown in (XVII) and (XVIII).

The synthesis of new large cryptand was also carried out. It was capable of encapsulating a benzene molecule in its cavity [60, 61] (XIX).

(XIX)

A novel photo-active anthracenyl crown showed reversible photo-cyclisation to form a cryptand photo switch [61] (see XIX). A new [K (222 cryptands)]$_2$ [μ = O] [μ = O$_2$ (Si(CH$_3$)$_2$ (GeSe$_2$)] was also prepared [62]. The molecular modelling studies on molecular recognition involving cryptands and cryptates provided a deeper insight into their conformational and recognising properties and binding selective [63] sites. The caged anions with special reference to prechlorate and perfluoro anion cryptands showed that they were completely encapsulated within the cryptand molecule. Among BF_4^- and SiF_6^{2-} the latter was favourably encapsulated [64]. The different products were obtained from HgSe$_2$ [K (222 cryptand)]$_2$ by electrochemical reduction [65].

2.7 Synthesis of Aza Crown Ethers

The interest in the synthesis of aza crown ethers continued as they formed complexes with stability ranging in-between the stability of crown ethers and the stability of nitrogen cyclams, i.e. cryptands. They had important use as synthetic receptor in molecular recognition process and good utility due to anion complexation properties. These were used as catalyst [66, 67] in nucleophillic substitution and redox reactions in preparing chromogenic ligands for s-block metals and useful in the chromatographic separations involving silica gel bonded aza crown ethers (see XX).

It was observed that crown ethers with ethylene bridges formed more stable complexes with metal cations than those with propylene bridge. The common methods used for synthesis involved template synthesis [68]. In synthesis of

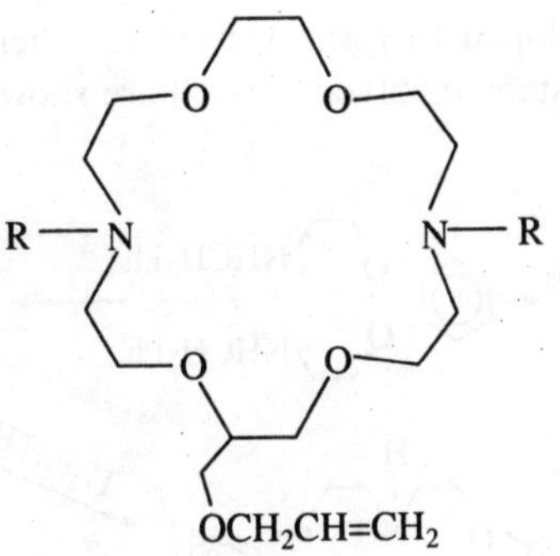

(XX)

18C6 from ditosylate and diol in the presence of [69] tertiary butoxide salt was enhanced when potassium ion was used as the template effect. Such effect was also observed during synthesis of aza crown ethers though less pronounced as softer N-donor atom formed weak complexes with s-block metals. High dilution technique was eliminated. The yield was increased with use of caesium carbonate due to formation of ring. Such enhancement in yield was called 'caesium effect'. This is not template effects as per Kellogg's [70, 71].

The aza crown ethers can be divided as monoaza, diaza, polyaza and benzoaza crown ethers. The method for synthesis is as follows (XXI):

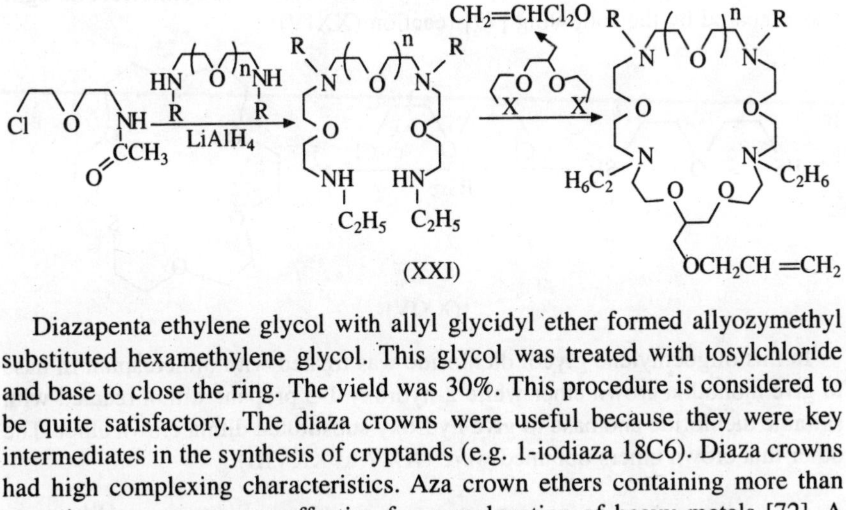

(XXI)

Diazapenta ethylene glycol with allyl glycidyl ether formed allyozymethyl substituted hexamethylene glycol. This glycol was treated with tosylchloride and base to close the ring. The yield was 30%. This procedure is considered to be quite satisfactory. The diaza crowns were useful because they were key intermediates in the synthesis of cryptands (e.g. 1-iodiaza 18C6). Diaza crowns had high complexing characteristics. Aza crown ethers containing more than two nitrogen atoms were effective for complexation of heavy metals [72]. A typical scheme for synthesis of polyazacrown ethers is shown in (XXII).

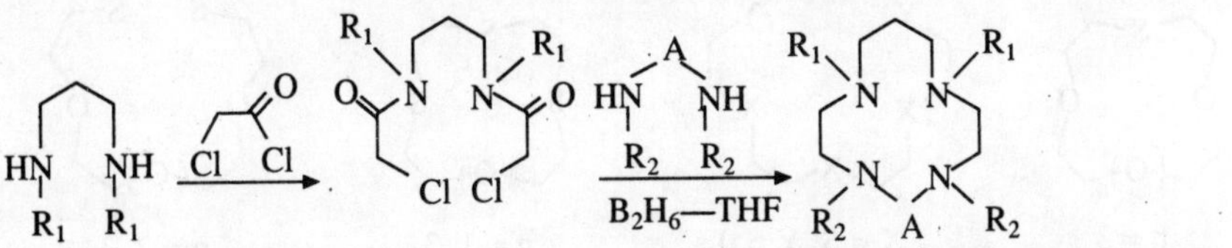

$A = O, NC_2H_6, R = CH_3, C_6H_5CH_2, C_2H_5$

(XXII)

A method was developed to form aza crown ether (polyaza crown ether) (XXII). The important steps involved [73-77] are shown in (XXIII):

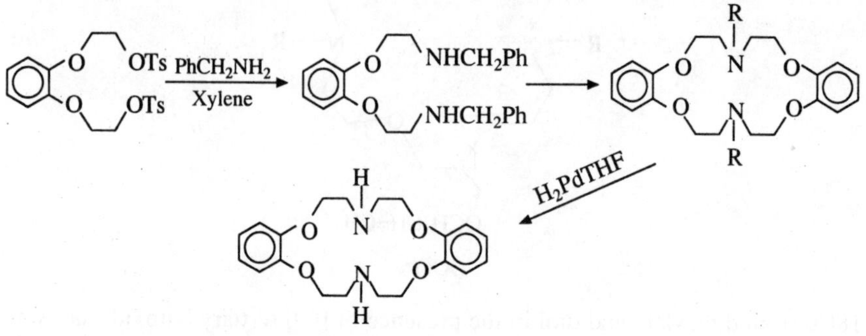

(XXIII)

As a rule polyaza crown ethers were prepared by reacting *p*-toluene sulphonamide with ditosylate derivative of oligoethylene glycol. Dibenzodiazo crown was prepared by three step synthesis [78].

2.8 Synthesis of Thia Crown Ethers and Related Compounds

Thia crown ethers are similar to crown ethers except that one or more of the oxygen atom of polyether was replaced by sulphur atom. The simplest analogue was prepared by the following [79] reaction (XXIV):

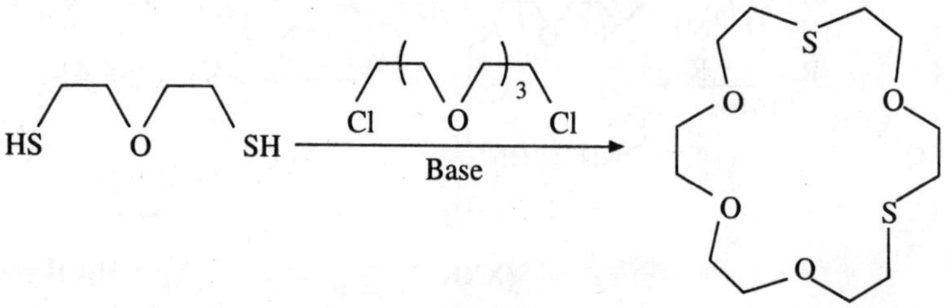

(XXIV)

In this oligoethylene glycol dichloride was treated with dimercaptan in base to give monothia crown ether while 2-hydroxy 1-3 propane-dithol reacted with suitable dichloride and base to give hydroxy substituted dithia crown ether. The other thia crown ethers obtained were (XXV to XXVIII):

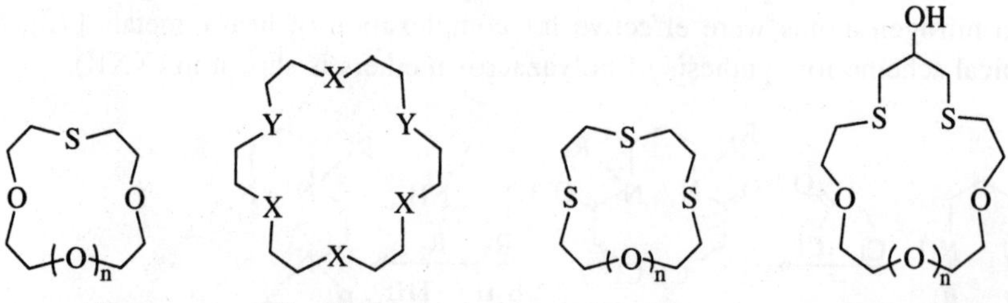

n = 3
(XXV)

X = S Y = O
(XXVI)

n = 1, 3
(XXVII)

n = 2, 3
(XXVIII)

2.9 Chiral Crown Ethers

The series of pyridino 18C6 ligands containing alkyl or phenyl substituents on two stereogenic centres in the macroring have been synthesised. They had structures [80-83] as shown in (XXIX) and (XXX):

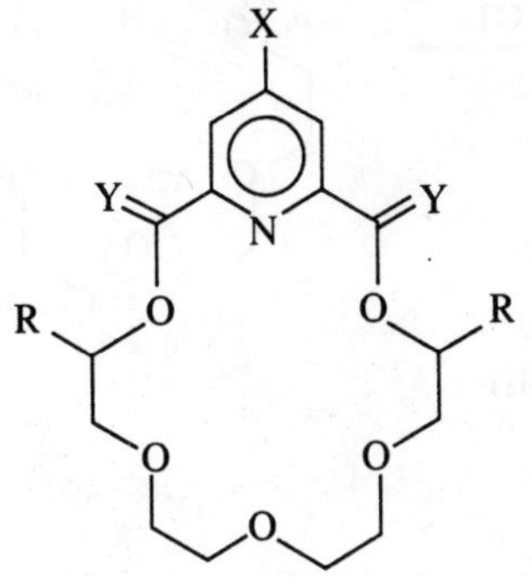

Y = O, X = H, Cl
R = CH$_3$

(XXIX)

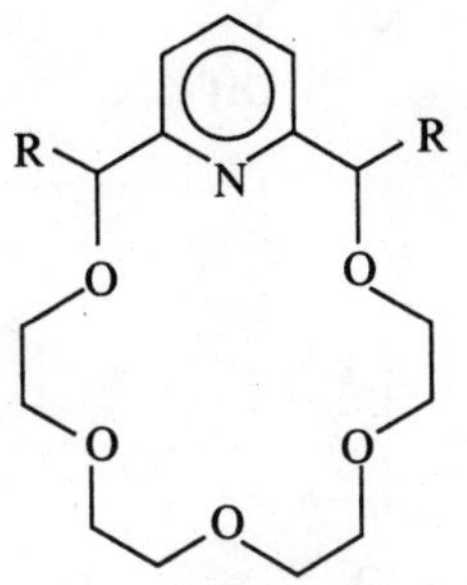

R = CH$_3$

(XXX)

These chiral ligands had good property to recognise the enantiomers of different chiral ammonium salts as determined by difference in stability. The 18C6 ligand exhibited enantiomeric recognisation to chiral ammonium salt.

2.10 Proton Ionisable Crown Ethers

Such compounds were synthesised to observe proton driven metal ion transport across liquid membrane. The compound 4-hydroxy pyridine proton ionisable crown was synthesised as follows (XXXI):

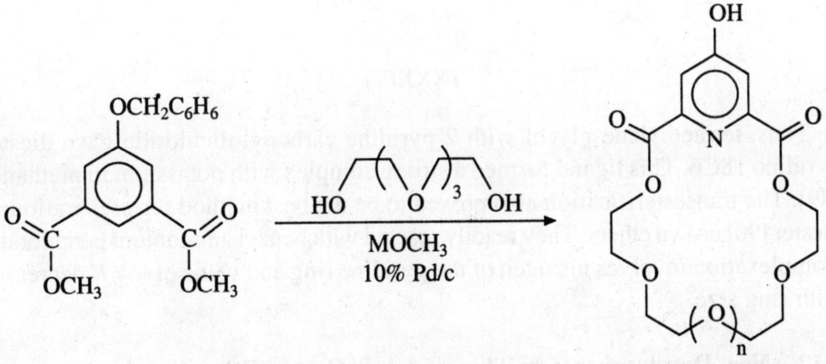

n = 1, R = H

(XXXI)

The 4-pyridino derivative was synthesised as follows [84-86] (XXXII):

2.11 Diester Crown Ethers

These were extensions of crown ethers and were easier to synthesise. A typical diester crown ether was synthesised [87, 88] as indicated in (XXXIII) and (XXXIV).

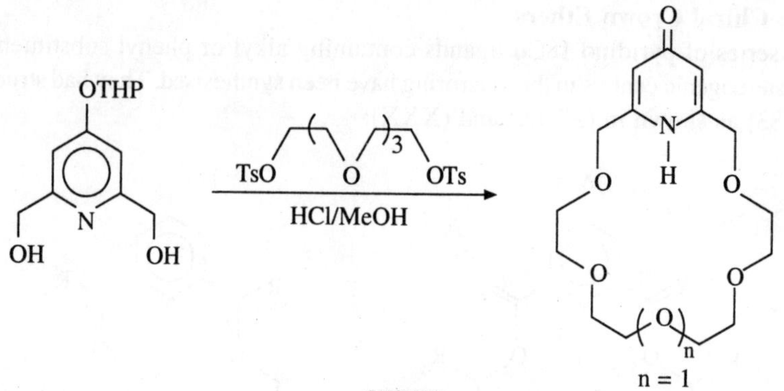

(XXXII)

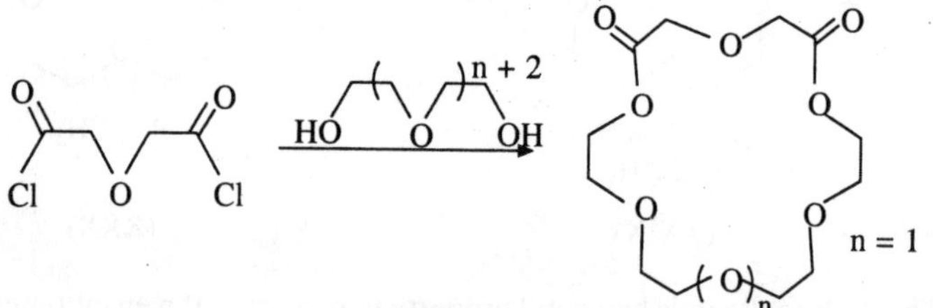

(XXXIII)

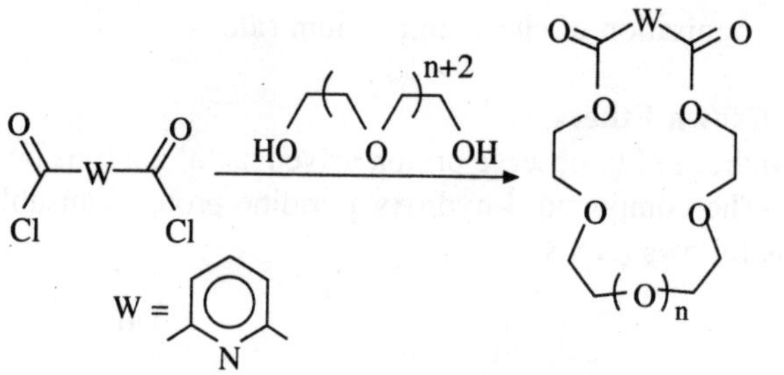

(XXXIV)

Thus, tetraethylene glycol with 2-pyridine carbonyl dichloride gave diester pyridino 18C6. This ligand formed a strong complex with potassium in methanol [89]. The transesterification also proved to be the best method for preparation of diester [90] crown ethers. They readily reacted with benzyl ammonium perchlorate. Complexation involves nitrogen of the pyridine ring and value of log K decreases with ring size.

2.12 New Developments in Thia and Aza Crown Ether Synthesis

A functionalised thia crown ether containing butyl unit was synthesised using caesium and diborane for the functionalisation of methylene group [91-92]. New macrocycle containing azoxy subunit was introduced as macroring subunit and extended their application in ion selective electrodes [93-94]. New crown thia ether complexes of soft acid metals like copper and silver were synthesised and their structure studied. It showed silver was tetrahedrally bound with four thio ether groups to form four symmetry related complexes and each ligand molecule was coordinated to four separate silver ions to give three dimensional network [95].

There was surge of activity to improve synthesis of aza crown ethers and cryptands. The cheap starting materials were employed and only one or two steps reactions were preferred in such synthesis BOC protected *m*-xylene diamine when treated with ethylene glycol ditosylate to form BOC protected diazabenzo 18 crown 6 was obtained leading to new synthesis technique. Diazobenzo 21 crown 6 and diazo 17C5 crowns were thus synthesised. The BOC protecting group was eliminated in methanolic hydrogen chloride to form the diaza crown ether (XXXV) [96]. A new synthesis of macrotricyclic ligand based on 1, 4, 8, 11-tetra-azacyclotetradecane was devised involving five steps starting from trans-dioxacyclam as diprotected macrocycle. The compound is in structure (XXXVI) [97]. A new macrocyclic poly-azacyclo alkane-poly N-carboxylate anion was developed as receptor for amino acid [98].

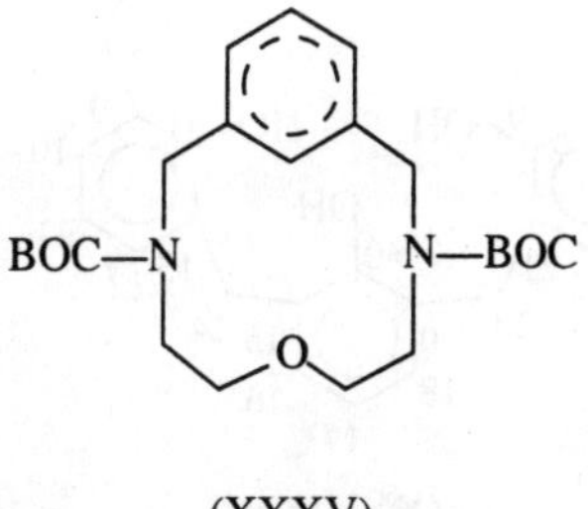

(XXXV) 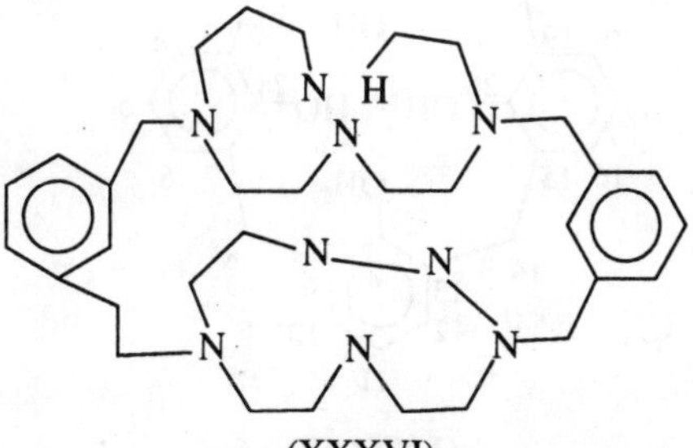   (XXXVI)

Finally, a pyrazole containing aza crown ether was synthesised by intramolecular cyclo additions. The medium and large ring annulated aza crown ethers were thus obtained [99].

2.13 Synthesis of Calix(n) Arenes

Zinke [100] was the first to detect the presence of cyclic tetramer (XXXVIIa) which was obtained by acid catalysed treatment of aldehydes and phenol to give tetramethyl calix [4] arene [101]. In subsequent compounds (CH_3) was replaced by R = C $(CH_3)_2$ CH_2CH_3 or – $C(CH_3)_2$ – CH_2C $(CH_3)_3$, C_6H_5. It was obtained by Niederl [100] as shown in (XXXVIIb).

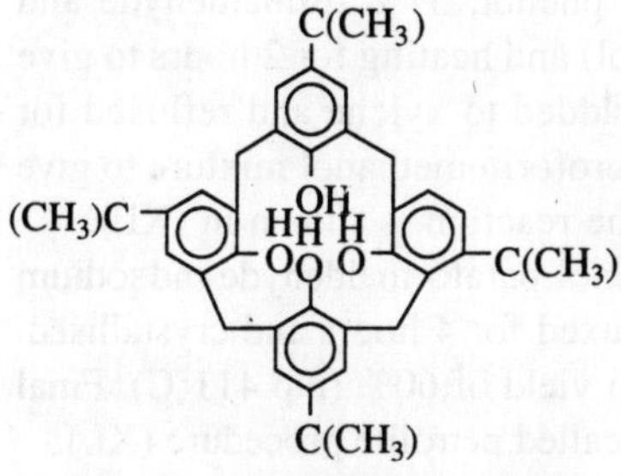

(XXXVIIa) 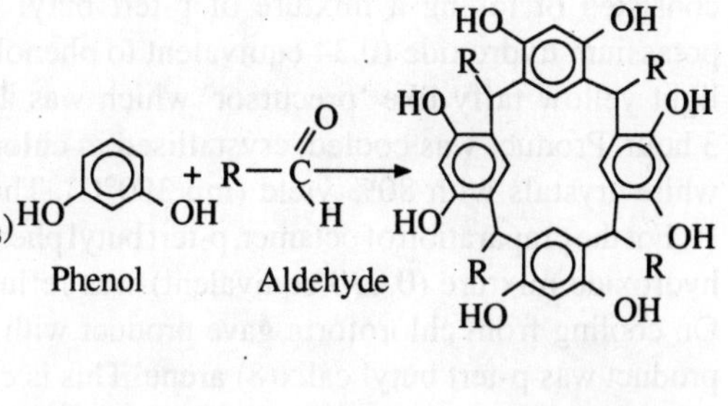   (XXXVIIb)

This was phenol-formaldehyde reaction. The name 'calixarene' was derived from 'calix' (Greek: *vase* or *chalice*) and 'arene' indicating the presence of aryl residues in macrocyclic structure. On examination of these space filling models, it showed the existance of very deep cavity capable of forming complexes with molecules of requisite size. Usually p-substituted phenols did not form calix(4) arene as the only product but was accompanied by other products. The reaction of p-tert butyl phenol and formaldehyde served as starting material for the synthesis of tetramer, hexamer XXXVIII or XXIX, respectively. A judicious selection of the solvents, bases, reactant ratios, gave best yield. The resorcinol derived calix(n) arene were also prepared. The IUPAC nomenclature was difficult [102].

(XXXVIII)   (XXXIX)

There were number of procedures in base catalysed reaction process. In Zinke and Cornforth [103-106] procedure, a mixture of p-tert butyl phenol, 37% formaldehyde and alkali hydroxide equivalent to 0.045 M of phenol was heated for 2 hours at 120°C to give 'precursor'. This was refluxed with diphenylether for 2 hours cooled and crystallised from toluene to give p-tert butyl calix(4) arene (mp 342°C). With variation in base higher homologue were obtained, e.g. 0.30 equivalent of potassium hydroxide gave hexamer. This proved that the amount of base in cyclo-oligomerisation process affected the end product. Lithium hydroxide gave tetramer; potassium hydroxide gave hexamer and sodium hydroxide gave octamer. A modified petrolite [107-109] procedure for synthesis of hexamer consisted of taking a mixture of p-tert butyl phenol, 37% formaldehyde and potassium hydroxide (0.34 equivalent to phenol) and heating for 2 hours to give light yellow taffy like 'precursor' which was added to xylene and refluxed for 3 hour. Product was cooled, crystallised in chloroform methanol mixture to give white crystals with 80% yield (mp 380°C). The reaction is shown in (XL).

For the preparation of octamer, p-tert butyl phenol, paraformaldehyde and sodium hydroxide mixture (0.030 equivalent) was refluxed for 4 hours and crystallised. On cooling from chloroform gave product with yield of 60% (mp 411°C). Final product was p-tert butyl calix(8) arene. This is called petrolite procedure (XL).

A procedure for p-substituted calixarene involving the substituent as p-tert

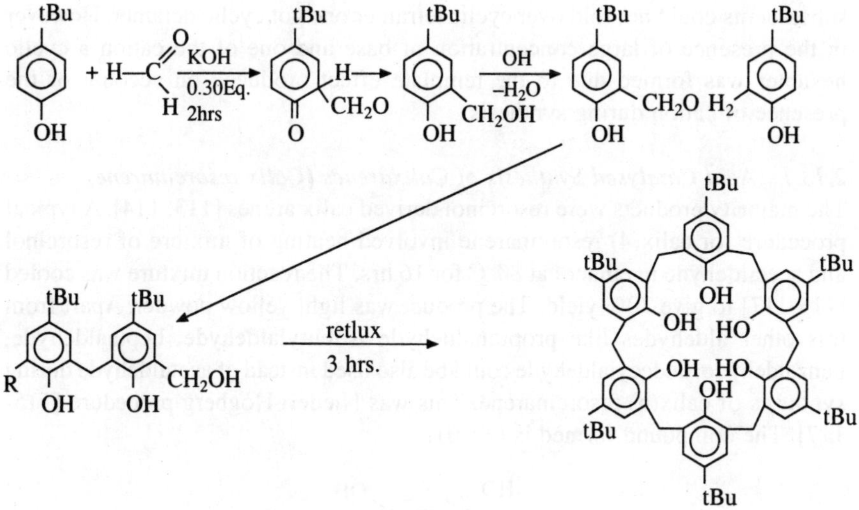

(XL)

pentylphenol, p-n-octyl phenol, p-m-nonyl phenol and p-n-dodecyl phenol was also synthesised but mixture of cyclic hexamer, heptamer and octamer were usually [110] encountered.

It is worthwhile to consider the mechanism of formation of calixarene. The pathway of based induced oligomerisation of phenols and formaldehyde has been studied. The steps were as [111, 112] per following process indicated in (XLI and XLII):

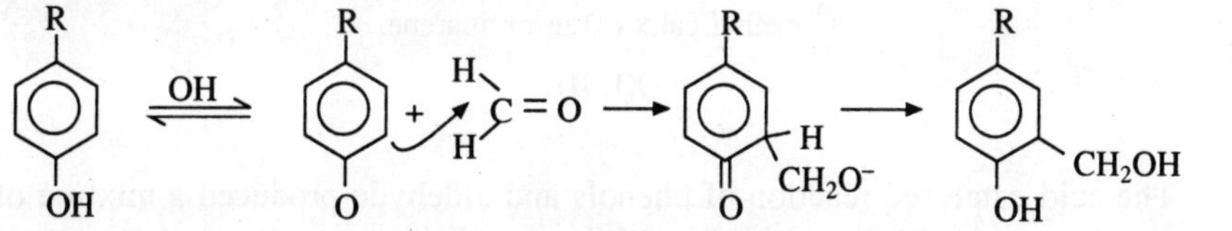

(XLI)

The reaction under vigorous conditions gave dioxylmethyl compounds as shown (XLII):

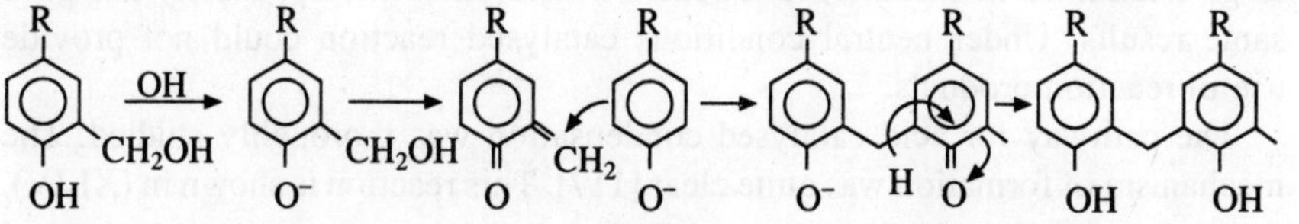

(XLII)

In base induced calixarene formation the conversion of the hydroxy methylated linear oligomer to cyclic oligomer involved the formation of hemicalixarenes in which forces of intramolecular and intermolecular hydrogen bonding played an important role, by removal of water and formaldehyde from hemicalixarene. In less base at 180°C tetramer had resulted and with phenol in which p-alkyl substituents could not fold over cyclic tetramer one got cyclic octamer. However in the presence of large concentration of base and one of the cation a cyclic hexamer was formed due to the template effect produced on account of the presence of cation during synthesis.

2.13.1 Acid Catalysed Synthesis of Calixarenes (Calix resorcinarene)

The majority products were resorcinol derived calix arenes [113, 114]. A typical procedure for calix(4) resorcinarene involved heating of mixture of resorcinol and acetaldehyde in ethanol at 80°C for 16 hrs. The reaction mixture was cooled [115-117] to give 70% yield. The product was light yellow powder. Apart from this other aldehydes like propionaldehyde isobutylaldehyde, heptaaldehyde, benzaldehyde, dodecylaldehyde could be also used instead of acetaldehyde during synthesis of calix(n) resorcinarene. This was Niederl Hogberg procedure [115-117]. The compound formed is (XLIII).

C methyl calix (4) resorcinarene.

(XLIII)

The acid catalysed reaction of phenols and aldehyde produced a mixture of linear oligomers with long length, with less tendency to cyclise in comparison with one, in base catalysed reaction for such purpose ketones were useless. Nonhydroxyl containing calixarenes were made by acid catalysed reaction of mesitylene and 1, 2, 3, 5 tetramethyl benzene with formaldehyde in acetic acid to give tetramer. A Friedel-Crafts reaction with chloromethyl mesitylene gave same results. Under neutral conditions catalysed reaction could not provide useful reaction products.

The pathway for acid catalysed condensation was thoroughly studied. The mechanism of formation was quite clear [117]. This reaction is shown in (XLIV).

2.13.2 Chiral Calixarenes

Calixarenes differed from cyclodextrins due to the absence of chirality. A p-tert butyl calix(8) arene ester was prepared and optically studied. The lower rim functionalisation helped to introduce chirality [118]. They are XLV, XLVI and XLVII structures. All these chiral calixarenes were incapable of resolution into entiomers under ordinary conditions as they were conformationally mobile.

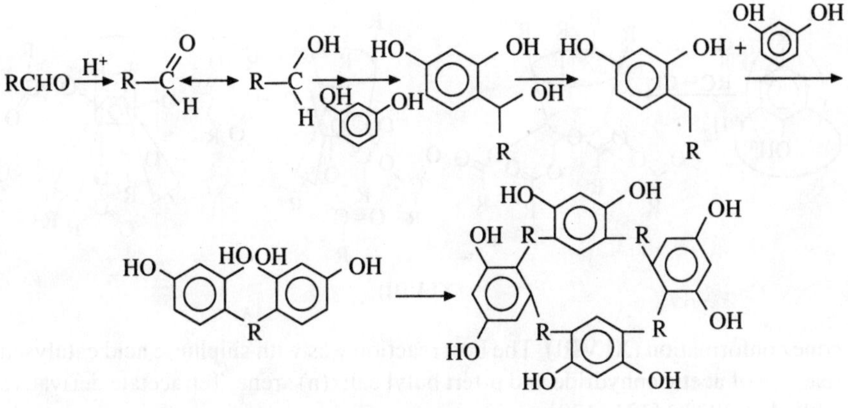

(XLIV)

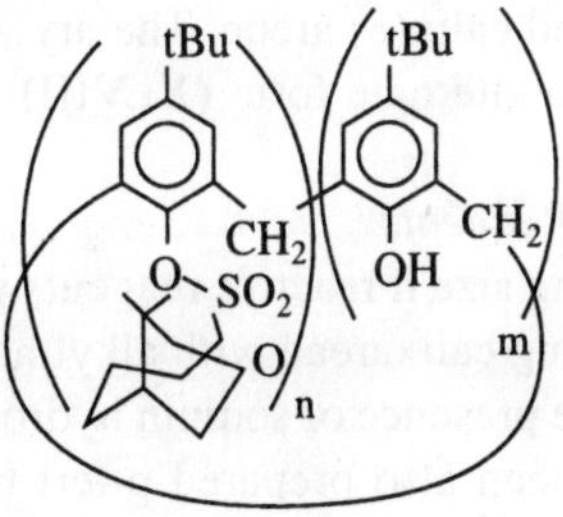

(XLV) 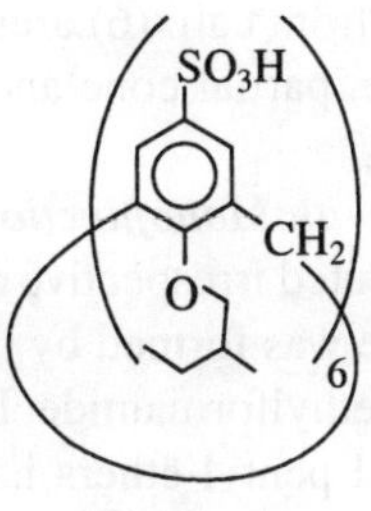 (XLVI) 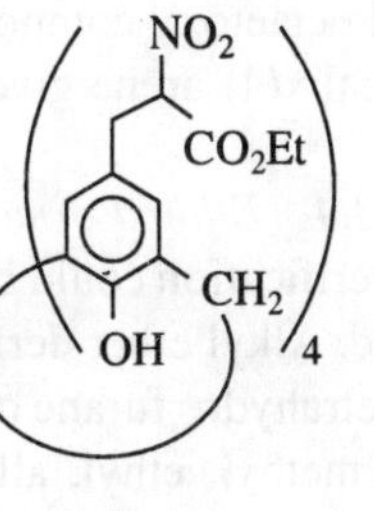 (XLVII)

There were homo calixarenes, oxacalixarene, homoxa calixarenes. Attention is paid only to those compounds possessing [1n] metacyclophane structures [119].

Thus one step methods for synthesis of the calixarene was most common. Subsequently linear stepwise methods and new convergent methods for calixarene synthesis were developed.

2.13.3 Introduction of Functional Groups

Calixarene proved to be promising complexing ligands for metals only on introduction of functional groups in their upper or lower rims. Several reactions involved substitution in lower rim. Ester formation of monofunctional reagents was one such example particular for base catalysed reaction while upper rim substitution was common for acid catalysed reactions. The acetate derivative was most commonly used in complexation. Such compound was more soluble with low melting point. With excess of acetylation agent the OH-groups were converted into ester group. Acetylation converted calix(4) arene into partial cone conformation (XLVIII). Aluminium trichloride catalysed and sodium hydride induced arylation of calix(4) areane, p-tert butyl calix(4) arene and p-allyl calix(4) arene using benzoyl chlorides with various parasubstituents [120] were thus obtained.

The ratio of 1-3 alternate/cone conformer was dependent on parasubstituent of the arylating agent. Low temperature and more arylating agents favoured

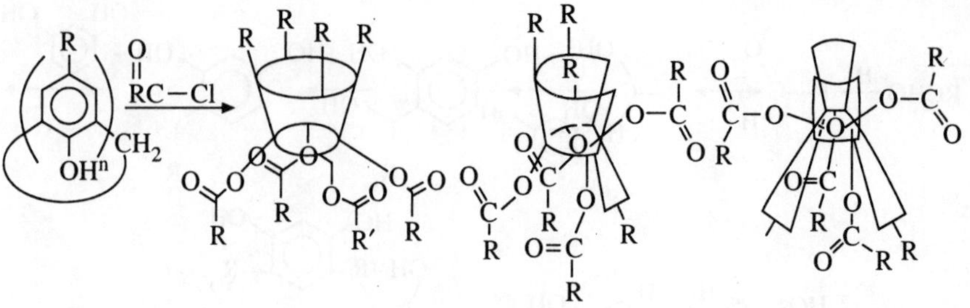

(XLVIII)

cone conformation (XLVIII). The best reaction was with sulphuric acid catalysed reaction of acetic anhydride and p-tert butyl calix(n) arene. Tetracetate derivative melted at 383°C [121, 122].

The larger calixarenes when acid catalysed on acetylation gave hexaetoxy and octaetoxy compounds from calix(6) arene and calix(8) arene. The arylation of calix(4) arene give cone, partial cone and 1, 3 alternate form (XLVIII).

2.13.4 Ether Formation with Monofunctional Reagents

Etherification could be effected irrespective of ring size if reactive reagents were used. Alkyl ether derivative was formed by treating calixarene with alkyl halide in tetrahydro furane or dimethylformamide. In the presence of sodium hydroxide the methyl, ethyl, allyl and pentyl ethers have been also prepared p-tert butyl calix(4) arene when treated with dimethylsulphate in the presence of BaO-Ba $(OH)_2$ in DMF gave trimethylether. By both esterification and etherification with reagents addition a new functional groups could be introduced into calixarenes, e.g. 2-4 dinitrophenyl ether or p-tert butyl calix(8) arene formed by reaction of p-tert butyl calix(8) arene with 2-4 dinitrochloro benzene in pyridine solution [123-124]. This is shown in structure (XLIX).

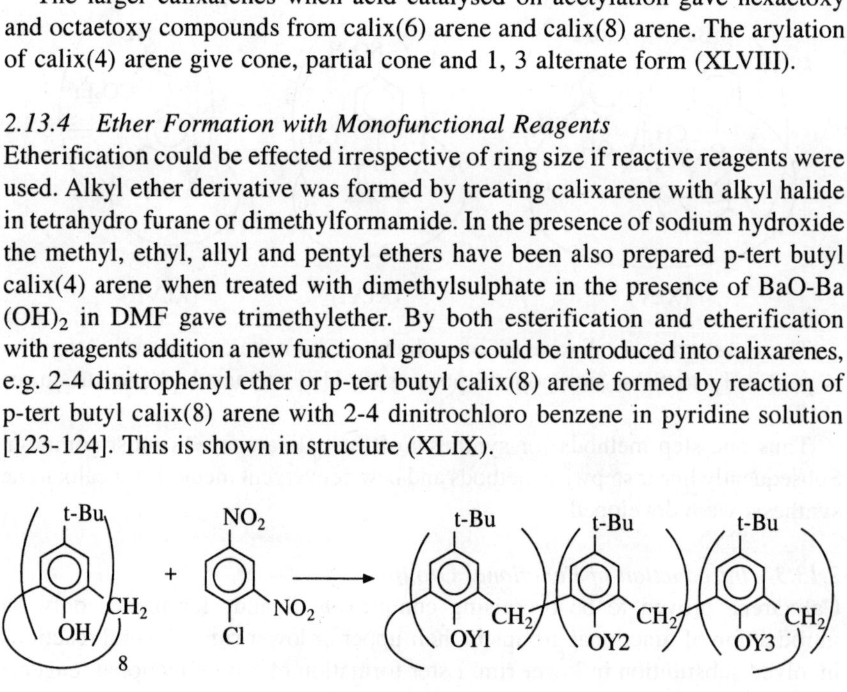

(XLIX)

The water soluble calixarene was obtained by treating p-tert butyl calix(4) arene with tert butyl α-bromoacetate and sodium hydroxide in THF solution to give tetra ester with 90% yield. The reaction of calix(4) arene with 3, 5 dinitrobenzoyl chloride in aluminium chloride to form diester was used to lower rim functionalised calix arenes [125-126]. In this structure the substituents were n = 4, 6 or 8, R_2 = t-butyl, t-octyl or hydrogen while R_1 = H, CH_2COOCH_3, $CH_2COOC_2H_5$, or $CH_2CONHBu$. These were all lower rim functionalised calixarenes as shown in (L) and (LI).

(L) structure with R₂, OR₁, CH₂, n

(LI) structure with t-Bu, OCH₂COOCH₃, 6

The compound hexa-acetyl calix(6) arene for instance had been extensively used in solvent extraction of thorium, chromium, manganese, iron, cobalt, palladium, copper, bismuth, thallium and lead.

2.13.5 Reactions at Upper Rim of Calixarene

It was interesting to note that the p-substituent in calixarene, viz. p-tertiary butyl, p-tertoctyl or p-tertiary pentyl group of the phenolic ring could be easily removed by reverse Friedel-Crafts reaction. This was called dealkylation of p-alkyl calixarenes. Removal from cyclic compound was relatively easier than such removal from linear oligomer. The unsubstituted para position was quite useful for subsequent work in synthesis (structure LII).

(LII) reaction scheme: t-butyl calixarene → (AlCl₃, Toluene (phenol)) → p-H calixarene

Such dealkylation was carried out in phenol-toluene mixture. More yield was obtained in hexa and octamer. Usually t-butyl group para to free OH-group was released easily [127, 128]. The tert butylation of calixarene gives 65% yield (LII).

2.13.6 Electrophilic Substitution

This was done on p-H calixarene with aromatic substitution reaction. Nitration for instance was first attempted. A p-sulphato calix(6) arene was synthesised with 75% yield by treatment with sulphuric acid at 100°C. Single step synthesis involving simultaneous dealkylation and sulphuration was also possible. Starting with p-sulpha calix(n) arene as starting material p-nitro compound was obtained by treatment with nitric acid at 5°C. The p-H-calixarene was also used for subsequent introduction of groups. The yield for sulpho derivative compound was 78% while for the nitrito derivative yield was only 20% (structure LIII).

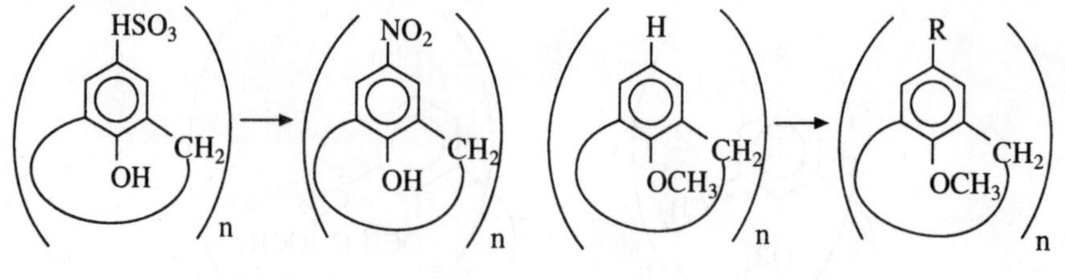

40 Analytical Chemistry of Macrocyclic and Supramolecular Compounds

(LIII) → (LIV)

Thus, substitution at the lower as well as at upper rim was also possible (LIV). Instead of direct nitration, usually route through sulphonation was preferred [129].

A better nitration yield was possible through treatment of calix(4) arene with nitric acid, acetic acid, benzene mixture or treatment of sodium hydroxide-nitrate in water followed by oxidation with dilute nitric acid [130]. The acetylation and benzoylation under Friedel-Crafts reaction using acetyl or benzoyl chloride and aluminium chloride gave ortho-substitute product and esters failed to [131] react further at para position. Methyl ether of calixarene (n=6, 8) gave 65% yield of p-acetyl calix(6) arene as shown here with. This was true with p-acetyl calix(8) arene. These compound were water soluble. Thus, calixarene ethers were better for electrophillic substitution than parent calixarene. The several reactions like halogenation involving bromination was carried out.

A great deal of interest arose in introduction of the aryl group in p-position but had limited success and applicability in analysis (structures LV, LVI, LVII, LVIII). Many problems encountered in eletrophillic substitution could be easily surmounted by use of para Claisen rearrangement to produce p-allyl calixarene p-toluenesulphonate was converted into various functionalised calix(4) arene. The route was also useful for large calixarenes but with poor yield [132]. The functionalisation via the p-Quinomethoides route was also explored by taking advantage of nucleophillic character of p-position of phenolates leading to "Mannich type reaction" [134]. The compounds were useful in complexation, where $Z = - N(CH_3)_2$, $-N(C_2H_5)_2$ and $-N(CH_2CH=CH_2)_2$ as (LVI).

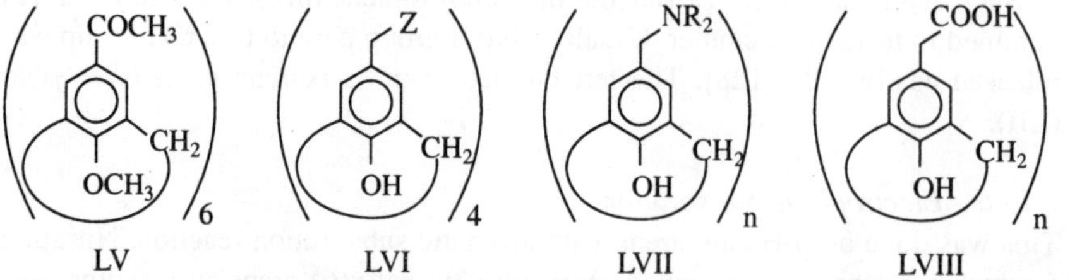

LV   LVI   LVII   LVIII

In fact no p-quinomethides were actually isolated [133] via this route but amino calixarene and carboxycalixarene were obtained. However, nucleophiles which were too weakly basic failed to effect reaction [134] in the p-chloro methylation route [133]. This proved as starting material for synthesis of p-alkyl calix(4)arene if R=H, CH_3, C_6H_5 etc.

2.14 Recent Developments in Synthesis of Calixarenes

In addition to conventional methods for the synthesis of supramolecular compounds number of new methods have come to forefront. They generally involved less number of steps in the synthesis or those of the reactions involving less number of reactants or one offering better yield of the end product. Most of these reactions were leading to formation of one starting material leading to the formation of calixarenes via different routes [135].

For instance a new synthetic receptor [136] was devised where in it permitted simultaneous binding of a cation with polyether crown ring and anion to its boron site. In fact many synthesis gave better understanding of some biological processes and indicated new routes [137]. The compounds with new ribbons and threads [138] contained H-bond dipole-dipole attractions and stackness of π-electrons. New compounds called 'Rotaxane' were developed by using template effect (structures LIX and LX). The rotaxane were one where in cyclic molecule wheel was threated on the separate axle with the axle then being capped at both ends with bulky groups to lock the wheel in place as in (LX):

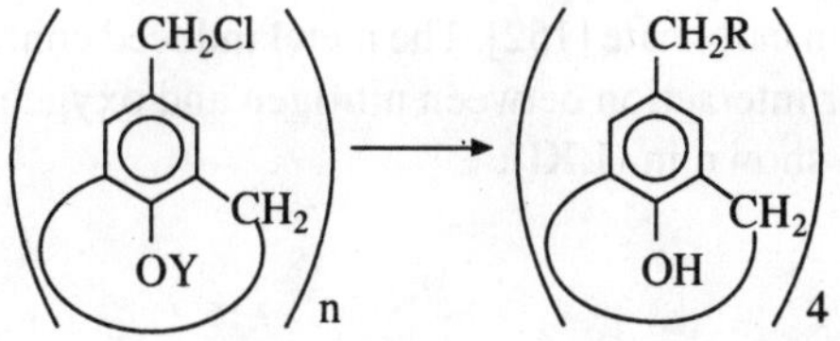

(LIX)

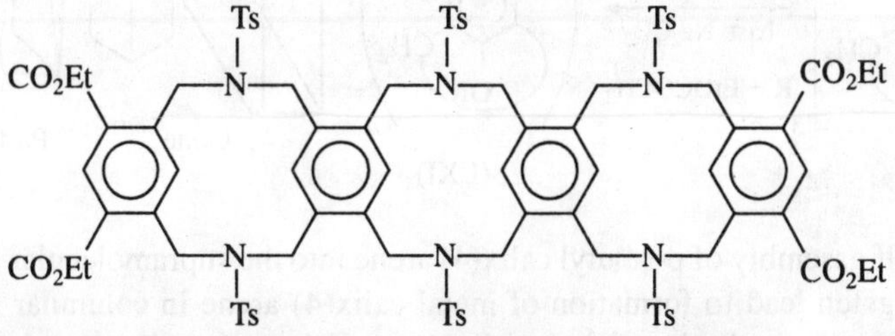

(LX)

The catalytic hydrogenation of calix(4) arene afforded the transderivative to produce [139] transmacrocyclic diether. The 1:1 p-butyl calix(8) arene complex of C-60 resulted due to encapsulation of aggregates of C-60 in double conformation [140]. The synthetic [141] supramolecular compounds involving non-covalent bonds were developed.

Just as chromogenic crown ethers were synthesised; similarly attempts were made to develop new chromogenic calixarene specially calix(8) arene [142]. Such compounds proved useful for detection of amines and metal ions in chemical analysis.

New chromogenic calixarenes were designed for detection of amines which proved potential diagnostic tool as reagents [143]. Apart from calix(8) arene a new type of calix(4) arene containing ester group was synthesised to obtain a compound containing quinone segment. The nitrated derivatives were used for complexing alkylamines [144]. Some time chiral calixarene [145] were obtained from methanone in three step synthesis. During synthesis of calix(6) resorcinarenes, a metacyclophane derived from resorcinol was accomplished by cyclocondensation of 2-alkyl resorcinol with 1, 3, 5 trioxene in ethanol-hydrochloric acid mixture to give tetra and hexamer [146] with brisk activity. In aza compounds of crown ether identical efforts were made to develop aza calixarenes. Hexahomotriaze calix(3) arene was obtained by reaction of 2-6 bis(hydroxyl methyl)$_4$ methyl phenol and benzylamine gave new product [147]. Simultaneously like crown imido [148] derivatives, metalimido calixarene complexes were also obtained. Heterocalixarenes featuring benzimidazol 2-one subunit were prepared from benzimidazol 2-one and 1-3 bis(bromo methyl) benzene [149].

Just as template effect was effective in synthesis of the crown ethers and cryptands it also proved to be of pronounced importance in calixarene conformation through host guest type of interaction [150]. Thus the selective synthesis of p-t-butyl calix(4) arene was effected by template effect in base where in 100% cone-selectivity was changed to 100% partial cone-selectivity in the presence of sodium carbonate or caesium carbonate [152]. The metal induced conformational changes in calix(4) arene for interaction between nitrogen and oxygen radical [151] were interesting and are shown in (LXI):

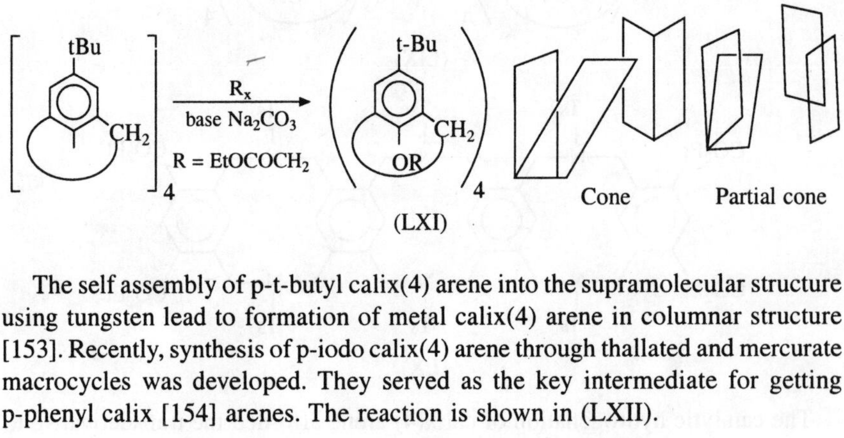

(LXI)

The self assembly of p-t-butyl calix(4) arene into the supramolecular structure using tungsten lead to formation of metal calix(4) arene in columnar structure [153]. Recently, synthesis of p-iodo calix(4) arene through thallated and mercurate macrocycles was developed. They served as the key intermediate for getting p-phenyl calix [154] arenes. The reaction is shown in (LXII).

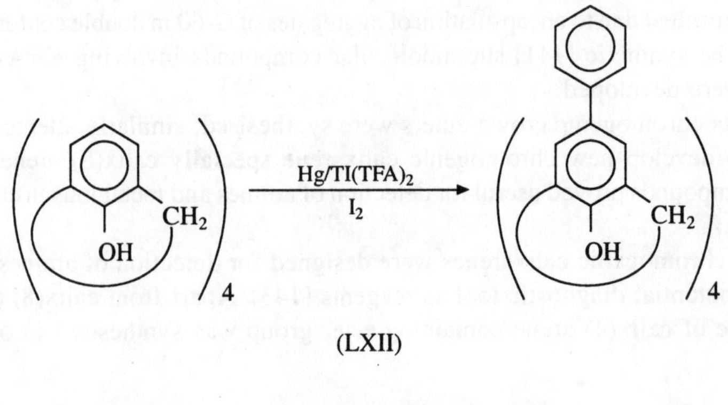

(LXII)

The different polyethylene glycol ditosylates in the presence of potassium carbonate in acetonitrile with p-tert butyl calix(4) arene led to the formation of double calix (4) arene crown ethers [155]. The last compound readily reacted with alkali metals-calix(4) arene based carcecomplexes were obtained from calix(4) arene by doped inclusion chemical modification [156]. Another new invotion in synthesis was of phosphorylated calixarene containing ($CH_2P(O)Ph_2$) entity using alkylating agent from t-butyl calix(4) arene [157]. It was interesting to note that synthesis of novel supramolecular structure by combining calix(4) and calix(6) arene with cyclodextrins or calix(4) resorcinarene or cyclotrivertrylene [158] was useful. A pronounced cavity effect was noticed in calix(4) arene in electrophillic aromatic substitution reactions between nonsubstituted benzene derivative and hydroxymethyl calix(4) arene [159]. A copolymer containing calix(4) arene disphenol was obtained containing variable amounts of macrocycle in the chain [160]. Unlike aza crown ethers, azobenzene modified 1, 3 calix(4) bis crown 1 (2, 2′) and 1 (4, 4′) were also obtained by the interaction of parent compound with one polyether crown-6 and one azobenzene modified crown-6, each attached to side of the calix(4) arene in 1, 3 from [161]. The higher homologus like calix(9, 10, 11, 12) arenes were also synthesised from cyclocondensation reaction mixture of linear phenolic oligomers in alkaline media [162]. An attmpt was made to get 1, 2 bridged calix(4) arene bis crowns in 1, 2 alternate conformation. Its stereochemistry depended on the nature of solvent used, length of polyether unit and base for deprotonation of calix(4) arene [163].

In such compounds two opposite p-position linked by aliphatic chain for its complexation properties [164]. A large number of calix(4) arene have been coupled to cavitands and cyclodextrin to form selectivity functionalised compounds chameleons [165]. Amino methyl calix(4) arene having properties of capturing insects by the molecular venus fly trap was obtained from calix(4) arene, formaldehyde and dimethylamine in THF [166]. A binaphthyl chromogenic calix(4) arene was synthesised from indophenol to give bisnaphthyl calix (4) arene [167]. They may prove to be optically active.

In comparison to calix(n) arene which were usually obtained by base catalysed reaction, the acid catalysed calix(n) resorcinarene were less studied. An evidence was furnished for the incorporation of alkyl group in polyhydroxyaromatic cavity of such compounds [168]. Another attempt was made to include the acetyl chloline trimethyl ammonium moiety in calix(4) resorcinarene [170] by multiple cation π-interaction. An intra-complex electron transfer H-bonded calixarene-porphyrin system was noticed [169]. The arylation and arylmethylation of calix(6) arene was effected with p-substituted aryoyl chloride and arylmethyl halides. The p-substituent influenced conformation outcome of the reaction [171]. A new macrocavitand was obtained by reaction (LXIII).

Thus, calix(8) arene if capped with tetra methoxyl p-tetra chloromethyl calix(4) arene a new product was obtained with different characteristics [172] (LXIII). The 2-hydroethoxyl substituted calixarene was prepared by the reduction of corresponding ethyl calixaryl acetate lithium aluminium hydride. Thus, the suitable derivatives of calix (6, 8) arene had been obtained [173]. Molecular recognisation of hexaaquo transition metal ions was effected by supramolecular compounds

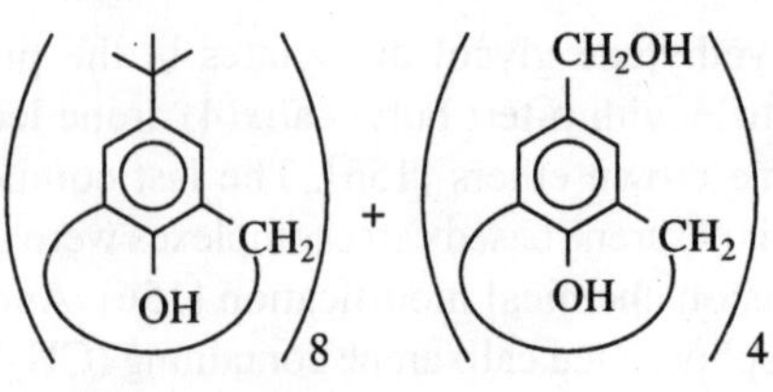

(LXIII)

[174]. Such metallo calixarenes with metallodi phosphine fragments in reaction taking place in cavity was studied [175].

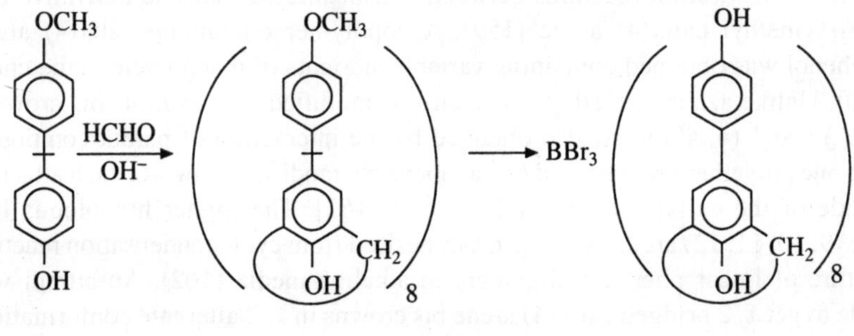

(LXIV)

A p-(1-4 hydroxyphenyl-1-methyl) calix(8) arene was obtained by the reaction as new product with deep cavity [176] to encapsulate metal cation. This is shown in (LXIV).

A simple synthesis of new supramolecular host containing the Trogers base moiety was reported with Cram's picrate extraction method where its cation binding properties were examined [177]. The calix(4) arene conformation properties were used for encapsulation of the metal ions. For instance 1, 3 alternate structure was best for entrapping potassium. The compounds was 1, 3 dialkoxy calix(4) arene crown 5 conformers. Crown 6 derivative was selective for caesium ion [178] during complexation reactions.

Conclusion
In summary we see that, in last two decades cyclodextrins and crown ethers have played a leading role in host guest chemistry. The former compound included organic molecules while latter compound bound metal cations and ammonium ions. Calixarene and their derivatives formed selective complexes with all kinds of organic or inorganic guest molecules. In fact calixarene could be called as the third host molecules next to cyclodextrins and crown ether [179]. They were metacyclophanes consisting of phenol and methylene bridge. Their method of preparation was simple, due to cyclic phenolic hydroxy group was useful for building up functionalised host molecule and they had many sized π-rich cavity for incorporating guest molecules compared to other two host molecule, calixarene could complex ionic as well as neutral guests [179]. These compounds provided new route for the synthesis of Buckiminister fullerene (C-60) in combination

with the calix(8) arene. The self assembly in calix(n) arene compounds had numerous merits over covalent bond driven organic synthesis [180]. The difficulty in characterisation was however a great hurdle in such studies but they had potential practical applications. Self assembly did not provide intermolecule species unless it was used in solid state [180].

In the next chapter an endeavour is made to provide a deep insight to knowledge of characterisation of these novel compounds. The most fascinating account will pertain to the metal complexation. As a matter of fact amongst the variety of organic ligands used in the analytical chemistry for complexation to quantitative analysis, these compounds found privileged place on account for their capacity to encapsulate metal cation in cavity in crown ethers or cryptands or to encapsulate not only metals but organic molecule in cavity of cone of calixarene compounds.

References

1. M. Hiraoka, Crown compounds their characteristics and applications, Elesvier Publications, Amsterdam (1982).
2. M. Hiraoka, Crown ethers and analogus compounds, Elesvier Publications, Amsterdam (1992).
3. G.W. Gokel, Crown Ethers and Cryptands Royal Chemical Society (1999).
4. S.M. Khopkar, Mayuri Gandhi, Crown Ethers and Cryptands in Solvent Extraction, Unpublished Monograph.
5. J.S. Bradshaw, Synthesis of multidentate compounds, Ed. R.M. Izatt and J.J. Christensen, Academic Press (1978).
6. C.J. Pedersen, Macrocyclic polyethers Dibenzo 18 Crown 6 polyether and Dicyclohexyl-18-Crown 6 polyethers Org. Synth. **52**, 66 1972).
7. L. Rossa, F. Vogtle, Synthesis of medio and macrocyclic compounds by high dilution principle techniques Top. Curr. Chem. **11B** 1 (1982).
8. S. Shinkai, Dynamic control of cation binding. In: Y. Ionoue, G.W. Gokel (Ed) Cation binding by Macrocycles-Marcel Dekker New York (1990) pp. 397.
9. C.J. Pedersen, J. Am. Chem. Soc. **89**, 7017 (1967).
10. E.L. Eliel, Sterochemistry of Carbon Compounds McGraw-Hill New York (1962).
11. H.K. Frensdorff, J. Am. Chem. Soc. **93**, 4684 (1971).
12. R.M. Izatt, D.P. Nelson, J.H. Rytting, B.L. Haymore, J.J. Christensen, J. Am. Chem. Soc. **93**, 1619 (1971).
13. M. Hiraoka, Yuki Gosei Kagaka (Japanese) **33**, 782 (1975).
14. M. Madan, D.J. Cram, Chem. Comm. 427 (1975).
15. E.P Kiba, M.G. Siegel, L.R. Sousa, G.D.Y. Sogan, D.J. Cram, J. Am. Chem. Soc. **95**, 2691 (1973).
16. J.S. Bradshaw, J. Hetrocyclic Chem. **10**, 1 (1973).
17. J.M. Timko, R.C. Helgeson, N. Newcomb, G.W. Gokel, D.J. Cram, J. Am. Chem. Soc. **96**, 7097 (1974).
18. C.J. Pedersen, H.K. Frensdorff, Angew Chem. Int. Ed. **11**, 16, (1972).
19. M. Chinquini; P. Tundo, Synthesis 516 (1976).
20. G.W. Gokel, D.M. Dishong, C.J. Diamond, J. Chem. Soc. Chem. Commun. 1053 (1980).
21. P.E. Stott, Chemical Eng. News **54** (37) 5 (1976).
22. T.H. Gouw, Chemical Eng. News **54** (44) 6 (1976).
23. P.E. Statt, Chemical Eng. News **54** (51) 5 (1976).

24. T.G. Egorova, O.P. Sheremet, A.V. Kopytin, Yu. N. Gusev V.N. Kobrina, Izv. sibold. Akad. Nauk SSSR Serkhim Nauk **9** 147 (1980), AA **40** 5C45 (1981).
25. L.F. Lindoy, Nature 356, 197 (1992).
26. A. Takadate, T. Masuda, C. Murata, T. Tanaka, S. Goya, Bull. Chem. Soc. Japan **68**, 3105 (1995).
27. J. Broos, I.K. Sakadinskaya, J.F.J. Engbersen, W. Verboon D.N. Reinhoudt, J. Chem. Soc. Chem. Comm. 255 (1995).
28. N. Fukada T. Ohtsu, M-Miwa, M. Mashino, Y. Takeda, Bull. Chem. Soc. Japan **69**, 1397 (1996).
29. A.M. Groth, I.F. Lindoy G.V. Mechan, J. Chem. Soc. Perkin Trans. **1** 1553 (1996).
30. B. Lakshmi, A. Prabhavathi Devi, M. Nagarajan, J. Chem. Soc. Perkin Trans **1**, 10, 1496 (1997).
31. A.M. Mathur, A.B. Scranton, Sepn Sci. and Tech. **32**, 285 (1997).
32. J.M. Lehn, Noble Prize Lecture 1987, J. Inc. Phenomena **6**, 351 (1988).
33. G.W. Gokel, S.H. Korzeniowski–Macrocyclic polyether Synthesis, Springer-Verlag (1982).
34. G.W. Gokel, Chem. Soc. Review **21**, 39 (1992).
35. T. Ueda, T. Adachi, K. Sumiya, T. Yoshida, J. Chem. Soc. Chem. Comm. 935 (1995).
36. J. Bradshw, R.M. Izatt, Acct chem. Res. **30**, 338 (1997).
37. S. Shinkai, K. Miyazaki, O. Manabe, J. Chem. Soc. Perk Tran, 1, **449** (1987).
38. K. Naemura, J. Fuji, K. Ogasahara, K. Hirose, Y. Tobe, J. Chem. Soc. Chem. Comm. 2749 (1996).
39. M. Montalti, R. Ballardini, L. Prodi, V. Belzani, J. Chem. Soc. Chem. Comm. 2011 (1996).
40. A.F. Reguillon. N. Dumont, B. Dunjic, M. Lemaire, Tetrahedran **53**, 1343 (1997).
41. K.E. Krakowiak, J. Incl. Phen. and Mol. recog. **29**, 283 (197).
42. B. Dietrich, Inclusion compounds J.L. Atwood, J.D. Davies, D.D. Mac. Nicol. (Ed). Academic Press London Vol. II p. 337 (1984).
43. B. Dietrich, J.M. Lehn, J.P. Sauvage, Tetrahedron letters 2885 (1969).
44. B. Dietrich, J.M. Lehn, J.P. Sauvage Tetrarhedron Letters 2889 (1969).
45. J.M. Lehn, A.H.Alberts, R.J. Annùniata, J. Am. Chem. Soc. **99**, 8502 (1977).
46. R.A. Bartsch, D.A. Babb, B.E. Knudsen, J. Incl. Phenom. **5**, 515 (1987).
47. K.E. Krakowiak, J.S. Bradshaw, N.K. Dalley, C.Y. Zhu, G.L. Yi., J.C. Curtis, D. Li. R.M. Izatt, J. Org. Chem. **57**, 3166 (1992).
48. K.E. Krakowiak, J.S. Bradshaw, R.M. Izatt, J. Heterocyclic Chem. **27**, 1011 (1990).
49. K.E. Krakowiak, J.S. Bradshaw, J. org. Chem **60** 7070 (1995).
50. K.E. Krakowiak, P.A. Krakowiak, J.S. Bradshaw, Tetrahedron **34**, 777 (1993).
51. K.E. Krakowiak, J.S. Bradshaw X.L. Koul, N.K. Dalley, Tetrahedron **51**, 1599 (1995).
52. K.E. Krakowiak, J.S. Bradshaw, X.L. Kou, N.K. Dalley, J. Hetrocydic chem. **82**, 931 (1995).
53. K.E. Krakowiak, J.S. Bradshaw, Israel J. Chem. **32**, 3 (1992).
54. K.E. Krakowiak, I.S. Bradshaw H. Art R.M. Izatt, Pure and applied chem. **65**, 5111 (1993).
55. K.E. Krakowiak, J.S. Bradshaw, R.M. Izatt, Syn Lett 611 (1993).
56. G.A. Melson-Coordination Chemistry of Macrocyclic Compounds Plenum, Press New York (1979)
57. R.W. Alder, R.B. Sessions, J.M. Mellor, M.F. Rowlins, J. Chem. Soc. Chem, Commun. 747 (1997).

58. P.H. Smith, M.E. Barr, J.R. Brainard, D.K. Ford, H-Freiser S. Muralidharan. S.D. Reilly, R.R. Ryan, L.A. Silks, W.H. Yu, J. Org. Chem. **58**, 7939 (1993).
59. K.E. Krakowiak, J.S. Bradshaw, R.M. Izatt, J. Hetrocyclic Chemistry **27**, 1011 (1994).
60. R.J.A. Janssen, L.F. Lindoy, O.A. Mathews, G.V. Meehan A.N. Soboler, A.H. White, J. Chem Soc. Chem. Comm. 735 (1995).
61. J.H.R. Tucker, H.B. Laurent, P. Marsau, S.W. Riley J.P. Desgrergne J. Chem. Soc. Chemical Commun. 1165 (1997).
62. D.M. Smith, C.W. Park, J.A. Ibers, Inorg, Chem. **36**, 3798 (1997).
63. G. Wipff, J. Coordinate Chem. **27**, 7 (1992).
64. G. Morgan, V. McKee, J. Nelson, J. Chem. Soc Chem. Com. 1649 (1995).
65. C.W. Park, D.M. Smith, M.A. Pell, J.A. Ibers, Inorg. Chem. **36**, 942 (1997).
66. A.K. Kauser, J. Chem. Soc. Perk. Trans. **5,** 227 (1953).
67. E. Weber, Kontakte (Merck) 38 (1983).
68. R.N. Green, Tetrahedron Letters 1793 (1972).
69. H.K. Frensdorff, J. Am. Chem. Sco. **93**, 600 (1971).
70. G. Dijkstra, W.H. Kruizinga, R.M. Kellogg, J. Org. Chem. **52**, 4230 (1987).
71. B.K.Vriesema, J. Buter, R.M. Kellogg, J. Org. chem. **49**, 110 (1984).
72. K.E. Krakowiak, J.S. Bradshaw, R.M. Izatt, D.J. Zamecka-Krakowiak, J. Org. Chem. **54**, 4061 (1989).
73. J.S. Bradshaw, K.E. Krakowiak, R.M. Izatt, J. Heterocyclic Chem. **26**, 1431 (1990).
74. J.S. Bradshaw, K.E. Krakowiak, R.M. Izatt, D.J.Z. Zamecka, Krakowiak, Tetrahedron Letters **31**, 1077 (1990).
75. K.E. Krakowiak, J.S. Bradshaw, R.M. Izatt, I. Org Chem. **55**, 3364 (1990).
76. K.E. Krakowiak, J.S. Bradshaw W. Jiang N.K. Dalley, G. Wu, R.M. Izatt, J. Org. Chem. **56**, 2675 (1991).
77. K.E. Krakowiak, J.S. Bradshaw, R.M. Izatt, Syn Letters 611 (1993).
78. C.J. Pedersen, M.H. Bromes, U.S. Patent 3847 949 (1974).
79. J.S. Bradshaw, J.Y. Hui, J.J Haymorel, J.J. Christensen, R.M. lzatt, J. Het Chem.
80. R.M. Izatt, C.Y. Zhu, N.K. Dalley, J.S. Curtis, X.L. Kou, J.S. Bradshaw, J. Phy. Org. Chem. **5**, 656 (1992).
81. R.M. Izatt, T.M. Wang, P. Huszthy, J.K. Hathaway, X.X. Zhang, J.C. Curtis, J.S. Bradshaw, C.Y. Zhu, J. Incl. Phenom and Mol. Recog. **17**, 157 (1994).
82. X.X. Zhang, R.M. Izatt, J.S. Bradshaw, P. Huszthy, Anal. Quim. **92**, 64 (1996).
83. X.X. Zhang, R.M. Izatt, P. Huszthy, J.S. Bradshw, Chem. Rev. Cited in J. Incl. Phen. **29**, 221 (1997).
84. J.S. Bradshaw, M.L. Colter, Y. Nakatsuji, N.O. Spencer M.F. Brown, R.M. Izatt, G. Arena, K. Tse B.E. Wilson, J.D. Lamb, N.K. Dalley, F.C. Morin, D.M. Grant, J. Org. Chem. **50**, 4865 (1985).
85. Y. Nakatsuji, J.S. Bradshaw, P.K. Tse, G. Arena, B.K. Wilson, N.K. Dalley and R.M. Izatt, J. Chem. Soc. Chem. Comm. 749 (1985).
86. J.S. Bradshaw, Y. Nakalsuji, P. Huszthy, B.E. Wilson, N.K. Dalley R.M. Izatt, J. heterocyclic Chem. **23**, 353 (1986).
87. R.M. Izatt, J.D. Lamb, G.E. Mass, R.E. Asay, J.S. Bradshaw J.J. Christensen, J. Amer, Chem Soc. **99**, 2365 (1977).
88. G.E. Mass, J.S. Bradshaw, R.M. Izatt, J.J. Christensen, J. Org. Chem. **42**, 3937 (1977).
89. M. Newcomb, G.W. Gokel, D.J. Cram, J. Am. Chem. Soc. **96**, 6810 (174).
90. J.S. Bradshaw, B.A. Jones, R.B. Nielsen, N.O. Spenser, P.K. Thompson, J. Heterocyclic Chem. **20**, 957 (1983).

91. J. Buter, R.M. Kellogg, F. Van Bolhuls, J. Chem. Soc. Chem. Comm. 282 (1990).
92. K.E. Krakowiak, J.S. Bradshaw, D.J. Zamecka Krakowiak, Chemical review **89**, 929 (1989).
93. J.S. Bradshaw, J. Incl Phenom. and Mol. Recog. 221 (1997).
94. J.F. Biernat, E. Luboch, A. Cygan, J. Cord. Chem. **27**, 215 (1992).
95. M. Manakata, L.P. Wu, M. Yamamoto, T.K. Sowa, M.K. Maekawa, J. Chem. Soc. Dalton Trans. 3215 (1995).
96. K.E. Krakowiak, J.S. Bradshaw, J. Heter. Chem. **32**, 1639 (1995).
97. F. Rabiet, F. Denat, R. Guildrd-Syn Comm. **27**, 929 (1997).
98. M. Kodama, Bull Chem. Soc. Japan **69**, 979 (1997)
99. G. Broggini, L. Garanti. G. Molteni, G. Zecchi, Tetrahedron **53**, 3005 (1997).
100. C. David Gutsche, Calixarenes, Royal Chemical Society Londan (1999).
101. Z. Asfari and J. Vicens, Tetrahedron Letters **29**, 2659 (1988).
102. A. Zinke, R. Kretz, E. Leggewie, K. Hossinger, Monatsch **83**, 1213 (1952).
103. A. Zinke, E. Ziegler, Ber B74, 1729 (1941). Ber **77**, 2644 (1944).
104. J.W. Cornforth P. D′ Arcyhart, G.A. Nicholls, R/J. W. Rees, J. A. Stock, Br. J. Pharmscol, **10**, 73 (1955).
105. C.D Gutsche, M. Iqbal, D. Stewart, J. Org. Chem. **51**, 742 (1986).
106. G.D. Gutlsche, M. Iqbal Organic synthesis in Calixarenes, Royal Chemical Society London (1988).
107. C.D. Gutsche, D. Stewert, Calixarenes, Royal Chemical Society London (1999).
108. C.D. Gutsche, B. Dhawan, K.H. No. K. Muthukrishnan, J. Am. Chem. Soc. **100**, 189 (1984).
109. B. Dhawan, S.I. Chem. C.D. Gutshce, Makromol. Chem. **188**, 921 (1987).
110. S.R. Izatt, R.T. Hawkins, J.J. Christensen, R.M. Izatt, J. Am. Chem. Soc. **107**, 63 (1985).
111. F. Ullman, K. Brittner, Ber. **42**, 2539 (1909).
112. A Zinke, R. Kertz, E. Leggewie, K. Hossinger, Montash Chem. **83**, 1213 (1952).
113. N. Caro, Ber. **25**, 1939 (11892).
114. A.G.S. Hogberg, J. Am. Chem. Soc. **102**, 6046 (1980).
115. A.G.S. Hogberg, Ph.D. Thesis, Royal Inst. Tech. Stockholm (1977).
116. A.G.S. Hogberg. J. Org. Chem. **45**, 4498 (1980).
117. D.J. Cram, S. Karbach, H.E. Kim, C.B. Knobler, E.F. Marerick, J.L. Ericeson, R.C. Helgeson, J. Am. Chem. Soc. **110**, 2229 (1988).
118. R.F. Hunter, C. Turner, Chem. and Ind. 72 (1957).
119. M. Tashiro, A. Tsuge, 12th Int Conference of Macrocyclic Chem. Hiroshima Abstract p. 95 (1987).
120. I. Iqabal, T. Mangia Oico, C.D. Gutsche, Tetrahedron **43**, 4917 (1987).
121. V. Bocchi, D. Foina, R. Pochini, R. Ungaro, G.D. Anderetta, Tetrahedron **38**, 373 (1987).
122. C.D. Gutsche, B. Dhawan, J.A. Levine, K.H. No., L.J. Bauer Tetrahedron **39**, 409 (1983).
123. M.A. Mckervey, E.M Seward, G. Ferguson, B. Ruhl, S.J.Harris, J. Chem. Soc. Chem. Comm. 388 (1985).
124. M.A. Mckervey, E.M. Seward, J. Org. Chem. **51**, 3581 (1986).
125. G. Ferguson, B. Kaitner, M.A. McKerrey, E.M. Seward, J. Chem. Soc. Chem. Com. 584 (1987).
126. S.K. Chang, S.K. Kwon, 1. Cho, Chem. Letters 947 (1987).
127. S. Shinkai, S. Mori, T. Tsubaki, T. Sone, O. Manabe, Tetrahedron Letter **25**, 5315 (1984).
128. S. Shinkai, K. Araki, T. Tsubaki, T. Arimura, O. Manabe, J. Chem. Soc. Perkin-1; 2297 (1987).

129. S. Shinkai, T. Tsubaki, T. Sone, O. Manabe, Tetrahedron Letters **26**, 3343 (1985).
130. K. No., Y. Noh, Bull Korean Chem. Soc. **7**, 314 (1986).
131. C.D. Gutsche, L.G. Lin, Tetrahedron **42**, 1633 (1986).
132. C.D. Gutsche, K.C. Nam, J. Am. Chem. Soc. **110**, 6153 (1988).
133. A. Arduini, A. Casnati, A. Pochini, R. Ungaro, 5th Int. Sympo Inclusion Phenomena pH13 (1988).
134. C.D. Gutsche, J. Alam, Tetrahedron **44**, 4689 (1988).
135. C.D. Gutsche, B. Dhawan, K.H. No, R Muthukrishnan, J. Am. Chem. Soc. **103**, 3782 (1981).
136. L.F. Lindoy, Nature **356**, 197 (1992).
137. P. Linnane, S. Shinkai, Chem. and Ind. 811 (1994).
138. L.L. Lindoy, Nature **376**, 293 (1995).
139. F. Grynszpan, S.E. Biali, J. Chem. Comm. 195 (1996).
140. C.L. Raston, J.L. Atwood, P.J. Nichols, B.N. Sudria, J. Chem. Soc. Chem. Comm. 2615 (1990).
141. J.F. Stoddat, Account. Chem. Res. **30**, 393 (1997).
142. H.M. Chawla, K. Srinivas, Tetrahedron Letter **35**, 2925 (1994).
143. H.M. Chawla K. Srinivas, J. Chem. Soc. Chem. Comm. **259**, 3 (1994).
144. Q.Y. Zheng, C.F. Chen, Z.T. Huang, Tetrahedron **53**, 10345 (1997).
145. J. Jauch, V. Schuring, Tetrahedron Assymmelry **8**, 169 (1997).
146. H. Konish, K. Ohata, O. Morikawa, K. Kobayashi, J. Chem. Soc. Chem. Comm. 309 (1995),
147. H. Takemura, K. Yashimura, 1.U Khan, T. Shinmyozu I. Inazu, Teterahedron letters **33**, 5775 (1992).
148. V.C. Gibson, C. Redshanw, W. Clegg, M.R.J. Elsegood, J. Chem, Soc. Chem. Com. 2371 (1995).
149. E. Weber, J. Trepte, K. Gloe, M. Piel, M, Czugler V.C. Kravstov, Y.A. Simonove, J. Lipkowski, E.V. Ganin, J. Chem. Soc. Perkin Trans. II 2359 (1996).
150. T. Arimura, M. Kubota, K. Araki, S. Shinkai, T. Matsuda, Tetrahedron Letters **30**, 2563 (1989).
151. N. Oradi, K. Araki, R. Nakamura, H. Otasuka, S. Shinkai, J. Chem. Soc. Chem. Comm. (1989).
152. K. Iwamoto, K. Fujimoto, T. Malsuda, S. Shinkai, Tetrahedron letters **31**, 7169 (1990).
153. A.Z. Gerosa, E. Solavi, L Giannin, C. Floriani A.C. Villa, C. Rizzoli, J. Chem. Comm. 119 (1996).
154. A. Arduini, A. Pochini, A. Rizzi, A.R. Sicuri R. Ungaro, Tetrahedron Letters **31**, 4653 (1990).
155. Z. Asfari, J. Weiss, S. Pappaelavdo, J. Vicens, Pure and Applied Chem. **65**, 585 (1993).
156. A.M.A. Wageninoen, J.P.M. Daynharen, W. Werboom D.N. Reinhoudt, J. Chem. Soc. Chem. Comm. 1941 (1995).
157. C. Dieleman, C. Loeber D. Matt, A.D. Cian, J. Fischer, J. Chem. Soc. Dalton. Trans. 3097 (1995).
158. A.M.A. Van Wageningen, W. Verboon, D.N. Reinhoudt, Pure and Applied Chem. **68**, 1273 (1996).
159. O. Struck, J.P.M. Duynhoren, W. Verboon, S. Harkema, D.N. Remhoudt, J. Chem. Soc. Chem. Comm. 1517 (1996).
160. A. Dondoni, C. Ghiglione, A. Marra, M. Scoponi, Chemical Comm. **7** 673 (1997).
161. M. Saadioui, Z. Asfari, J, Vicens, Tetrahedron letters, **38**, 1187 (1997).
162. I. Dumazet, J.B.R. Vains, R. Lamartine, Synt Comm. **27**, 2547 (1997).

163. A. Aruduini L. Domiano, A. Pochini, A Sechi R. Ungaro, F. Ugozzoli, O. Struck, W. Verboom, D.N. Reinhoudt, Tetrahedron **53**, 3767 (1997).
164. V. Bohmer. W. Vogt, Pure and Applied Chem. **65**, 403 (1997).
165. E.V. Dienst, W.I. Bakker, J.F.J. Engbersen, W. Verboom D.N. Reinhoudt, Pure and Applied Chem. **65**, 387 (1993).
166. J.L. Atwood, G.W. Orr, S.G. Bott, K.D. Robinson, Angewante, Chemi, **32**, 1093 (1993).
167. Y. Kubo, S. Maruyama, N. Ohhora, M. Nakamura, S. Tokita, J. Chem. Soc. Chem. Comm. 1727 (1995).
168. Y.K. Kuchi, Y. Aoyama, Bull Chem. Soc. Japan **69**, 217 (1996).
169. T. Arimura, C.T. Brown, S.L. Springs, J.L. Sessler, J. Chem. Soc. Chem. Comm. 2293 (1996).
170. K. Murayama K. Aoki, Chem. Comm. 119 (1997).
171. S.K. Reddy, C.D. Gutsche, J. Org. Chem. **57**, 3160 (1992).
172. A. Aruduini, A. Pochini, A. Secchi, R. Ungaro, J. Chem. Soc. Chem. Comm. 879 (1995).
173. J.K. Moran. E.M. Georglier A.T Yardonov, J.M. Mague D.M. Roundhault, Org. Chem. **59**, 5990 (1994).
174. U. Auerbach, O Stockheim, T. Weyer muller, K. Wieghardt, B. Nuber, Angew Chem. **32**, 714 (1993).
175. C. Wieser, D. Matt, J. Fischer, A. Harriman, J. Chem. Soc. Dalton Trans. **14**, 2391 (1997).
176. C.H. Tung, H.F. Ji, J. Chem. Soc, Perkin trans II, 185 (1997).
177. A. Manjula, M. Nagrajan, Tetrahedron. **53**, 11859 (1997).
178. R. Ungaro, A. Arunduini, A. Casnati, A. Pochini F. Ugozzoli, Pure and Applied Chem. **68**, 1213 (1996).
179. M. Takeshita, S. Shinkai, Bull Chem. Soc. Japan **68**, 1088 (1995).
180. D.S. Lawrance, T. Jirang, M. Levett, Chemical Review **95**, 2229 (1995).

3
Characterisation and Metal Complexation

3.1 Introduction

The utility of any complex in analytical chemistry depends upon the information of its characteristics. Thus, the characterisation of metal complexes is of utmost importance. Many such complexes were studied for metals which included β-diketones, oxines and oximes, naphthols, azonaphthols and thiosubstituted derivative of various ligands. In the chemistry of macrocyclic and supramolecular compounds on account of their interesting stereochemistry and conformations, such study attains greater importance from the view point of the characterisation of not only of the ligands but also of the metal complexes.

Therefore, this chapter attempts to trace early work dealing with thermodynamic data regarding the solution chemistry of such complexes. The factors influencing mechanism of complexation and nature of the complex formed are studied. The discovery of macrocyclic effect led to explanation of several anomalies in the chemical reaction. It considered entropy and enthalpy of complexes and reactions. And also included the study of ionic size of the metal, stability constant of complexes in aqueous solution and analogues parameters influencing complex formation.

The complexation of cryptands with metals was also thoroughly investigated. They were mainly characterised by analytical absorption spectroscopy. The same technique was extended to the characteristics of complexes with calixarenes. The technique of UV, visible and IR spectroscopy threw a light on the nature of bonding. In addition to absorption spectroscopy, the NMR methods were dependable and rapid to elucidate the structure of these complexes. The ESR technique was also resorted to understand the nature of metal ligand bonding and existance of free radical. Apart from proton NMR, C^{13} NMR was also extensively used. Large amount of data on characterisation of metal complexes with crown ethers, aza and thia crown ethers, and calixarene, collected.

The study of metal complexes with calixarene gave fascinating information on stereochemistry and conformation. The phenomena of ion transport, extent of extraction, chromogenic behaviour was well understood. Such complexes were characterised by NMR spectroscopie methods. Since calixarene gave three dimensional structures, such studies with NMR proved beneficial to understand nature of bonding.

Mass spectrometry was less frequently used for the characterisation of complexes as well as new ligands. Information on triple cation binding and identification of such complexes was done by mass spectrometry. Mostly alkali metals and few lanthanide complexes were studied by this method. Tandem mass spectrometry was utilised as the complimentary tool.

Studies involving UV, visible IR spectroscopy, ESR, NMR and mass spectrometry gave clear insight on nature of host guest chemistry. Much information on cavity size of crown ether, or annular space of basket like compounds such as calixarene or three dimensional networking in the case of complexes with cryptands was procured by judicious choice of these techniques.

A clear insight of the synthesis of these novel ligands, their characterisation and mechanism of metal complexation and elucidation of structure by X-ray, TGA, DTA etc is of utmost importance to develop newer methods of quantitative analysis in analytical chemistry.

Crown ethers, cryptands and calixarenes came to forefront mainly because of their ability to form complexes not only with metals but also with anions and organic neutral molecules. The primary reason for such complexation was ease with which the conformation of these compounds could be altered. It was possible to change the cavity size in calixarenes or to introduce the functional group at appropriate position to facilitate the complexation of these hydrophobic ligands specially in the organic solvent. Therefore, this chapter endeavours to consider various parameters which govern the process of complexation with these compounds. On account of their special application in analytical chemistry, a due emphasis is led on the understanding of metal complexation and characterisation of the ligands.

3.2 Metal Complexes with Crown Ethers

Crown complexes are unusually stable. The binding species, viz. metal, ammonium or organic ionic compounds were guest attached to crown compound which acted as host due to the presence of donor atoms in its ring: donor being oxygen, nitrogen or sulpur. Such complexation depended upon relative size of the diameter of the cavity of crown compounds as well as the diameter of the cation. The charge of cation, type of the donor atom and basicity of ligands influenced the complexation process. Such complexes were soluble in various organic solvents specially nonpolar solvents due to the presence of hydrophobic group present on crown. These complexes were formed by ion-dipole interaction between cation and negatively charged oxygen atoms on the polyether ring. The structure had

DB18C6 + M^{n+} ⇌ M^{n+}

(I)   (II)

cation trapped in a crown ether, cavity size of which had matched with ionic size or diameter of cation, one pairs of electrons of oxygen atom were directed inside the ring. This oxygen atom was located at equidistance from cation. A regular arrangement conformed enthropy on crown ether compounds with organic ionic compounds as primary ammonium salts or aromatic dizonium salts and molecular complexes with bromine. The HBr and oxonium ions were also identified. The reaction observed by Pedersen is shown in (I) and (II).

Thus a structure was made of cation $[M^{n+}]$ trapped in the crown ether cavity that was of same size as of cationic diameter of the ion. The stability of such complexes ascertained in terms of stability constant or complexation constant. A maximum complexation occurred if ionic size could fit well the cavity of crown ether. The solubility of such complexes in nonpolor solvents was attributed to the presence of hydrophobic group present on crown ether ring, specially outside the ring. The counter anions like picrate were immensely active as they existed in solvent as 'naked' anions incapable of being solvated by solvents.

3.2.1 Early Work

Several reports have appeared on properties of crown ethers [1, 2] by Pedersen, Hiraoka [3] and Takagi [4] but Lehn [5] ably characterised cryptands while Christensen [6] reviewed cyclic polythioethers and polyamines by providing useful thermodynamic data [7, 8] pertaining to solution chemistry of such complexes.

3.2.2 Mechanism of Complexation

The complex formation usually proceeds by binding of cations caused by electrostatic ion-dipole interaction between cations and negatively charged oxygen-donor atoms arranged regularly in the polyether ring. The UV, visible spectra showed that the aromatic crown ethers absorbed at 275 nm while alicyclic crown ether absorbed at 220 nm. On complexation of aromatic crown ether there was a bathochromic shift by 6 nm towards red region.

Such complexes could be characterised by IR, NMR, X-ray and mass spectrometry. The electroanalytical methods including conductivity, polarography were also used on extensive scale. Such techniques would be discussed in subsequent sections. For instance solvent extraction of picrate complexes of alkali metals [9] with 18C6 or DB18C6 gave definite evidence of complexation. It also facilitated quantitative analysis of the metal concerned. The factors which usually affected the stability of the [10] complex include following parameters:

(a) The size of cation and cavity of crown ether (Table 3.2.1)
(b) Number of oxygen atoms in ring (more oxygen atom gave better stability)
(c) Arrangement of oxygen atom (plannar structure gave more stability)
(d) Symmetrical arrangement of oxygen atom (more stable was the complex)
(e) Basicity of donor atom like oxygen (more basic better was the stability of complex)
(f) Steric hindrance in polyether ring (it was not conducive for stability)
(g) Electrical charge of the cation (of utmost importance in stabilising)

Table 3.2.1 Cavity size of crown ether and cryptands [10]

Compound	Cavity Diameter (Å)	Compound	Cavity Diameter (Å)
14C4	1.2–1.5	Cryptand-1, 1, 1	1.0
15C5	1.7–2.2	Cryptand-2, 1, 1	1.6
18C6	2.6–3.2	Cryptand-2, 2, 1	2.2
21C7	3.4–4.3	Cryptand-2, 2, 2	2.8
24C8	> 4.0	Cryptand-3, 2, 2	3.6
		Cryptand-3, 3, 3	3.6
		Cryptand-3, 3, 2	4.8

3.2.3 Nature of Complex

Usually a 1:1 complex was formed when a cation diameter had very well fitted with the cavity size of the complex, it became stable but if diameter was larger than cavity size then the complex was not stable. Further a 2:1 or 3:2 complex formed as a sandwich structure, as cation was located slightly apart from the plane in which oxygen donors on the crown ring were arranged. When a cationic diameter was smaller than the cavity it gave a configuration where oxygen atom was located at the shortest distance from cation to give 1:2 complex. The determination of stability constant was possible by various methods that included conductance measurement [11–14] titration calorimetry [15–18] and NMR spectroscopy [19]. The last method depended upon the chemical shift or the width difference between complexed and uncomplexed species and was used to detect ligand or cation. One could also use ion selective electrodes [20, 21]. The hole size concept ·vas important for ion fitting in cavity, e.g. 15C5 had hole size 1.7–2.2 Å. Sodium with diameter 1.9 Å formed readily complex, but hole size concept would be most true for rigid system. 18C6 binded potassium more strongly with stability constant 6.08. The hole size concept was better represented by cryptands, e.g. cryptand 2, 2, 1 with cavity size 2.2 Å fitted sodium with diameter 1.7–2.2 Å while cryptands 2, 2, 2 with cavity size 2.8 Å fitted potassium with same size, with stability constant of 10.0.

3.2.4 Macrocyclic Effect

It is important to know macrocyclic effect' [22] at this stage from the following examples: Potassium formed complex with compound (IV) which was less stable as compared to complex of potassium with compounds (III).

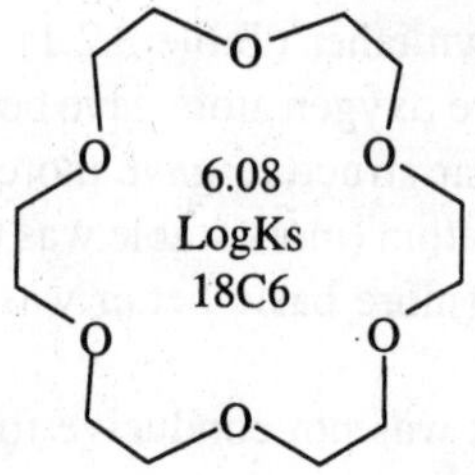

6.08 LogKs 18C6

(III)

2.3 LogKs

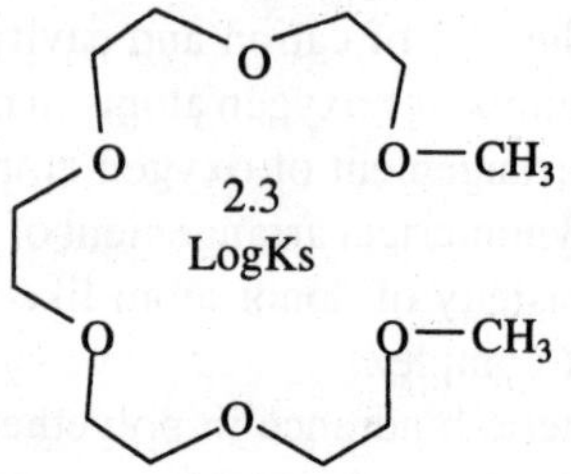

(IV)

The macrocyclic effect of stabilisation was 6000 fold for (III). Similarly, cryptand 2, 2, 2 complexes of potassium (V) with log K_f = 9.75 was stable, in comparison to the open chain compound (VI) with log K_f = 4.8. This showed marocyclic effect is extended to 90.000 fold excess.

(V)   (VI)

3.2.5 Entropy Consideration

The enthalpic terms was important. Inductive stabilisation and steric congestion were enthalpic. Such interpretation predominated in host guest chemistry. A placement of bezo group in 15C5 decreased sodium binding as 1 sp^2 hybridised oxygen atom were less basic than sp^3 hybridised oxygen atom. DC 18C6 was better than DB 18C6 as complexing ligand as former had increased hydrophobicity. Usually nitrogen-exhibited decrease in affinity for s-block metals. As a rule the crown ethers were more flexible than cryptands as former were planner while latter were three dimensional structures. The complexation rate was faster in crown ethers than in cryptands.

3.2.6 Complexes with Organic Compounds

Apart from metals organic molecules could also form complexes. Guanidinium was complexed with 21C7. A diazonium ion could be complexed by circular and polar cavity of crown ether. A hydronium ion complex was formed with DC18C6 in presence of perchlorate anion. These were all stoichiometric complexes where positively charged species had interacted by a pole-dipole mechanism or by H-bonding. The molecular inclusion complexes were those whose macrocycles peak in the lattice with voids was filled by smaller molecular species. Anions could be trapped as cryptands were better. The anions were trapped in cryptands by protonation. Ionic diameter of important cations which were generally used for the complexation are worth considering. Table 3.2.2 gives such data for important metals. Stability constants for metal complexes are given in Tables 3.2.3 and 3.2.4.

Let us consider large crown complexes of cation, e.g. sodium complexed with 12C4 with cavity of 1.5 Å, while sodium diameter was 2 Å. The 12C4 provided 4-donor group, but sodium preferred six positions as the coordination number enabling solvation. In B15C5, it was coordinated to six oxygen atoms, five from crown and one from water. If large cation was dumped on small crown a sandwich structure originated—18C6 and caesium with thiocyanate as counter anion. Now consider crown complexes of cations which were too small. It was also possible that 18C6 and sodium can form complex or sodium with 24C8 or potassium with

Table 3.2.2 Ionic size of some metals

Metal	Size (Å)	Metal	Size (Å)
Li	1.2	Zn	1.48
Na	1.90	Cd	1.94
K	2.66	Hg^+	1.52
Rb	2.96	Hg^{++}	2.20
Cs	3.34	Fe	1.28
Ca	1.98	Co	1.48
Sr	2.26	Ni	1.38
Ba	2.70	Pb	2.40

30C10 formed complexes. Such structures were investigated by X-ray diffraction studies. Generally, transition metal ions formed stable complexes with ligands having N- or S-atoms as donors which were soft bases. The affinity of crown ethers to soft base was weak due to the presence of O-atom as the donor. Further transition metals had large solvation energy.

Table 3.2.3 Stability constants of some complexes in water [13, 17]

Crowns	Na	K	Cs	Ag	Tl	Sr	Hg^{2+}
15C5	0.70	0.74	0.80	0.94	1.23	1.95	1.68
18C6	0.80	2.03	0.99	1.50	2.27	2.72	2.42
DC18C6	0.69	1.63	–	1.59	–	–	4.43

Table 3.2.4 Stability constant of complex of potassium in methanol or water [2, 10, 57]

Crown ether	K_f-constant	Cryptands	K_f-constant
DC14C4	1.30	Cryptand-2, 1, 1	< 2.6
18C6	6.10	Cryptand-2, 2,1	3.95
DC18C6	6.01	Cryptand-2, 2, 2	5.4
DB18C6	5.00	Cryptand-3, 2, 2	2.2
24C8	3.48	Cryptand-3, 3, 2	< 2.0
DB24C8	3.49	Cryptand-3, 3, 3	< 2.0
21C7	4.40	Cryptand-2B 2, 2	4.0
DB21C7	4.30		
DB30C10	4.60		

Usually the stability constant indicated the degree of the complex formation such stability of resulting complex in solution is important [23-25]. The stability constant was measured in water and methanol. The K_f values increased for small cations in low polar solvents as there was less competition between solvation of the solvent and coordination of the crown ring resulting from weak solvation ability of the low polar solvents to the cation. As regards thermochemical aspect such as ΔH, ΔG, ΔS was concerned all of them changed on complexation. If calorific value was large and entropy change was there, more stable, complex were formed [26-28].

Table 3.2.5 shows significant variation in ΔH (enthalpy change), ΔS (entropy change), ΔG (free energy) calculated from value of K_f. The calorific value and entropy change were of significance during complexation reaction [29-30]. The reaction of s-block metals proceeded rapidly but in transition metal complexes the rate of dissociation of the complex decreased rapidly when the ligands were multidentate and were macrocyclic ligands [31-32].

Table 3.2.5 Thermal constants on complexation for potassium [6]

Crown ether	log K_f	ΔH (kcal/mole)	ΔS (cal/deg mol)	Solvent comp. % (wt. methanol)
B15C5	1.20	−1.80	−0.50	20
18C6	4.30	−9.68	−12.70	70
DB24C8	2.42	−8.54	−17.60	70
DB27C9	2.86	−9.50	−18.80	70

The crown complexes were soluble in solvents with high dielectric constants, but solubility depended also on the polarity of the solvent. Medium type polarised solvents supplied effective solvation energy. Usually aromatic hydrocarbons, halogenated, hydrocarbon, ethers, amides DMSO were good solvents for crown complexes. In that context aliphatic saturated hydrocarbons were useless. The solubility also depended upon anions involved, it increased with increasing diameter of the anion. Hard anions like fluoride sulphate were useless but soft anions like iodide, thiocyanate, picrate were good for dissolution. Crown ethers and cryptands increased the solubility of sodium, potassium, caesium in ethers and they were solubilised by crown compounds in benzene, toluene [33-36]. It was interesting to note some striking features of metals. The ion dipole interaction between donor atom and cation declined with negative charge of donor and complex forming tendency decreased. However, tendency for complexation of transition metals increased significantly [20].

3.3 Metal Complexes With Cryptands

These were steric cage type macrobicyclic crown compound with two-nitrogen atoms on the bridge. The complexes were termed as cryptates. They were more stable than crown complexes [37-39]. The steric configuration of cryptands with bridge nitrogen atom were exo-exo (out-out), exo-endo (out-in) and endo-endo (in-in) type [40]. These are shown in structures (VII) to (IX).

The bicyclic cryptand complexes were 1:1 type (endo-endocent) in cryptand 2, 2, 2 while forming complexes with alkali metals sometimes cavity size was enlarged by torsion of polyether chain between two N-atom. The anion cryptates were also possible to form with anions like halides, nitrate chlorate etc. In electroanalytical studies of the cryptate complex, the reduction potential of cryptate was negative as compared to metal ions, the dissociation of complex proceeded during reaction of bielectron reaction [41-44].

The topological inferences by cryptands in solvation reaction was of significance for alkali and alkaline earth metals with small ionic size as steric binding promoted solvation, while for transition elements like gold covalent bonds with N donor participated in complex formation. On account of cage type structure cryptates were stable in enthalpy and ΔH increased with increasing diameter of complexing cation [45-50]. As a rule thermodynamically cryptates were more stable than the crown ether complexes. Cryptands promoted solubilisation of s-block metals in organic solvents like. THF, ether, alkylamines. Crown ether complexation was more faster than cryptand [33] complexes. Similarly, as compared to crown ethers, cryptands were more effective in complexing anions by protonation, although crown ether and cryptand complexation appear parallel each other. However the hole and ion cavity size concept was applicable to cryptates also, e.g. cryptand 222 had cavity of 2.8 Å matching with ionic diameter of potassium, viz. 2.60 Å hence it formed complex with it, in contrast, sodium did not form complex with ionic size of 1.95 Å. The macro-tricyclic cryptands were ditopic receptors with two crown ether rings bound together.

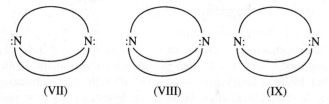

(VII) (VIII) (IX)

3.4 Characterisation of Complexes by Absorption Spectroscopy

Amongst spectral methods the extensively used technique was UV-visible and IR spectroscopy for the characterisation of metal complexes of crown ethers, cryptands and calixarenes. In fact large number of chromogenic crown ethers were developed and studied by UV-visible spectral methods.

The summary of UV-visible IR spectral methods generally used for the characterisation of various metal complexes are summarised in Table 3.4.1. In few instance laser Raman spectroscopy was also used for such characterisation [51, 52]. This was specially true of complexes of lithium with various crown ethers [54] along with infrared spectroscopy [55]. It was possible to separate charge transfer complexes by colour measurements [56]. All aza crown complexes were invariably studied by visible spectroscopy, e.g. aza 15C5 complex of sodium was thus studied [57]. The structure of calcium with substituted 18C6 or 24C8 was best studied by IR spectral methods and its structure was elucidated [60].

In study of inner transition metals the spectra gave definite information on size of ring, donor atom and type of substituent which ordinarily influenced f and d transition metals absorption spectra [63]. A typical antisymmetrical and symmetrical $-\overset{|}{\underset{|}{C}}-O-$ bond stretching was studied for the manganese crown ether complexes [65].

Similar to the study of crown ether complexes, attempts were made to study complexes of cryptands as well as those of the calixarene by absorption spectroscopy, to understand host guest interactions [68]. The spectra of heteroditopic cryptand with the copper was studied at 875 nm in acetonitrile and tetrahedral coordination complex was identified.

The most extensively studied complexes of metals were with thia and aza crown ethers. The shift of OH-stretching bond to lower frequency during complexation of s-block metals with aza crown ethers was studied [71] and it furnished useful information.

The study of the complexes of calixarene was greatly facilitated by absorption spectroscopy. For calixarene unusually low frequency at stretching vibration of the OH group occurred in IR region of 3150 cm^{-1} due to presence of strong intramolecular hydrogen bonding. FTIR helped to [75] elucidate intramolecular bonding. Usually low frequency were observed for linear phenol formaldehyde oligomers. In finger print region, i.e. 1500-900 cm^{-1} all peaks appeared identical with peak at 693 and 571 cm^{-1}.

The cyclic as well as linear oligomers had peak at 280 nm. Although reactions were monitored by other technique, UV-visible-IR continued to be important supporting tool. A three centred bond in the methylated p-tertiarybutyl calix(6) arene was ascertained by FT-IR method [76]. A significant bathochromic shift to red region was observed for alkali picrate interaction with calix(n) arenes, with nature of the complex as 1:1 [78]. The stabilisation of deprotonated form of calix(s) arene complex with europium was best studied by absorption spectroscopy. The details of inferences drawn on study of complexes with these crown compounds with absorption spectroscopy are summarised in Tables 3.4.1 through 3.4.4.

3.5 Characterisation of Crown Complex by NMR Spectroscopy

Like absorption spectroscopy, NMR spectroscopic methods have been popularly used to ascertain nature of complexes of crown ethers or cryptands with various metals. The toxicity of compounds like 12C4 was ascertained by NMR. The proton and C^{13} NMR showed shift at 3.65 ppm and 70.621 ppm, respectively. This was correlated with data on mass spectra. For instance NMR of DB18C6 showed upper multiplet 4.11 ppm down field from tetramethyl silane (TMS) were ratio of 2:2 lower singlet 6.92 ppm down field from TMS at ratio of 1.0. While NMR of DC18C6 had upper multiplets at 1.50 ppm down field from TMS were at ratio 1:17 lower singlet 3.67 ppm down field from TMS were ratio 1:17 lower singlet 3.67 ppm downfield from TMS were ratio 1:00.

The NMR and ESR spectroscopy had proved to be very beneficial for characterisation and structural analysis of metal complexes with crown ethers, aza or thia crown ethers, cryptands and calix(n) arenes. A large number of papers have appeared in this field, dealing with structure of alkali metal, alkaline earths, transition elements and some main group elements. A brief survey of such complexation is worth considering.

Since picric acid was used as auxilliary complexing ligand the H' NMR of 4-substituted B15C5 were studied and association constants were evaluated. On basis of substituents change in chemical shift was noted [81]. The influence of dilution shift effect on proton NMR and split lattice relaxation of benzoaza crown ethers in deutrated chloroform was noted. This is shown in structures (X) and (XI):

The H-bond formation between crown and solvent was not responsible for the dilution shift. The reaction mechanism was also explained [82]. On account

Table 3.4.1 Characterisation of crown complexes by absorption spectroscopy

Elements	Characterisation of Crown Complex
Alkali metals	Cyclic heptacosa (oxyethylene) (cyclo-E_{27}, 81C27) was studied by Laser Raman spectroscopy as one of the method and was compared with oxyethylene dimethylether [51]. Thiol derivative of DB18C6 was complexed with sodium and potassium and was studied by surface enhanced Raman spectroscopy indicating band at 1355 cm^{-1} due to complexation with potassium, rubidium, sodium and caesium ions, binding of these ions was strengthened by thiocyanate [52]. Alkyl phosphoric acid armed crown ethers were studied for complexation with alkali metals, it showed specific transport of alkali metals which fit the cavity size of the ring; structure also influenced selectivity [53] of complexation.
Li (I)	Complexes of 12C4, 15C5 and 18C6 were studied by Fourier deconvolation of C-H stretching spectra by Raman spectroscopy in aqueous solutions. Such complexation led to unperturbed stretching frequency with formation of 2 : 1 and 1 : 1 complex with potassium and sodium with 15C5 and 2:1 with sodium with 12C4, [54] resp. Infrared spectra of LiBr $(CH_2CH_2O)_5H_2O$ was plotted at 400-4000 cm^{-1}, a splitting of the OH-band due to two different modes of dentation of water was noted [55].
Na (I)	Complexes of 4'- substituted benzo 15C5 with sodium picrate was studied by UV-spectroscopy. Charge transfer complexes were also isolated with different colors [56]. Two cationic styrl dyes containing aza 15C5 were complexed and ligand complex ratio obtained [57] unsymmetrical annealated tetrathiaful valence-S_2O_4 crowns were complexed. UV-visible studies of sodium crown tetrathiaful showed 1:1 complex but for bis form 1 : 2 complex was formed [58].
K (I)	Crown ethers containing dicinamoyl groups were complexed with potassium e.g. 1-10 bis (p-2 chloroformyl) phenyl 1, 4, 7, 10 tetrahydroxdecane was used with UV-light it gave intramolecular photocyclo adduct, they were selective for extraction of potassium and rubidium [59].
Ca (II)	With long wavelength IR spectra of the calcium complexes with 15C5, 18C6, DB18C6, DB24C8 were characterised and structures elucidated [60].
Sr (II)	Crown ethers containing 1-8 dioxyxanthones, e.g. 1, 8 (3, 6, 9, 12, 15 pentaoxaheptanedecane-1, 17 diyl) dioxyxanthone when complexed with strontium or barium and characterised by UV-measurements showed shift with added metal due to complex formation by placing metal close to carbonyl oxygen [61]. The complexation of crown ether spiropyran with metal showed enhanced aggregation of marocyanine moiety due to photoisomerisation [62].
U (III)	The spectras of uranium (III) with 12C4, 15C5, 18C6, DB18C6, DC18C6, and diaza-18C6 were examined for f-d transition. It showed size of ring, nature of the substituents and donor atoms influenced the absorption spectra [63].
Sc (III)	The [$ScCl_2^+$] oxo-crown species was reacted with dibenzo derivative of 18C6, or 24C8 or 30C10 in the presence of antimony chloride to give stabilised complex as [$ScCl_2$ (DB18C6) $SbCl_6$] [64].

(Contd.)

Characterisation and Metal Complexation

Elements	Characterisation of Crown Complex
Mn (II)	12C4, 15C5 and 18C6 were reacted with manganese chloride to give 1:1 complex and were studied by FT-IR and Raman spectroscopy to note antisym and sym C–O–C bond stretching at 800-1200 cm^{-1} and MnO and MnCi$_2$ stretching at 2400 cm^{-1} [65].
Sb (V)	Hydronium ion-crown ether antimonate salts, e.g. [(H$_3$O)$^+$ (18C6)] [SbCl$_6^-$]^0 were formed with 18C6, 15C5 or 12C4 in acetonitrile. In complex with 15C5 redox formation of the neutral complex was encountered [66]. With iodine as acceptor charge transfer spectra of crown complexes were studied by UV-spectroscopy [67].

Table 3.4.2 Characterisation of complexes with cryptand by absorption spectroscopy

Elements	Characterisation of Complex
Fe (III)	^{57}Fe containing ferrocene cryptands complexes were studied by Mosbaur spectroscopy to study host guest interactions [68].
Cu (II)	New heteroditopic cryptand with naphthyl side arm was synthesised and was reacted with copper and nickel to give mononuclear cryptates, the former accepts H$_2$S to indicate Cu-SH$_2$ bond [69] heteroditopic cryptand formed complex with copper and zinc, the blue copper complex showed band at 875 and 745 nm of ligand field and ligand to metal charge transfer at 275 nm in acetonitrile for pseudo tetrahedral CuN$_4$ coordination complex [70].

Table 3.4.3 Characterisation of complexes with thia and aza crown ethers

Elements	Characterisation of Complex
Alkali metals	NN' bis (4-hydroxy 3, 5 dimethyl) 1, 4, 10, 13 tetraoxa-7, 16 diaza cyclodecane formed complexes with alkali metals. Their FT-IR spectra showed OH-stretch band of phenolic OH group was shifted to lower frequency ca 290 cm^{-1} compared to host compound [71].
Ba (II)	The complex with benzo thiazolium steryl aza crown and its derivatives were spectroscopically studied on picosecond and kilosecond time scales to indicate rotation around both the olefinic >C=C< bond and adjacent –C–C–bonds [72]. A cation specific light controlled transient chromoionophore based on a benzothazolium steryl aza crown ether decomplexed with barium in acetonitrile was studied by UV-Vis absorption spectroscopy to show that photoisomerisation leads to stabilisation due to dual intramolecular complexation of barium [73].
Hg (II)	Styrl dyes containing benzodithia 18C6 were complexed with mercury and silver. They were studied by Raman spectroscopy and assignment of bands were made arising out of changes in spectra by complexation or change in solvent polarity [74].

of intermolecular interaction in two aryl rings of crown ether and adjacent C$_{60}$ in DB18C6 KC$_{60}$ indicated unusual air stability with negative chemical shift [83]. The NMR spectra of macrocyclic ether bislactone showed chemical shifts of

Table 3.4.4 **Characterisation of complexes with calixarenes by absorption spectroscopy**

Elements	Characterisation of the Complex
Na (I)	The IR and Raman spectra of p-tert butyl calix(4) arene was recorded at 3800–400 cm^{-1} to show the presences of intramolecular H—bonds [75]. The H—bond structure of methylated p-tert butyl calix(6) arene were examined by FT-IR indicating the presences of three centred bond [76]. Oxime and Schiff base containing aza crown ethers were synthesised they interacted strongly with sodium and potassium in methanol [77]. Alkali picrates on interaction with calix(n) arene esters showed a bathochromic shifts in absorption maxima for 1:1 complex [78]. A novel chromogenic pyridinium derivation of calix(4) arenes was complexed with alkali metals. The spectra in different solvent showed solvato and halochromism formed in site on complexation [79].
Eu (III)	p-tert butyl calix(5) arene in acetonitrile complex with lanthanides indicated inter and intramolecular stabilisation of deprotonated form, a deformation of cone conformation due to complexation and quenching of metal luminescence [80].

(X) (XI)

CH_2 next to nitrogen atom increasing with size of molecule and [84] solvent polarity. Crown ethers obtained from 2–7 dihydroxyacridine and 2–7 dihydroxy-acridine-9-one were characterised by proton and C_{13} NMR technique a chemical shifts were related to tautomerism of acridine-9-one. Former formed no complex while latter complexed readily with iron, copper and silver [85]. A new compound bis (1, 11) or thiocyclophanonic 18C6 was synthesised from DB18C6 and studied by NMR [86]. Biscyclic crown ether obtained from biphenyl was studied for complexation with mercury [87] metallocrown and metallocalixarene platinum bistriflates were reacted with bipyridinc and bidendate ligands to give tetranuclear macrocycles as observed by multinuclear NMR studies [88] with end objective of use in membrane transport studies.

Stability of methylbenzo 18C6 complex with sodium was best summarised by this technique [89]. The spin relaxation time was noted during NMR study of complexes of s-block metals with the various crown ethers [91]. A structural change due to the complexation of crown ethers with alkali metals was ascertained by NMR technique [94].

The exchange kinetics was also studied by NMR spectroscopy [97]. In such investigations Na^{23}-NMR gave very valuable information. Low temperature C^{13}

NMR has helped the chemist to show rapidly exchanging structures for different complexes of potassium [100].

Presence of close conformation for rubidium and caesium with bis crown ether derivative was monitored by proton NMR [103]. The exchange kinetic of 18C6 complex with lanthanide series of elements was studied by NMR [106] indicating molecular pathway.

Phosphocryptand complex of potassium was discovered for first time with reference to conformational changes for the complexation by NMR methods [110].

C^{13}-NMR proved to be useful tool for the study of new cryptand and aza crown complexes [114]. The plannar structure of barium with aza-15C5 (1:1) was ascertained by this technique [119].

Thus, it would be noticed although NMR spectroscopy was useful to ascertain nature of coordination complex. It never proved useful for quantitative analysis of either the metal or the complex. The details of various findings on characterisation with NMR spectroscopy are summarised in Tables 3.5.1 to 3.5.3.

Table 3.5.1 Characterisation of crown complexes by NMR

Elements	Characterisation of Crown Complexes
Alkali metals	Methyl benzo 18C6 when reacted with sodium formed complexes which was studied by Na^{23}-NMR, it showed that stability was related to number of methyl group on crown [89]. Naphthalene crown ether complexes of alkali metals on studying by C^{13} NMR in CD_3OD solution showed that chemical shifts resulted at the limit of complete 1:1 complexation conformational difference arose for deviation from limiting shift [90]. The conformational changes for alkali metal complexes with various crown ethers when studied by NMR indicated spin lattice relaxation time [91]. The structural change for DB16C5 oxyacetic acid on complexation with alkali metals during extractive separations showed that proton ionisable lariat ethers with formaldehyde led to the formation of new ion exchange resin containing polyether binding site [92]. The electronic effect and steric hindrance ought to be considered while studying complexing ability of macrolactones with alkali metal cations [93]. The alkali metal complexes of 18C6, B18C6, DB18C6, were studied by NMR to understand structural changes due to complexation, for electron distribution and binding energy [94].
Li (I)	Crown ether complexes in molten salts were studied by Li^7-NMR to obtain stability constant, such constant increased from 18C6 to B15C5 and was least with 15C5 [95]. The NMR studies of DB14C4 lariat ether complex with sodium and lithium showed that the orientation of side arm depended on nature of atom linking side arm to ring. Further if it was attached through macrocycle ring it was oriented to cavity [96-98].
Na (I)	Exchange kinetics of sodium complex with 18C6 in THF were studied by Na^{23}-NMR spectra of sodium complex with 12C4, 15C5, B15C5, 18C6, DC18C6 and DB18C6 it showed 1:1 complex with best stability in 15C5 and least stability in 12C4 [99].

(Contd.)

Elements	Characterisation of Crown Complexes
K (I)	Bis(benzocrown) ether derivatives of potassium were studied by low temperature C^{13}-NMR, it showed rapidly exchanging structure for different complexes of 1:1 composition. It also showed the most extreme behaviour indicative of stronger complexation [100].
Cs (I)	Nuclear over häuser effect spectroscopy (NOES) and C^{13}-NMR of complexes of caesium with DB24C8, DB18C6, B15C5 showed reduction in distance between CH-α and -CH_2 groups due to fluttering motion of benzogroup and elongation of crown moiety [101].
Rb (I), Cs^{133}	The complexes of rubidium and caesium with 18C6, DC21C7, DC24C8 in acetonitrile were studied with Cs^{133}-NMR techinque, they formed 1:1 and 2:1 complex, rubidium complex was more stable [102]. Bis crown ether derivative of triaminotriazine was examined by proton NMR. Rubidium and caesium were effective in promoting formation of closed conformer while half open or fully open form due to rotational barrier of C (triazine)-N bond was seen for [103] potassium. The luminescence spectra in crystalline supramolecular (Rb18C6 $MnBr_4$) $(TlBr_4)_2$ complex when studied showed that $MnBr_4^{2-}$ ions were effective as the luminescent probes [104].
Ca (II)	Complex with benzo crown ether was studied by NMR involving C^{13} and C^{17} nuclei [105].
La (III)	Exchange kinetics of 18C6 complex with lanthanum, calcium lead and barium in methanol when studied by NMR showed binuclear path way for lanthanum (III) [106].
Hg (II)	Tetraseleno crown ether complexes of merury (II) were studied by NMR, they showed complexing ability vai their selenium atoms [107].

Table 3.5.2 Characterisation of the cryptand complexes by NMR

Elements	Characterisation of Complexes
NH_4^+ radicle	Macrocyclic compound with two catecholate group when reacted with boric acid and ammonia in ethanol gave ammonium complex as pseudo cryptand with highest selectivity [108].
Li (I)	The unusual coordination distorted square planner geometry in interior of cryptand-helicate was stabilised due to steric effects and H-bonds between two water molecules attached to lithium [109].
K (I)	Bis (phosphotriester) macrobicyclic polyether cryptand called phosphocrypt viz. O=P $[O(CH_2)_2O (CH2)O(CH_2)_2 O]_3$ P=O formed stable complexes in water nitrogen bonded cryptand was less stable than P-bonded cryptand due to cation-dipole interaction and better conformal changes for complexation [110].
Cu (II)	Heteroditopic cryptand imposing distorted geometry was prepared for complexation with copper to give inclusion complex, bound anion readily accepted thiocyanate and azide anions [111].
Ni (III)	Heteroditopic cryptand complex so formed was octahedrally coordinated with equitonal sites occupied by 3-secondary amino nitrogen atom of cryptand and N-atom of solvent [112].
Ag (I)	Octaazo stiff base cryptands were complexed with it to give polynulceating and extended coordination nature of the cryptand compound [113].

Characterisation and Metal Complexation

Table 3.5.3 Characteristics of complexes with thia and aza crown ethers with NMR

Elements	Characterisation of Complexes
General	New cryptands and bis(aza12C4) ether were synthesised and studied by C^{13} NMR [114]. H'-NMR and C^{13}-NMR spectra of N-substituted dibenzoaza crown ether was recorded in relation to charge distribution of crown ethers [115]. H'-NMR and C^{13}-NMR spectra of N-acyl substituted benzoazacrown ethers was recorded to study effect of substituents on structure and shift [116] N N' bis (1-pyrenyl methyl)-1, 4, 10, 13 tetraoxa, 7, 16 diazacyclo-octadecane a derivative of Diaza 18C6 when synthesised showed photophysical properties in the presence of complexing metal cation, and enhanced emission intensity [117]. Azophenolic crown ethers were prepared and complexed with chiral amines [118].
Ba (II)	The study of aza 15C5 ether dye complex with barium and silver showed plannar structure with 1:1 complexation indicating binding of hard barium and soft silver ion [119].
Co (II)	Bis substituted amide linked aza crown ether cobalticinium compound was prepared and studied by NMR showed cooperative enhancement of the strength of anionic complexation [120].
Pd (II)	Mixed azathioether crown 2, 5, 8 trithia (2, 9) 1-10 phenanthraline formed complex with ligand adopting folded conformation. Similar complexes for platinum and rhodium [121] was noted.
Hg (II)	Troponoid thiacrown complex was prepared, cavity size was important in extraction and transport of mercury through membrane in preferred to transport of copper [122].

3.6 Metal Complexes with Calixarenes

Calixarenes are increasingly used for complexation of metal ions. So far limited metal complexes of calixarenes have been prepared, but many more will be prepared in future. The main reason for prefering calixarenes as the complexing ligand were:

1. Calixarene framework had two rims which could be easily modified to suit complexation reaction.
2. Cavity size for intercalation could be easily modified by changing (n), i.e. nos of calixarene rings or phenylrings.
3. Ligand group could be varied to accommodate the metal ion of any size with typical coordination properties.
4. Calixarene had sufficient flexibility to adopt to particular cavity size and shape.
5. Chemical modifications of each rim could easily include a legitating group on the rim to improve its solubility in solvents.
6. They could incorporate different functional groups on each rim to prepare new calixarenes with better selectivity.
7. Non-centrosymmetric structure of calixarene made it attractive for various ligand applications.

The upper rim without group/groups is called 'naked' calixarenes (see structure XII to XV). This position could be easily substituted by sulphate, phosphonate ligands to improve its solubility. One could also readily prepare chloromethylated derivatives useful as the precursors.

The sulphonation of calixarenes proceeds as per following procedure (shown in structures from XVI to XVIII).

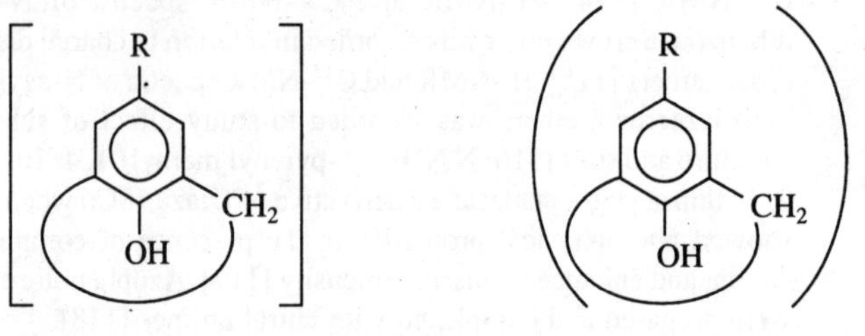

(XII)   (XIII)

A concentrated sulphuric acid was passed at 80°C for 4 hours [123-125]. On treating with octylchloromethyl ether and stannous chloride and triethylphosphite one could get arylphosphonic derivative. Calixarenes with chloromethyl group on upper rim could be also used as a synthetic precursors for other compounds. (XVI to XVIII).

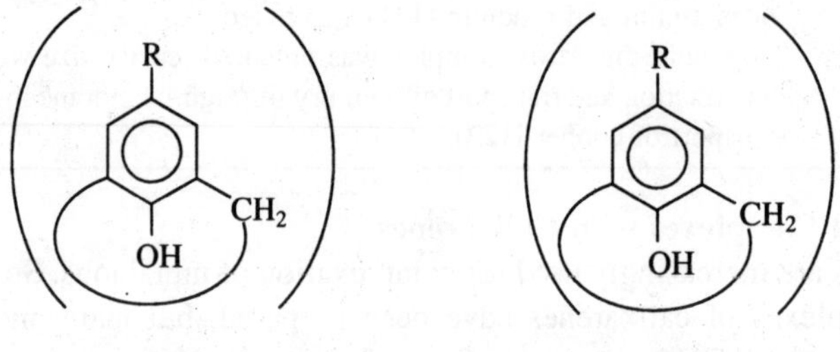

(XIV)   (XV)

The group was reactive to nucleophillic reagents. By Mannich reaction [126] broad range of different groups could be introduced in the upper rim of the calixarene. This was also possible to carry out modification of lower rim. Each rim could be substituted at lower phenolic positions.

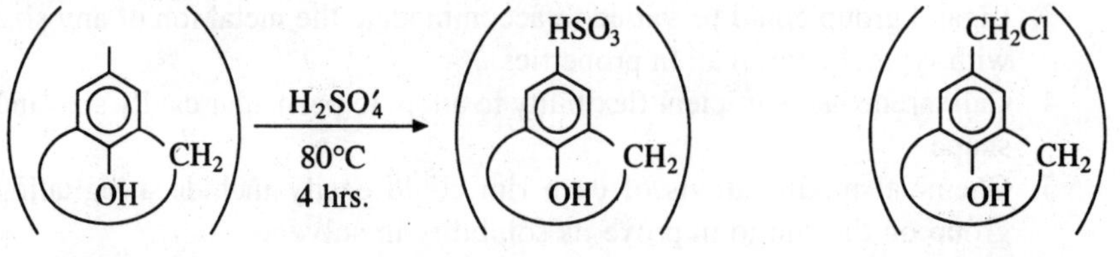

(XVI) $\xrightarrow{H_2SO_4,\ 80°C,\ 4\ hrs.}$ (XVII)   (XVIII)

Such lower rim could accommodate ester, ether, phosphinite or carboxylate.

Calixarene rims could be modified to attach them to polymeric support for immobilisation useful in extraction chromatography [127]. They could be used for molecular recognisation by incorporating binding groups in cavity. They had

H-bonding donors or acceptors with net result the molecule tended to associate intramolecularly giving closed receptor as shown in structure (XIX) to (XXII) and (XXIII), if M = metal, G = ligand bonding site (XXIII).

Calixarenes readily formed complexes with alkali metal, the inner transition elements and few transition group elements. They were used to transport cations in membrane technology. The significant advantage of calixarenes was to accommodate metals of any size, its rim could be adjusted using n = 4, 6, 8 membered ring. The unsubstituted calixarenes themselves were inefficient as carriers of the alkali metals. Like crown ethers cavity size and cation diameter determined the complexation possibility. For calix(4) arene cavity was 2.4 Å for calix(6) arene it was < 4.8 Å while for alkali metals ionic size varied from 1.52–3.40 Å indicating possibility of interaction (Table 3.6.1) with them.

Table 3.6.1 Complexation possibility and transport of cations (mol/S × 10^5) in base

Cation	Diameter Å	Calix(4) hole Å	Calix(6) hole Å
Li	1.52	–	10
Na	2.04	'2	22
K	2.76	< 0.7	13
Rb	3.04	6	71
Cs	3.40	260	810

As a rule calix(n) arenes which were substituted at lower phenolic rim could be used for complexation of s-block metals. The lower substituted group might be ester, ketone, ether groups attached to the lower rim. The complexation was due to donor oxygen atom. Generally water soluble calix(n) arenes had sulphonate substitution in para position of the upper rim to show proton-transfer reactions in the presences of sodium ions [128]. Tetramers showed preference for extraction of sodium, the tetraketones were effective extractants for lithium, while hexamers ketones and ester preferred larger cations. Finally octamers showed low level phase transfer effectivenes of s-block metals.

Usually small calixarenes were useful for the complexation of alkali metals. Tetramer for sodium, hexamer for potassium rubidium and caesium, but octamer is useless for alkali metal while none could complex with lithium. As regards ionic radii of sodium and calcium ion was similar hence both could complex with calix(4) arene [129]. Calix(4) arene could be isolated in cone and partial cone conformation which in turn exhibited different selectivity during complexation. In solvent extraction cone form was more effective but not partial cone conformer. This was evident from Table 3.6.2.

Table 3.6.2 Extent of extraction (%)

Calix(4) arene	Sodium	Potassium
Cone-1	0.8	1.9
1–3 alternate-1	4.6	16.2
1–3 alternate-2	8.8	56.8
1–3 alternate-4	0	88.8
Cone-2	3.9	3.9
Cone-3	0	9.0

This showed for 1–3 alternate calix(4) arenes, π-donor effect of phenyl ring was important in improving selectivity and stability constant of metal ion binding to calixarenes [130]. The charge effect could also effect ion binding [131]. The carboxylate substituted calixarenes were used for complexation of s-block metals and their protonates and stability constants were evaluated. This showed substitution at lower rim had enhanced stability constant showing better selectivity [132]. It was further noted the compound with R = ethyl in lower rim under went interconversion between the cone and partial cone forms. Generally cone form showed better extractibility.

The complexation readily influenced calixarene properties. Calixarenes esters had monolayer typical of ring size. Unsubstituted derivatives did not form monolayer. The monolayer derivatives were reactive with monovalent cations by showing expansion on complexation with them [133].

Chromogenic calixarene had been developed as ion sensors. In case the ion bound complex had different conformation than uncomplexed form emissive properties of chromophore could be changed to design ion sensors. They were prepared by including chromophoric functional groups on the lower rim enabling them to bind metal ion [134]. The adjacent compound had high selectivity for lithium with sharp band for absorption at 550 nm (XXIV).

The three dimensional shape of calixarene was good for complexation of metal ion. The metal could be encapsulated in cavity but it could escape out also to prevent that the cap was chemically attached to prevent escape of metal ions. Such caped compounds were called as 'calix spherands'. A crown ether metallocycle was introduced to give calix crowns [135]. Molecular dynamics with thermodynamic method was used to understand binding of alkali metals to calix spherands.

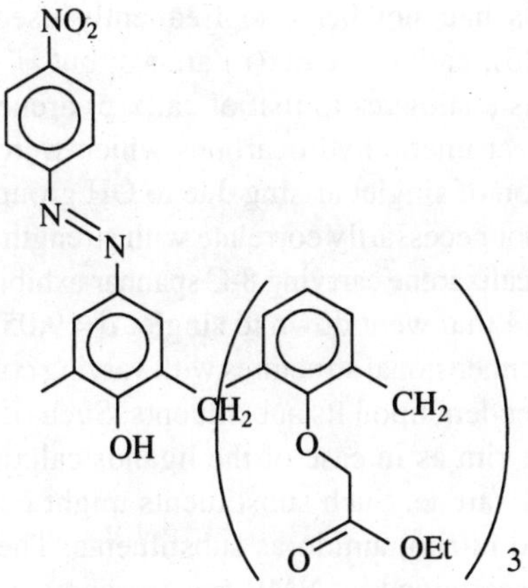

(XXIV)

Apart from alkali metals, calixarenes readily formed complexes with *f*-block metals. Europium (III) had been extensively studied. Here also stability of complex could be increased by lower rim substitution, e.g. ethyl phosphonate [136] was best substituted for lanthanides. The coordination of water molecule made such complexes with gadonium a promising for magnetic resonance imaging (MRI) in medicine. The carboxylate or hydroxame substituted calix(6) arene was best for uranium [137]. If hydrophobic group or terbutyl or hexyl was introduced in para position instead of sulphonate they promoted extraction of uranium (VI) from aqueous phase [138].

For transition metals chemically modified calixarenes were useful. Calixarenes with phenolic group on their lower rim could bind metal by deprotonation action of phenolate ligand. They could form water soluble complexes also, e.g. iron (III) at pH 2.5–5.5 formed violet complex with calix(6) arene [139]. As opposed to this upper rim substituted amino calixarene were best for complexing with nickel copper, cobalt, iron, palladium [140]. Such complexes could be used for molecular recognisation. Some of the complexes could be used for synthesis of liquid crystals. Phosphorous donor atom attached to lower rim of tetramer could be used as donor ligands for transition metals [141].

3.7 Characterisation of calix(n) arene complexes by NMR spectroscopy

The NMR spectroscopy became one of the most effective tool available to chemist to characterise calixarene complexes. In fact ^{13}C-NMR which had advantage over H'-NMR helped chemist to differentiate between linear and cyclic oligomers of p-tert butyl calix(4) arene. The latter had simple spectra due to cyclic nature. It had four resonance lines due to the aromatic C-atoms and one from methylene carbon, and two from tert butyl carbons. The reasonance from aromatic protons, tert butyl protons and hydroxyl protons were singlet and that from CH_2 protons was doublets. The calix(4) arene and calix(8) arene were thus differentiated by the fact that in pyridine former retained a part of doublets at low temperature while latter showed presence of signlet at very low temperature.

The NMR methods had not been so frequently used for odd number of calixarenes like calix(5), calix(7), calix(n) arenes; but H'-NMR spectra of odd number calixarene was analogues to that of calix(6) arene in nonpolar solvents such as chloroform or aromatic hydrocarbons which were frequently used.

As a rule the position of singlet arising due to OH group varied with ring size of calixarene and did not necessarily correlate with strength of hydrogen bonding. The upper rim bridged calixarene carrying 8-C spannar exhibited two OH-resonance at $\delta = 9.26$ and $\delta = 9.14$ that went down to singlet $\delta = 9.05$ at room temperature. Calixarene were three dimensional structures with varying cone size. The selectivity of calixarene was dependent upon its substituents. Such of the substituent might be present in the outer rim as in case of the ligands calixresorcinarene or inner rim was present in calixarene. Such substituents might consist of acetyl group, sulphonate group, keto nitrito, amido as substituents. The exact desposition of such group could be ascertained by NMR spectroscopy. A good amount of light was specially thrown on the nature of the complex formed with the metal ions by NMR technique.

Calix(4) arene formed tetracyanoethylene complexes in dichloromethane in ratio of 1:1 species with absorption maxima at 498 nm; but proton NMR showed that four alkyl groups were turned out word and was sterically stable. In such complexes, complexation occurred in conjunction with out-in displacement [142]. A novel calixarene was prepared from 5-t-butyltetrahydro-1, 3, 5 triazine-2 (1H) one [143]. Conformational characterisation of p-tertbutyl calix(6) arene ethers was investigated by NMR, it showed cone conformation for derivatives like tribenzyltrimethyl ether while tetrabenzyldimethyl ether had 1, 2, 3 alternate conformation [144] 5, 11, 17, 23, 29 penta-t-butyl calix(5) arene 31, 32, 33, 34 pentol was better synthesised by o-alkylation, it showed butyl group did not inhibit but octyl group did inhibit annulus rotation [145] tetra sulphonate derivative of calix(4) resorcinarene formed 1:1 complex with hydrophillic guest molecules; to indicate multiple host guest interactions, complexation favoured by electron withdrawing residues and electron donating substituents. It favoured enthalpy change, the NMR showed that for complexation driving force was C-H-π interaction between CH-bonds of the guest as soft acid and benzene ring of host as soft base [146]. The pulse gradient spin echo (PGSE) NMR technique was used to evaluate the diffusion coefficients of p-tert butyl calix(n) arene [147]. Silica bonded calixarene phase were obtained to give calix(4) arene tetra-amide and calix(6) arene hexaester phase, NMR showed calixarene bonding to the silica surface [148]. The four conformers of tetra-acetoxy p-tert butyl calix(4) arene was determind by H'-NMR to show four conformers were not flexible with varying stability as 1, 3 alternate > 1, 2 alternate > partial cone > cone (most stable form) [149].

Tetrasulphonated calix(4) arene 1, 3 dicarboxylic acid had significant effect on acid base and inclusion properties of the ditopic receptor [150] o-tert butyl calix(n) arene (where n = 3, 4) with 3, 4, 5 exohydroxy groups were synthesised, and when studied by NMR showed annular flexibility in deutrated chloroform and presence of intramolecular H-bonds between proximal OH-groups [151] (structure XXV).

$$n = 2, 3, 4$$
(XXV)

In supramolecular complex of C-60 based calix(5) arene the Van der Waal's interactions between host and guest molecule were important in complexation [152]. Extracted calix(4) arene and doubly bridged biscalix(4) arene were synthesised. The NMR studies showed for new compounds had two different pinched cone conformation [153].

Water soluble p-sulphano calix(n) arene with stilbene dyes were studied from view point of molecular association (XXVI).

(XXVI)

The (XXVI) compound reacted in 1:1 ratio with dye to form complexes, the absorbance of such complex was decreased by the process of association, the substitution with surfactants released free dye molecule and formal 1:1 complex. This was used to analyse surfactant [154]. Four methylene bridges in p-tert butyl calix(4) arene were replaced by sulphur bridges to give thia calix(n) arenes which were more flexible conformationally and served as inclusion host for organic molecules [155]. Calix (aza) crown capped by diamide bridges in 1, 3 position on lower rim were synthesised and were alkylated and studied by NMR spectroscopy [156] calix(4) arene in 1, 3 alternate conformation containing hard and soft ion binding sites were synthesised [157].

The significant applications of NMR spectroscopy for the characterisation of calixarene complexes with various metals are best summarised in Table 3.7.1.

One significant feature of this technique was it had been extensively used to study particularly complexes of sodium, ruthenium, cobalt, platinum and silver and few anions with calixarenes [158-159]. Some outstanding features worthy of mention are taken in to considerations. The oscillations of sodium between two calix(4) arene ionophores was noted by NMR method [158]. Most of such studies were carried out in deutrated acetonitrile or chloroform as solvents.

The complex of cobalt (II) with calix(6) arene hexasulphonate was studied to

Table 3.7.1 Characterisation of calixarene complexes by NMR

Element	Characterisation of Complex
Na (I)	Bis calix(4) arenes doubly bridged with oxyethylene chains at lower rim had been prepared, they were selective for sodium and potassium, these ions oscillated intramolecularly between two calix(4) arene ionophore sites [158]. Calix(4) arene tetraester was complexed with sodium and studied by H'-NMR and Na^{23}-NMR in deutrated acetonitrile and chloroform, a 1:1 complex was formed [159].
Ru (II)	A cyclic macrocyclic and calix(4) arene Ru-bipyridyl receptor molecule was prepared a new class of anion receptor to sense an ionic guest species was available, H-bonding was relevant for anion complexation [160].
Co (II)	Calix(6) arene hexasulphonate formed stable complex (1:1) with cobalt (II) sepulcohrate in water and 2:1 complex formed in solid state [161] condition.
Pt (IV)	A platinum hydride was obtained by interaction of trans [P-H (Cl) (PPH3)$_2$] with 5, 11, 17, 23 tetratertbutyl 25–27 bis (diethylcarbamoylmethoxyl) (26-28) bis (diphenylphosphonomethoxy) calix(4) arene. Hydrido was directed in hemispherical cavity [162].
Fe (III)	Inclusion complex of ferrocene derivatives by water soluble calix(6) arene was prepared H^1 NMR studies showed that binding interaction between calixarene host and guest were analogues to one in organic compounds by cyclodextrin or cyclophane [163].
Ag	Binaphthyl bridged calix(4) arene in cone conformation were synthesised and were then complexed with sliver, to give 1:1 complex. Here silver ion encapsulated in polar cavity. They were useful for chiral recognition and catalysis [164] chiral mono-alkylated p-tert butyl (1, 2) calix(4) crownether complexes with silver were formed, in a ratio of 1:1:1 with picrate with 25% extraction in methylene chloride [165].
P (V)	p-tert butyl calix(4) tetrakis (diphenyl phosphinite) and tetrakis (dimethyl phosphinite) acted as phosphorous binding surfaces for nonmetallic and heterodi metallic species [166].
Anions	π-metalated calixarenes in their cavity bound anions, presence of transition metal centre enhance acidity of -OH at lower rim of calixarene, in higher homologues anionic guest were embedded in molecular cavity [167].
Anions	Polyaza crownether derivatives of calix(4) arene as anion receptor was investigated. The ammonium derivative, viz. 25, 27 [2, 2′ (2, 2,′ (2, 5, 8 triammonium)nonyl) diphenoxyl diethyl] p-tert butyl calix(4) arene trichloride formed complexes with Cl$^-$, NO$_3^-$, CO$_3^{2-}$ and AsO$_2^-$ [168].
Xe	Xe was enclathrated in p-tert butyl calix(4) arene cavity which behaved as cavity probe as well as guest by ^{129}Xe-NMR studies showed that Xe occupied two cavity sites of the calixarene molecule [169].

ascertain the nature of the species. It was 2:1 and 1:1 complex [161] in solid form as well as in water respectively.

The ferrocene complexes were very widely studied. For instance inclusion complex with calix(6) arene was prepared and when studied by H'-NMR showed that binding interaction between calixarene host and ferrocene guest were similar to organic compounds like cyclodextrin or cyclophane [163].

As discussed earlier calix(n) arene formed complexes with metal, organic molecules and the anions. The anions were bound in cavity of calixarenes as embeded species [167]. The calix(n) arene acted as anion receptor when capped with polyaza crown ethers. Such complexes were formed with nitrate, chloride, carbonate and arsenite [168].

Finally an interesting phenomena was observed, viz. the entrapment of Xe-in p-tertiary butyl calix(4) arene by NMR spectroscopy. Such studies revealed that Xe-occupied two cavity sites of calixarene molecule [169].

3.8 Characterisation of Calixarene Complexes by Mass Spectrometry

The determination of molecular weight of calixarene was possible by use of mass spectrometry technique. Initially it was done by classical method of depression in freezing point or the elevation of boiling point. Then osmometers were also used for study of insoluble calixarenes, but now-a-days no single technique was comparable with mass spectrometry. A calixarene derived from resorcinol and acetaldehyde was studied. A signal at m/e 656 proved presence of four aromatic rings and gave conformation of cyclic tetramer.

Then investigation for phenol formaldehyde derivatives were latter initiated. In few instances it gave wrong information, e.g. calix(8) arene with signal at m/e 648 was for tetramer but latter a weak signal at m/e 1872 was recorded for octamer.

The large amount of data was available on mass spectra of p-alkyl calixarenes. Usually cyclic oligomers loose methyl or tert butyl groups and retain ring structure in beam, while linear oligomer disintegrates to phenolic units.

However mass spectrometry could not play decisive role in assigning structure of calixarene metal complexes as was possible by XRF or NMR methods.

Amongst spectroscopy techniques; UV, visible, IR, Raman, NMR were extensively used for the characterisation of metal complexes with crownethers cryptands and specially calixarenes. However, very positive information was obtained from mass spectrometry though it is not spectral method of characterisation. In fact gas chromatography coupled with mass spectrometry not only provided information on the nature of complexes formed with metals, but also they provided methods for the individual separations of one metal complex from other. In fact mass spectrometry was used for solution of problems in synthesis and isolation of crown ethers [170]. Mass spectrometry provided an answer to question whether calix(n) arenes really existed in discrete monomers in solution. In fact it did exist as discrete monomer due to strong intramolecular H-bonding. The higher homologues attained pinched ring conformation due to destabilisation of intermolecular H-bonding [171].

The first evidence for triple cation binding by multiring macrocyclic polyether was provided by mass spectrometry. Three crown balamphiphillic system was present in such system [172] much depended on technique. For instance fast atom bombardment mass spectra of bis-(crownether) was effective for ionisation of compounds [173] while electron impact mass spectrometry of macrocyclic polyether lactones, confirmed presence of hydrogenated oxixane ions [174]. Finally the laser microprobe mass spectrometry of crowns and cryptands complexes in larger silicates showed presence of cationized or protonated molecules produced in microplasma. The cation-ligand association and hydrogen exchange of cryptand for acidic copper (II) was demonstrated [175] in such complexes.

Some synthetic studies throw light on complicated problem of identification of complexes. The deuterium labelled crown ethers specially lactone type complexes showed uncommon double hydrogen rearrangement [176]. The capping of calix(4) arene derivatives which were chiral ligands were synthesised, a new bis calix(4) arene derivative as callicrown with connected lower rims was prepared and studied by mass spectrometry [177] and were isolated by thin layer chromatography. The ultraviolet laser desorption ionisation mass spectrometry of calixarene dissolved in chloroform or acetonitrile was carried out [178]. It was observed that as the length of acyl group was increased in crown ethers or benzo crown ethers like B15C5 or B15C6 or its 4-acyl derivatives then the McLafferly rearrangement was also increased [179]. The mass spectral investigations of polyether compounds showed the presence of acidic and basic sites in molecular ions and neutral molecules respectively [180]. The pyrolysis gas chromatography and pyrolysis mass spectrometry investigation of polymeric crown ethers was used to prepare new polymers and gave information on composition and breakdown mechanism [181]. Recently new methods like the fast atom bombardment mass spectromerty (FAB-MS) and field desorption mass spectrometry (FD-MS) were used to study dizonium ion crown ether molecular complexes in gas phase. In such studies the ArN_2^+ crown 1:1 complex was encountered with charge transfer phenomena as the mode of complexation mechanism [182].

Some of the interesting observation related to the application of mass spectrometry used to study complexes of metals like sodium, potassium, magnesium, lanthanum and antimony or bismuth are summarised in Table 3.8.1.

Some of the significant observations from these tables are:

The sensitivity and hydrophobicity of coordination complexes of s-block metals with important crown ethers was noted by mass spectral investigations [183]. The secondary ionisation mass spectrometry proved to be beneficial to draw conclusions. A field desorption mass spectrometry was also used [184] to study hydrated and unhydrated complexes. The gas phase reactions were also studied by this technique. The collision induced dissociation reaction were investigated by autospace Q-tandem mass spectrometry. The fragmentation pattern of magnesium crown complex was studied in greater details [190]. A definite relationship between intensity of ions in gas and solution phase was investigated for crown ethers lanthanide complexes [191].

Conclusion

In conclusion one can see that although absorption spectroscopy including UV, visible, or IR are definite tools for characterisation of complex, some times impure compounds gave wrong signals for presence of unknown moieties.

The NMR spectroscopy had proved to be most valuable tool specially in

Table 3.8.1 Characterisation of complexes by mass spectrometry

Element	Characterisation of Calixarene Complexes
Alkali metals	The desorption of crown ether alkali metal complexes in liquid secondary ion mass spectrometry (SIMS) showed large difference in the relative abundances in complexes. At low concentration better sensitivity and at higher concentration, significant hydrophilicity was significant for distinguishing between mixed complexes from those having surface activity and limited action with solvents [183].
Na (I)	Hydrated and unhydrated complexes of sodium, and other alkali metals with 15C5, 18C6, were detected by field desorption mass spectrography [184]. FAB-MS was used to observe metal cation selectivity in complexation reaction. The complexes of 15C5 with barium, zinc silver, iron, magnesium, calcium, mercury, lead in presence of trifluoromethyl sulphonate were studied for abundance measurements [185]. Ion formation in fast bombardment mass spectrometry was used to study gas phase reactions, $(M+A)^+$ ions were produced, such of the formation depended on voltality of crown ethers [186].
K (I)	Autospace Q-tandem mass-spectrometer was used to study collision induced dissociation experiments, it showed relative bond strength inside activated host guest complex in gas phase [187]. The voltile organic solvents were used in mass spectrometry with field evaporation of ions from solution, at low temperature with water addition to improve emission, for sodium complex with crown ether, like 15C5, 18C6 in acetone showed the ions Na^+N solvated with H_2O [188].
	Laser induced liquid beam mass spectral studies of weak noncovalent interaction between B15C5, K complex showed ion selectives which was in agreement with stability constants of the complexes [189].
Mg (II)	Crown ethers containing this metal, when studied by mass spectrometry showed fragmentation pattern of (2-benzyl 1-3 xylene) 15C4 which was similar to that of 1-methylanthracene, it involved consecutive 1, 5, 4 transfer to generate an open chain polyether group [190].
La (III)	FAB-MS of lanthanide crown ether complexes showed a definite relationship between intensities of ions in gas phase and those of the ions involved in reactions in solution phase [191].
Sb/Bi (III)	Synthesis of crown ether esters with (benzyloxy methyl side arms using templates as Sb Ph_3, $BiPh_3$) was done. The key step was formation of complex with $BiPh_3$ and (benzyl oxy) methyl substituted diols [192].

study of calixarene complexes. The presence of cyclisation and absence of linear oligomer could be definitely confirmed by use of NMR. Not only H' NMR but in many instances ^{13}C NMR and in few instances ^{23}Na-NMR had been used extensively to characterise such compounds.

As regards mass spectrometry it was useful for characterisation of crown ether and calixarene compounds. Tandem mass spectral methods came to the rescue of chemist in latter part of the century.

In addition of nature of complex which could be easily characterised by spectral methods much consideration was required to study structure of such complex. This was particularly important as cavity-hole consideration in host guest chemistry for molecular recognisation was of great significance. An attempt will be made in next chapter to see how such structures were elucidated by XRF, thermal and radio-chemical methods of characterisation and analysis.

References

1. C.J. Pedersen, J. Am. Chem. Soc. **89**, 7017 (1967).
2. C.J. Pedersen, J. Am. Chem. Soc. **92**, 391 (1970).
3. M. Hiraoka, Yuki Gosei Kagaku **33**, 782 (1975).
4. M. Takagi, T. Matsuda, Kagaku-no-Ryoiki **31**, 208, 348 (1977).
5. J. M. Lehn, Structure and bonding, Vol. 16, p. 1, Springer-Verlag (1973).
6. J.J. Christensen, J.O. Hill, R.M. Izatt, Science **174**, 459 (1971).
7. R.M. Izatt, D.J. Eatough, J.J. Christensen, Structure and bonding, Vol. 16 p. 113, Springer-Verlag (1973).
8. J.J. Christensen, D.J. Eatough, R.M. Izatt, Chem. Review. **74** 351 (1974).
9. B.S. Mohite, Solvent extraction of alkali and alkaline earths with crown ethers Ph.D. Thesis I.I.T. Bombay (1986).
10. C.J. Pedersen, K. Frensdorff, Angew Chem. **11**, 16 (1972).
11. D.F. Evans, J. Solution Chem. **1**, 499 (1972).
12. E. Schori, J. Jagurgrodzinskii, Isral. J. Chem. **11**, 243 (1973).
13. R. Ungaro, B.E. Haj, J. Smid, J. Am. Chem. Soc. **98**, 5198 (1976).
14. N. Matsura, K. Umemoto, Y. Takeda, A. Sakaki, Bull Chem. Soc. Japan **49**, 1246 (1976).
15. R.M. Izatt, J.H. Rytting D.P. Nelson, B.L. Haymore, J.J. Christensen, J. Am. Chem. Soc. **93**, 1619 (1971).
16. S.M. Khopkar and Mayuri Gandhi, Crown Ethers and Cryptands in solvent extraction (unpublished work) 1994.
17. R.M. Izatt, D.E. Terry, B.L. Haymore, L.D Hansen, W.K. Dalley, A.G. Avondet, J.J. Christensen, J. Am. Chem. Soc. **98**, 7620 (1976).
18. R.M. Izatt, D.E. Terry, D.P. Nelson, Y. Chan, D.J. Eatough, J.S. Bradshaw, L.D. Hansen, J.J. Christensen, J. Am. Cham. Soc. **98**, 7626 (1976).
19. V.M. Timko, R.C. Helgeson, M. Newcomb, G.W. Gokel, D.J. Cram, J. Am. Chem. Soc. **96**, 7097 (1974).
20. H.K. Frensdorff, J. Am. Chem. Soc. **93**, 600 (1971).
21. G.A. Rechnitz, E. Eyal, Anal. Chem. **44**, 370 (1972).
22. G.F. Smith, D.W. Margarum, Chem. Comm. 807 (1975).
23. I. Tabushi, Sakiyu Gakkaishi **18**, 732 (1975).
24. I. Tabushi, Yuki Gosei Kagaku **33**, 37 (1975).
25. J.M. Timko, S.S. Moore, D.M. Walba, P.C. Hiberty D.J. Cram, J. Am. Chem. Soc. **99**, 4207 (1977).
26. S.M. Khopkar and Mayuri Gandhi, J. Sci. and Ind. Res. **55**, 139 (1996).
27. M.M. Shemyakin, Yu. A. Orchinnikov, V.T. Ivanov, V.K. Antony, E.I. Vinogradova, A.M. Shkrov, G.G. Malenkov, A.V. Evstratov, I.A. Laing, E.I. Melnik, I.D. Ryaboya, J. Memb. Biolog. **1**, 402 (1969).
28. W.L. Duax, H. Hauptman, C.M. Weeks, D.A. Norton, Science **176**, 911 (1972).
29. M. Kodama, E. Kimura, Bull. Chem. Soc. Japan **49**, 2465 (1976).
30. U. Takaki, T. Hogen Esch, J. Smid, J. Am. Chem. Soc. **93**, 6760 (1971).

31. N.F. Curtis, Cord. Chem. Rev. **3**, 3 (1968).
32. L.F. Lindoy, D.H. Busch, Preparative inorganic reactions, Vol. 6, John Wiley (1971).
33. J.L. Dye, C.W. Andrews, S.E. Mathews, J. Phy. Chem. **79**, 3065 (1975).
34. J.L. Dye, Electrons in Fluids p. 77 Springer Verlag (1973).
35. J.L. Dye, J. Chem. Ed. **54**, 332 (1977).
36. J.L. Dye, Science 62 (1977).
37. C. Kappenstein, Bull Soc. Chim. Fr 89 (1974).
38. M.R. Truter, C.J. Pedersen, Endeavour **30**, 142 (1971).
39. M.R. Truter, Structure and bonding Vol. 16 p. 71 Spring Verlag (1973).
40. H.E. Simmons, C.H. Park, J. Am. Chem. Soc. **90**, 2428 (1968).
41. P. Peter, M. Gross, J. Electroanal Chem. **53**, 307 (1974).
42. P. Peter, M. Gross, J. Electroanal Chem. II **61**, 245 (1975).
43. J.P. Gisselbrecht, P. Peter, M. Gross, J. Electroanal Chem. **74**, 315 (1976).
44. J.P. Gisselbrecht, M. Gross, J. Electroanal Chem. **75**, 6307 (1977).
45. B. Dietrch, J. M. Lehn, J.P. Sauvage, J. Blanze, Tetrahedron **29**, 1629 (1973).
46. B. Dietrich, J.M. Lehn, J.P. Sauvage, Tetrahedron **29**, 1647 (1973).
47. B. Dietrch, J.M. Lehn, J.P. Sauvege, Tetrahedron letter 2889 (1969).
48. V.M. Loyola, R.G. Wilkins, J. Am. Chem. Soc. **97**, 7382 (1975).
49. E. Kauffmann, Helv Chem. Acta. **59**, 1099 (1976).
50. J.M. Lehn, J.P. Sauvage, Chem. Comm. 440 (1971).
51. Z. Yang, G.E. Yu, J. Cooke Z.A. Adib, K. Viras, H. Matsura, A.J. Ryan, C. Booth, J. Chem. Soc. Faraday Trans. **92**, 3173 (1996).
52. J.B. Heyns, L.M. Sears, R.C. Carcoran, K.T. Carron, Anal Chem. **66**, 1572 (1944).
53. Y. Habata, S. Akabori, Coord. Chem. Reviews **148**, 625 (1996).
54. V. Zhelyazkov, G. Georgiev, Zh. Nikolov, M. Miteva, Spectro Chem. Acta **54A** 625 (1989).
55. M. Zhao, Q. Zhang, Yingyong Huaxue **6**, 48 (1989); CA **111**, 473819 (1989).
56. Wu. Y, An, T. Jingchao, J.S. Bradshaw, R.W. Izatt, J. Incl. Pheno. Mol. Recog. Chem. **9**, 267 (1990).
57. L. Antonov, N. Mateeva, Talanta **41**, 1489 (1994).
58. D. Reinhold, V. Morisson, A.J. Moore, L.M. Goldenberg, M.R. Bryce, J.M. Raoul, M.C. Petty J. Garin, M. Saviron, I.K. Lednev, R.E. Hester, J.N. Moore, J. Chem. Soc. Perkin Trans. **21**, 587 (1996).
59. S. Akabori, S. Tsuchiya, Bull Chem. Soc. Japan **63**, 1623 (1990).
60. O.A. Raevskii, S.V. Trepalin, V.E. Zubareva, Kord Kim. **14**, 1188 (1988); CA **109**, 243002a (1988).
61. S.B. Roy, B.G. Cox, O.S. Mills, N.J. Mooney, C. Ian F. Watt, D. Kirkland, D. Martin, J. Chem. Soc. Perkin Trans. **2**, 2091 (1996).
62. K. Kimura, M. Sumida, M. Yokoyama, J. Chem. Soc. Chem. Commun. 1417 (1997).
63. S.A. Kulyukhin, A.N. Kamenskaya, N.B. Mikheev, Radio Khimiya **33**, 35 (1991); CA **115**, 265750u (1991).
64. M.D. Rudd, G.B. Drew, J. Chem. Soc. Dalton Tran. 811 (1995).
65. H. Li, T. Jiang, I.S. Butler, J. Raman spectroscopy **20**, 569 (1989).
66. G.R.Willey, M. Ravindran, Inorg. Chim. Acta **183**, 167 (1991).
67. R. Salman, A. Marsumi, M. Sabri, Spectro Chem. Acta Part A **49A**, 435 (1993).
68. C.D. Hall, I.P. Danks, P.J. Hammand, N.W. Sharp, M.J.K. Tomas, J. Org. Metallic Chem. **388**, 301 (1990).
69. K.G. Ragunathan, P.K. Bharadwaj, Proc. Ind. Acad. Sci. **107**, 519 (1995).
70. P. Ghosh, P.K. Bharadwaj, J. Chem. Soc. Dalton trans. **15**, 2673 (1997).

71. Y. Hobata, S. Akabori, J. Chem. Soc. Dalton trans, 3871 (1996).
72. I.K. Lednev, T.Q. Ye, R.E. Hester, J.N. Moore, J. Phy. Chem. A **101**, 4966 (1997).
73. I.K. Lednev, R.E. Hester, J.N. Moore, J. Am. Chem. Soc. **119**, 3456 (1997).
74. M.V. Alifimov, Y.V. Fedorov, O.A. Fedorova, S.S Gromov, R.E. Hester, I.K. Lednev, J.N. Moore, V.P. Oleshko, A.I. vedernikov, J. Chem. Soc. Perkin Trans. **2**, 1447 (1996).
75. Y. Xia, J.C. Hu, Z.H. Chen, J.P. He, Fenxi Kexue Xuebao **12**, 10 (1996); Anal. Abst. **58**, 9D163 (1996).
76. G. Janssen Rob, V. Willem, T.G. Bertlutz, J.H. Vandermaas, M. Maczka, J.P.M. Van Duynhaven, D.N. Reinhoudt, J. Chem. Soc. Perkin Trans. 1869 (1996).
77. A.V. Bordunov, J.S. Bradshaw, V.N. Pastushok, X.X. Zhang X. Kou, N.K. Dalley, Z. Yang, P.B. Savage, R.M. Izatt, Tetrahedron **53**, 17595 (1997).
78. T. Arimura, M. Kubota, T. Matsuda, O. Manobe, S. Shinkai, Bull. Chem. Soc. Japan **63**, 1674 (1989).
79. I. Bitter, A. Grun, L. Toke, G. Toth, B. Balazs, I.M. Ziegler, A. Grofesik, M. Kubinyi, Tetrahedron **53**, 16867 (1997).
80. L.J. Charbonniere, C. Balsiger, K.J. Schenk, J.G. Bunzli, J. Chem. Soc. Dalton Trans. 505 (1998).
81. A. Haoyan, W. Yangjie, S. Lianfang, Y. Hanzhen, Q. Jianging-Youji Huaxue, 275 (1986); CA **107**, 216982 x (1987).
82. Z. Jingcheng, S. Lianfang, L. Tianbao, W. Chengtai, Wuli Huaxue Xuebao **4**, 244 (1988); CA **110**, 94373 p. (1989).
83. J. Chan, F.F. Cai, Q.F. Shao, Z.E. Huang, S.M. Chen, J. Chem. Soc. Chem. Commun. 1111 (1990).
84. A. Zhang, S. Wang, S. Huang, S. Qin, Bopuxue Zazhi **7**, 225 (1990), CA **114**, 23312g (1991).
85. A. Vachet, A.M. Patellis, J.P. Goly, M. Goly, J. Barbe, J. Elguero, J. Org. Chem. **59**, 5156 (1994).
86. H. Hara S. Watanabe, M. Yameda, O. Hoshino, J. Chem. Soc. Perkin Trans. I 741 (1991).
87. A.M. Costero, M. Pitarch, C. Andreu, L.E. Ochando, J.M. Amigo, T. Debaerdemacker, Tetrahedron **52**, 669 (1997).
88. P.J. Stang, D.H. Cao, K. Chen, G.M. Gray, D.C. Muddiman R.D. Smith, J. Am. Chem. Soc. **119**, 5163 (1997).
89. Q. Luo, C. Li, X. Feng, J. Wei, H. Hu, A. Dai, Gaodeng Xuexia Xuebao **8**, 5 (1987).
90. M.R. Johanson, C.A. Colburn, S.J. Ganion, B. Son, J.A. Mosbo, I.R. Sousa, Mag Res. Chem. **26**, 197 (1988), CA **108** 211317V (1998).
91. E. Eleinpeter, S. Stoss, M. Gaebler, W. Schroth, Magnetic Resonance Chem. **27**, 676 (1989); CA **112**, 90045e (1990).
92. R.A. Bartsch, J.S. Kim, U. Olsher, D.W. Purkiss, V. Ramesh, N. Kent, D.T. Hayshita, Pure and Applied Chem. **65**, 399 (1993).
93. A.M. Costero, C. Andreu, M. Pitarch, R. Andreu, Tetrahedron **52**, 3683 (1996).
94. M.J. Wilson, R.A. Pethrick, D. Pugh, M. Saiful Islam, J. Chem. Sco. Faradays Trans. **93**, 2097 (1997).
95. E.M. Eyring, D.P. Cobranchi, B.A. Garland, A. Gehard, A.M. Highley, Y.H. Huang, G. Konya, S. Petrucci, R.V. Eldik, Pure and Applied Chem. **65**, 451 (1993).
96. Z. Chen, P.J. Moehs, R.A. Sachlenben, J. Chem. Soc. Perkin II 2549 (1996).
97. P. Szczygiel, M. Shamsipur, K. Hallenga, A.I. Popv, J. Phy. Chem. **91**, 1252 (1987).

98. A. Göcman, M. Bulut, C. Erk, Pure and Applied Chem. **65**, 447 (1993).
99. E. Karkhaneei, A. Afkhami, M. Shamsipur, Polyhedron **12**, 1989 (1996).
100. M.N.S. Hill, J.C. Lockhart, D.P. Mousley, J. Chem. Soc. Dalton Trans. 1455 (1996).
101. S.V.S. Mariappan, S. Subramanian, L. Gomathi, Magnetic Resonce Chem. **29**, 656 (1991); CA **115** 149052 t (1991).
102. R.T. Streeper, S. Khazaeli, Polyhedron **10**, 221 (1991).
103. J. Otsuki, K.C. Russell, J.M. Lehn, Bull Chem. Soc. Japan **70**, 1671 (1997).
104. N.S. Fender, F.R. Fronczek, V. John, A. Kahwa, G.L. McPherson, Inorg. Chem. **36**, 5539 (1997).
105. V.P. Kazachenko, V.O. Zavelskii, I.I. Bulgak, O.A. Raeuskii, Izv Akad. Nauk. SSSR Serkhim 532 (1988); CA **110**, 38966 x (1989).
106. N. Alizadeh, M. Shamsipur, J. Chem. Soc. Farady Trans. **92**, 4391 (1996).
107. A. Mazouz, P. Meunier, M.M. Kubickii, B. Hanquet, R. Amardeil, C. Barnet, A. Zahidi, J. Chem. Soc. Dalten Trans. **6**, 1043 (1997).
108. E. Graf, M.N. Hasseini, R. Ruppert, A.D. Cian, J. Fischer, J. Chem. Soc. Chem. Comm. 1505 (1995).
109. M. Albrecht, S. Kotila, Angew want chemi. **35** 1208 (1996).
110. C.B. Allan, L.O. Spreer, J. Org. Chem. **59**, 7695 (1994).
111. D.K. Chand, P.K. Bharadwaj, Inorg. Chem. **35**, 3380 (1996).
112. P. Ghosh, S. Sengupta, P.K. Bharadwaj, J. Chem. Soc Dalton **6**, 935 (1997).
113. S.Y. Yu, Q.H. Luo, B. Wu X.Y. Huang, T.L. Sheng, X.T. Wu, D.X. Wu, Polyhedron **17**, 453 (1997).
114. K.E. Krakowiak, J.S. Bradshws, J. Org. Chem. **56**, 3723 (1991).
115. S. Jiangao, W. Chengtai, X. Kurifang, S.Q. Fen, H. Yuzhen, Y. Hanzhen Q. Jianquing, and S. Lian fung, Bopuxue Zuzhi **2**, 173 (1985); CA **106**, 4454F (1987).
116. Zong Jingcheng, Shen Lianfang, Lu. Tiantao, Wu. Chengtai, Fenxi Huaxue **14**, 925 (1986); CA **107**, 197 385 m (1987).
117. K. Kubo, N. Kato, T. Sakurai, Bull Chem. Soc. Japan **70**, 3041 (1997).
118. K. Hirose, J. Fuji, K. Kamada, Y. Tobe, K. Naemura, J. Chem. Soc. Perkin Trans. **2**, 1649 (1997).
119. M.V. Alifimov, A.V. Churakov, Y.V. Fedorov, O.A. Fedrova, S.P. Gromov, R.Ę. Hester, J.A.K. Howard, L.G. kuzmina, I.K. Lednev, J.N. Moore, J. Chem. Soc. Perkin Trans. **2**, 2249 (1997).
120. P.D. Becr, A.R. Graydon, J. Org. Metallic Chem. **466**, 241 (1994).
121. F. Contu, F. Demartin, F.A. Devillanova, F.I. Saiai, A. Garau, V. Lippolis, A. Salis, G. Verani, J. Chem. Soc. Dalton Trans. 4401 (1997).
122. A. Mori, K. Kubo H. Takeshita, Cordination Chem. Review **148** 71 (1996).
123. S. Shinkai, S. Mori, T. Tsubaki, T. Sone, O. Manabe, Tetrahedron Letters **25**, 5315 (1984).
124. S. Shinkai, K. Araki, T. Tusbaki, T. Arimura, O. Manabe, J. Amer. Chem. Soc. Perking Trans. **1**, 2297 (1987).
125. J.L. Atwood, D.L. Clark, R.K. Juneja, G.W. Orr, K.D. Robinson, R.L. Vincent, J. Amer. Chem. Soc. 114, 7558 (1992).
126. C.D. Gutsche, I. Alam, Tetrahedron **44**, 489 (1988).
127. S.J. Harris, G. Barrett, M.A. Mckervey, J. Chem. Soc. Comm. 1224 (1991).
128. G. Arena, R. Cali, G.G. Lombardo, E. Rizzarelli, D. Scotto, R. Ungaro, A. Casnati, Supermol. Chem. **1**, 19 (1992).
129. F. Armaud-Neu, M.J. Schwing-Weill, K. Ziat, S. Cremin, S.J. Haris, M.A. Mckervey, New J. Chem. **15**, 33 (1991).

130. A. Ikeda, S. Skinkai, Tetrahedron Letters **33**, 7385 (1992).
131. M. Gomez-Kaifer, P.A. Reddy, C.D. Gutsche, L. Echegoyen, J. Am. Chem. Soc. **116** 3580 (1994).
132. F. Arnaud-Neu, G. Barrett, S.V. Harris, M. Owens, M.A. Mckerrey, M.J. Schwing weill, P. Schwine, Inorg. Chem. **32**, 2644 (1993).
133. Y. Ishikawa, T. Kunitake, T. Matsuda, T. Otsuka, S. Shinkaj, J. Chem. Soc. Chem. Comm. 736 (1989).
134. H. Shimizu, K. Iwanoto, K. Fuji moto, S. Shinkai, Chem. Letter 2147 (1991).
135. G. Deng, K. Sakaki, K. Nakashima, S. Shinkai, Chem. Letters 1287 (1992).
136. B.M. Furphy, J.M. Herrowfield, M.I. Ogden, B.W. Skelton, A.H. White, F.R. Wilner, J. Chem. Soc. Dalton Trans. 2217 (1989).
137. A.M. King, C.P. Moore, K.R.A.S. Sandanayeka, I.O. Sutherland, J. Chem. Soc. Chem. comm. 582 (1992).
138. S. Shinkai, Y. Shirahama, H. Satoh, O. Manabe, J. Chem. Soc. Perkin Trans. 2, **2**, 1167 (1989).
139. J.P. Scharff, M. Mahjoubi, R. Perrin, New J. Chem. **17**, 793 (1993).
140. C.D. Gutsche, K.C. Nem, J. Chem. Soc. **110**, 6153 (1988).
141. C. Floriani, D. Jacoby, A. Chiesi-Villa, C. Guastinii, Angew Chem. **28**, 137 (1989).
142. A. Ikeda, T. Nagsaki, K. Araki, S. Shinkai, Tetrahedron **48**, 1059 (1992).
143. P.R. Deve, G. Doyle, T. Axenrod, H. Yazdekhasti, H.L. Ammon, Tetrahedron Letters **33**, 1021 (1992).
144. S.K. Reddy, C.D. Gutsche, J. Org. Chem. **59**, 3871 (1994).
145. K. Iwamoto, K. Araki, S. Shinkai, Bull Chem. Soc. Japan **67** 1499 (1994).
146. T. Fuji Moto, R. Yanagihara, K. Kaboyashi, Y. Aoyama, Bull Chem. Soc. Japan **68**, 2113 (1995).
147. O. Mayzel, O. Aleksiuk, F. Grynszpan, S.E. Bialiyoohen, J. Chem. Soc. Chem. Comm. 1183 (1995).
148. R. Brindle K. Albert, S.J. Harris, C. Troltzsch, E. Horne, J.D. Glennon, J. Chromatography **731**, 41 (1996).
149. S. Akobri, H. Sonnone, Y. Habata, Y. Mukoyama, T. Ishil, J. Chem. Soc. Chem. Comm. 1467 (1996).
150. G. Arena, A. Casnati, L. Mirone, D. Sciotto, R. Ungaro, Tetrahedron Letters **38**, 1999 (1997).
151. G. Sartori, C. Portu, F. Bigi, R. Maggi, F. Peri, E. Marzi, M. Lanfranchi, M.A. Pellinghelli, Tetrahedron **53** 3287 (1997).
152. T. Haino, M. Yanase, Y. Fukazawa, Angew chem. **36**, 259 (1997).
153. M. Jorgensen, M. Larsen, P.S. Larsen, W.B. Petersen, H. Eggert, J. Chem. Soc. Perkin Trans. **1**, 2851 (1997).
154. M. Nishida, D. Ishii, I. Yoshida, S. Shinkai, Bull Chem. Soc. Japan **70**, 2131 (1997).
155. T. Sone, Y. Ohba, K. Moriya, H. Kumach, K. Ito, Tetrahedron **53**, 10689 (1997).
156. I. Bitter, A. Grun, G. Toth, B. Balazs, L. Toke, Tetrahedron **53**, 9799 (1997).
157. W. Aeungmaitrepirom, Z. Asfari, J. Vicens, Tetrahedron Letters **38**, 1907 (1997).
158. T. Ohseto, S. Shinkai, J. Chem. Soc. Perkin Trans. **2**, 1103 (1995).
159. Y. Israeli, C. Detellier, J. Phy. Chem. B **101**, 1897 (1997).
160. F. Szemes, D. Hesek, Z. Chen, S.W. Dent, M.G.B. Drew, A.J. Goulden, A.R. Graydan, A. Grieve, R.J. Mortimer, T. Wear, J.S. Weightman, P.D. Beer, Inorg. Chem. **35**, 5868 (1996).
161. R. Castro, L.A. Godinez, C.M. Criss, S.G. Bott, A.E. Kaifer, J. Chem. Soc. Chem. Comm. 935 (1997).

162. C. Wieser, D. Matt, L. Toupet, H. Bourgeois, J.P. Kintzinger, J. Chem. Soc. Dalton Trans. 4041 (1997).
163. L. Zhang, A. Macias, T. Lu, J.I. Gordon, G.W. Gokel, A.E. Kaifer, J. Chem. Soc. Chem. Comm. 1017 (1993).
164. E. Pinkhassik, I. Stibor, A. Casnati, R. Ungaro, J. Org. Chem. **62**, 8634 (1997).
165. F.A. Neu, S. Caccumese, S. Fuangswasdi, S. Pappalardo, M.F. Parisi, A. Petringa, G. Principato, J. Org. Chem. **62**, 8041 (1997).
166. M. Stolmer, C. Floriani, A.C. Villa, C. Rizzoli, Inorg, Chem. **36**, 1694 (1997).
167. M. Staffilani, K.S.B. Hancock, J.W. Steed, K.T. Holman, J.L. Atwood, R.K. Juneja, R.S. Burkhalter, J. Am. Chem. Soc. **119**, 6324 (1997).
168. T. Rojsajjakul, S. Veravong, G. Tum charern, T. Tuntulani, R.S. Magee, Tetrahardon **53**, 4669 (1997).
169. E.B. Brouwer, G.D. Enright, J.A. Ripmeester, J. Chem. Soc. Chem. Comm. 939 (1997).
170. R.V. Poponova, G.V. Vasilchenko, M.S. Chapakhin, O.V. Ivanovo, Vysokochist Veshcehestova-200 (1987) Zhukhim 19GD (1987).
171. F. Inokuchi, S. Shinkai, J. Chem. Soc. Perkin Trans. **2**, 601 (1996).
172. K. Wang, X. Han, R.W. Gross, G.W. Gokel, J. Chem. Soc. Chem. Comm. 641 (1995).
173. T. Zhao, Y. Jiang, L. Rang, J. Yang, Q. jiang, C. Zhu, D. Wang, H. Hu, Huaxue Xuebao **44**, 830 (1986); CA **196**, 119126 t (1987).
174. P. Gianni, C. Luciana, A. Carlo, P. Beatrice, T. Pietro, Org. Mass Spectro **22**, 162 (1987); CA **107**, 175340 r (1987).
175. B. Casal, H.E. Ruiz, L. Vanveck, F.C. Adams, J. Incl. Pheno and mol recog. **6**, 107 (1988).
176. Z. Huang, J. Chen, C. Wang, K. Cao, Wuhan Daxue Xuebao Zirankexueban 79 (1988); CA **111**, 193 944C (1989).
177. E. Pinkhassik, I. Stibor, V. Harlicek, Coll Czech Chem. Comm. **61**, 1182 (1996).
178. K. Linnemayr, G. Allmaier, Europian Mass spectroscopy **3**, 141 (1997).
179. D.K. Aslanova, A.K. Tashmukhamedova, R.R. Razakov, Khim Geterotskii Secdin, 898 (1987); CA **108**, 1308539J (1988).
180. P. Gianni, C. Luciana, M. Annamaria, T. Pietro, Org. Mass spectro **21**, 395 (1986); CA **106**, 83851j (1987).
181. E. Blasius, M. Keller, Z. Anal. Chem. **318** 390 (1984), AA **47**, 6C66 (1985).
182. K. Laali, R.P. Lattimer, J. Org. Chem. **54**, 496 (1989).
183. D. Giraud, I. Scherrens, M. Lever, O-Laprevote, B.C. Das, J. Chem. Soc. Perkin Trans. **2**, 901 (1996).
184. N.B. Zolotoi, G.V. Karpov, V.E. Skurat, Teo-Eksp Khim **24**, 239 (1988); CA **109**, 102430 m (1988).
185. R.A.W. Johnstone, M.E. Rose, J. Chem. Soc. Chem. Comm. 1268 (1983).
186. J.M. Miller, K. Balasanmugam, A. Fulcher, Org. Mass Spectro **24**, 497 (1989); CA **111**, 105005 u (1989).
187. B. Kralji, O.S. Timofeer, D. Zigon, T.I. Kirchenko, J. Marsel, Rapid Comm. Mass spect. **10**, 850 (1996); AA **59** 1D152 (1997).
188. N.B. Zolotoè G.V. Kavpov, Doklad Akad Nauk SSSR **303** 381 (1988); AA **51**, 5J117 (1989).
189. E. Sobott, W. Kleinekofort, B. Bratschy, Anal Chem. **69** 3589 (1997).
190. G.M. Gruter, B.M. VanBaar, T.J. Gerrits, O.S. Akkerman, F. Bickelhaupt, A. Barkow, D. Kuck, J. Chem. Soc. Perkin Trans. **2**, 925 (1996).
191. M. Yuan, Y. Zhu, D. Tang, F. Zhang, S. Liu, Yinggung Huaxue **6**, 30 (1989); CA **111**, 67201 V (1989).
192. Y. Habata, F. Fujishiro, S. Akabori, J. Chem. Soc. Perkin Trans. **1**, 953 (1996).

4
Metal Complexes and Their Structures

4.1 Introduction

The structure and binding of new crown compounds was usually determined by size of crown ring and guest; nature of host guest interaction and finally the orientation on the ligation. It is worth considering cation binding and structure of the crown compounds from the viewpoint of host guest chemistry

The donor groups in crown ether decides the selectivity of complexation. Such host guest complexation depends upon the interactions, viz. H-bonding, charge transfer force, hydrophobicity interaction. The molecular interaction is of atmost significance. A rigid crown ring offered strong selective binding on basis of cavity size selectivity consideration but flexible crown ring led to dynamic complexation and acted as effective carrier. An introduction of strong donor group on a crown ring skeleton provides specific site for host guest complexation. The electrochemical process could be coupled with crown guest interaction [1]. For calixarenes the introduction of carboxylate groups in to each benzene ring furnished pseudoplanar hexacoordinated structure in calix (6) arene [2, 3]. They had high stability constant for the size of guest cation (see structure I).

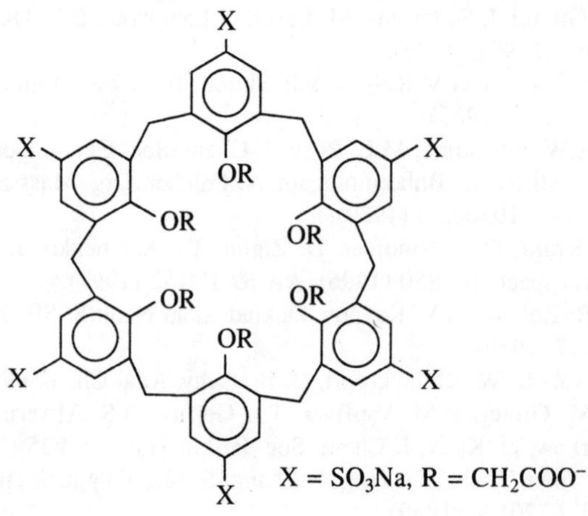

$X = SO_3Na$, $R = CH_2COO^-$

(I)

The small crown ethers showed greater tendency to form complex with small cation, while large cavity size crown compounds preferred large size cations. The conformation of flexible crown compounds were easily adjustable for complexing metal and molecular cations of various sizes. The complexes had now stability. Large size crown complexes generally were less stable [4] in comparison with small size crown complexes. The rapid progress had facilitated the design of crown ethers specific for uncharged guest compounds. Some had enforced cavities lined with convergent H-bonding sites. Armed crown compounds were made to combine the properties of rigid and stable structured crown ether with those of more flexible and labile open chain podand. By incorporating additional ligating groups into crown, three dimensional complex was formed, influencing their solubility in inorganic solvents. Such complexes had significance in solvent extraction procedures.

Cryptand-222 offered a three dimensional arrangement of donor atoms for cation binding, with high stability constant. The single armed crown ethers called 'lariat ethers' had side chain structure. Adjusting the size of of crown ring and varying arrangement of component heteroatoms made it possible to influence the binding strength and selectively of metals like s-block elements (see structure II) [5]. The enhanced binding was due to the presence of ether oxygen on the side arm of the lariat ether [2]. Double armed crown compounds were used to complex such cation which permitted flexibility and binding dynmics [6]. When two side arms were attached to a crown ring, complexation was increased by ligation from opposite sides of crown in the ring [6]. Such complexation was useful for alkali and alkaline earth's [7]. In armed crown compounds characteristics of guest species could be changed, with change in solubility or spectral characteristics [8], for instance redox properties of metal ion complexed with armed crown compounds usually varied astonishingly. One could thus alter ligating arm and parent macrocycle so as to accommodate guest species [9] Multiarmed polyamine macrocycles were known to act as host molecules for alkali metals. They had several ligating arms. Finally synthetic cyclic podands formed complexes of low stability in comparison to crown and cryptands. complexes.

(II)

The synthetic acyclic polyethers did not have intramolecular cavities for guest binding but could form 'pseudo cavity' for cation.

The open chain cryptands were (IV) synthesised, but for them stability constant were smaller than double armed compounds [10] as (III).

Thus cryptands and polycyclic crown compounds were most selective and challenging cation binders. In addition ditopic crown type compounds showed

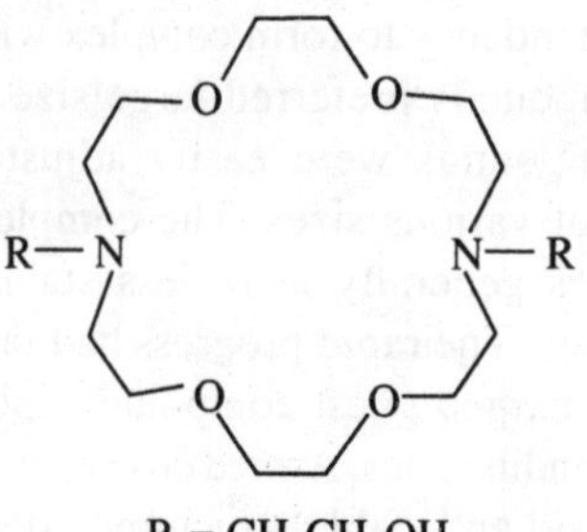

R = CH₂CH₂OH
(III)

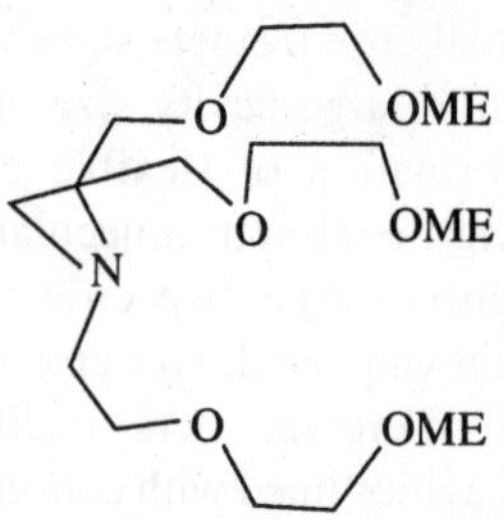

ME = Methylene blue
(IV)

good binding and recognition function. In such compounds two subunits bound guest species. [11] while polycyclic crown compounds were best binders for *d*-block elements [12]. The ditopic compounds formed dinuclear complexes [13] as they were themselves tricyclic compounds in nature.

It was interesting at this stage to examine the structure of cryptates mostly assigned by X-rays. In this compounds the sulphur atom if replaced by oxygen at opposite end of molecule, S-atoms turned outward and oxygen atoms turned inward to occupy cavity void. Further if S-atom were separated by two carbon atoms (Thia 18C6) then ethylene unit occupied the void. Crystal structures for diaza 18C6 was obtained. It showed empty space which existed in crown ether if adopted a symmetrical conformation with all donor turned inward situation was unfavourable. Hence macrocycle accepted the conformations where in space was occupied by a methylene or hydrogen. The structure was symmetrical showing only complexes of idealized size correspondence and symmetry could form, but it was not always true.

A general observation was that the bond distance were not same in 18C6 complex. In BI5C5 complexes (V) difference in M^+-O interaction distance were more pronounced. Na^+-O distance was shorter than K^+-O distance (VI). These distances varied with the coordination numbeer of metal ion, e.g. formed hexacoordinated complex with 18C6.

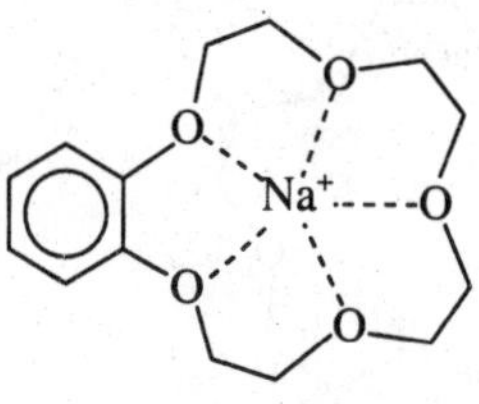

(V) 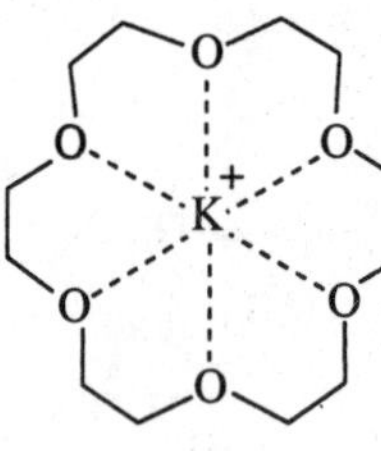   (VI)

When action was large, two crown molecules often participated in a sandwich type complex but when action was too small, more than the action it might be bound by ligand or ligand might wrap around the cation to accommodate required bond length and coordination number. Generally crown ethers and cryptands used donor atoms to stabilize cations or other dipolar species when complexation took place. N or O would usually act as a donor to a positively charged or polarised guest but protonated nitrogen or Lewis acid generally coordinated to a negative dipole of charge. These compounds were capable of structural adjustment so that the variety of guest molecules could be accommodated even when size shape correspondence was not favourable for the purpose of complexation reactions.

As a rule structure of solid complexes of metals with coordinating ligands like crown ethers and cryptands were usually examined by X-ray crystallography. Although such solid complexes had limited applicability in analytical chemistry, we would be considering few structures in order to understand the relationship between structure and the binding. Therefore, some attempt is made to consider few structure in solution and solids.

Usually crystal structure of 1:1, 2:1, 3:2, and 1:2 complexes formed according to size of cavity of crown ether and ionic diameter of cation. Apart from X-ray, thermal and radiochemical methods were also used for such purpose for elucidation of the structures. The crystal structure [14, 15] of RbSCN (VIII) with DB18C6 was studied by Pedersen. The complex was 1:1; rubidium was coordinated by six coplannar oxygen atoms equidistant from cation. The Rb-O distance 2.91 Å was equal to sum of Radius of Rb (1.48 Å) and Vander Waal's radius of oxygen (1.40 Å). Thiocyanate formed an ion pair by coordination of N- to Rb (Rb-N distance 2.94 Å). The structure of complex was as [16] shown in (VII). While the structure of parent compound DB18C6 was as shown in (VIII).

⊙ = Rb, O = O, ● = C

(VII) (VIII)

In contrast B15C5 formed 2:1 complex having sandwich structure with K- whose ionic diameter was larger than Na$^+$ (IX). Here in crystal K$^+$ was on centre symmetry so its O-neighbour was lying in two planes 3.34° apart from a pentagonal antiprism [17] and B15C5 with Na I [18] (X).

B15C5 formed 1:1 complex with Na$^+$, it was also pentagonal pyramid with Na$^+$ 0.75 Å of the plane of five O-atoms towards water molecule at appex H-bonding from water molecule held the complex cation to the anions in chains (O—. 3.51 and 3.47 Å). While B15C5 complex formed 2:1 complex having a sandwich structure with K$^+$ whose ionic diameter was larger than Na$^+$. In the crystal K$^+$ ions were on centre of symmetry so its ten O-neighbours were lying in two planes 3.34° apart from pentagonal antiprism. The change in two structures was due to torsion angles about two C-O bonds. The other structures DC18C6 with Na Br [19, 20], DB24C8 with KSCN [21-23] and DB30C10 with KI [24] had been extensively studied by X-ray crystallographic techniques [25].

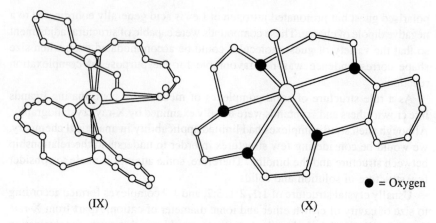

● = Oxygen

(IX)   (X)

Table 4.1 summarises the alkali metal oxygen distance for these complexes while above figures viz. IX and X are crystal structures and conformations of typical complexes of alkali metal with substituted crown ethers like B15C5.

Table 4.1 Metal oxygen distance [25]

Complex-Ligand	M-O (aliphatic)	M-O (aromatic)	M-O (water)	M-N(SCN)
Na I, B15C5·H$_2$O	2.427	2.372	2.285	–
NaSCN; DB18C6	2.735	2.783	–	3.322
NaBr, DB18C6	2.660	2.640	2.310	–
NaBr, DC19C6	2.673	–	2.349	–
NaSCN, 18C6 H$_2$O	2.450	–	–	–
KI (B15C5)$_2$ (2:1)	2.955	2.821	–	–
KSCN 18C6	2.770	–	–	–
KI, DB30C10 (1:1)	2.850	2.895	–	–
RbSCN DB18C6 (1:1)	2.900	2.913	–	2.913

Amongst complexes with alkaline earth the crystal structure of 18C6 with Ca (SCN)$_2$ was determined [26-27] so also that of DC18C6 with Ba (SCN)$_2$ was analyzed [28-29]. In such structures SXO-atoms were lying in same plane and the Ba-O distance and Ba-N distance were 2.77 ~ 2.29 Å and 2.89 Å respectively.

The complexes of crown ethers with lanthanides were also studied [30]. It was noted that the light lanthanide, viz. lanthanum, cerium, prosodyium formed metal complexes of type [Ln crown]$^{3+}$ (NO$_3^-$)$_3$ with B15C5 in acetone while heavy lanthanides like terbium, europium, gadolinium and lutetium formed complexes of type [Ln crown]^3 (NO$_3^-$)$_3$ their stability decreasing with increasing atomic number. Generally heavy lanthanides did not form stoichiometric complexes.

Actinide series of elements complex as UO$_2^{+2}$ and Th^{4+} were not having structure where ion was enclosed into the cavity of the crown ether but such structure was adduct type. The complex of UO$_2^{2+}$ with 18C6 [31-32] were examined as shown in structure (XI).

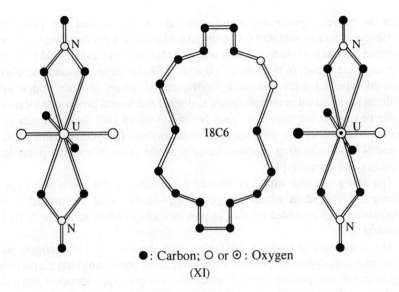

●: Carbon; ○ or ⊙: Oxygen
(XI)

This structure showed that UO_2^{++} ion was not present in crown ring but that neutral molecule $UO_2(NO_3)_2$ was linked to 18C6 by H-bond between water molecules and crown ether. The O-atom of linear UO_2^{++} was linked to uranium by coordination and that sixth-coordination in the equatorial plane of UO_2^{++} was occupied by two molecules of water and two bidendate NO_3^- as shown in structure (XI).

4.2 Structure in the Solution

Just before concluding the discussion on structure and metal complexes it is worthwhile to discuss structure of few complexes in solution. For complexes in solution the lattice forces were absent and such structures were influenced by polarisation of cation-anion behaviour of the solvent molecules and by the structure and behaviour of the ion pair. Some discussion on use of H^1NMR^1, $C^{13}NMR$ had been already covered in the previous chapter, e.g. complexes of B15C6, DB18C6, DB30C10 with sodium, potassium, caesium and barium with counter anions as $I^-SCN^-ClO_4^-$ respectively with each crown ether were studied by NMR technique [33]. The results showed that conformation was more or less same as in crystal lattice, however conformation of DB18C6 before complexation was different. The structure of complexes of DB30C10 was identical to one seen in solid state. On the whole structure and behaviour of the ion pairs of crown ether complexes in solution resembled closely to various uses as solubilisation and ion transport. The Cs^+-DB18C6 complex in solution was sandwich structure in which one $Cs^±$ was bound between two rings of the crown ether moiety.

It is worthwhile to consider few examples for the crystal structures of cryptates. Cryptands were macrobicyclic compounds with two nitrogen atoms, formed complexes with metal ions by strong binding of the ions into space lattice [34-35]. The complexes of cryptand 222 with cobalt (II) [36], or copper [37] and complexes of uranyl with cryptand 2 1 1 and cryptand 2, 2 1 were extensively studied [38]. The tertiary N-atom located on the bridge head were protonated to form the bicyclic quaternary ammonium salt. X-ray showed that protonated diazabicyclo alkane enclosed Cl⁻ in the space lattice to form the complex. It was clathrate compound where in anion was included in cyclic compound [39]. The structures are shown in the figure (XII and XIII). All these cryptand complexes were of type 1:1 with endo-endo configuration where in both bridge head N-atoms participated in complexation as donors and bound cation was located in centre of cage equidistant from each Na and O-donor [40]. In complexes of s-blocks metals cavity size was expanded by the torsion of polyether chains between N and N with increasing cationic diameter in the order of sodium, potassium, rubidium and caesium [41].

The anion cryptate would be formed by binding of the anion into cage of spherical cryptand in which bridge head N-atoms were protonated. Some information was provided on this in previous chapter (see structure XIII) for cryptand 222.

The complexes of cyclic polythiaether (see structure XII) with copper were called blue copper protons [42-43] which were metal containing enzyme for living organism (cf. iron proton) useful in electron transfer oxygen transport and redox catalyst in cell respiration metabolism. Its structure was studied [44] and analogy was drawn with structure of copper with polythiaether [45]. The results showed that there was the coordination of the S-of thioether in methionine residue.

It is worthwhile at this stage to consider structure of solid state complexes of calixarenes with metals. They also form solid complexes with organic solvents [46-48]. Tenacity of holding guest molecule varied from member to member. Cyclic tetramer and hexamer were best for the purpose. These molecular complexes were also called as cryptocavitate clathrato complexes.

The interaction of calixarene with inorganic compounds gave entities which were called 'complexes' or compounds containing oxygen, metal oxygen bonds. They were lower in bridging compounds. The interaction of transition metals with calixarene was studied [49]. A titanium calix(4) arene complex was obtained (XIV) and was studied by X-ray. Here two titanium atoms were sandwiched between a pair of cyclic tetramar as (XV).

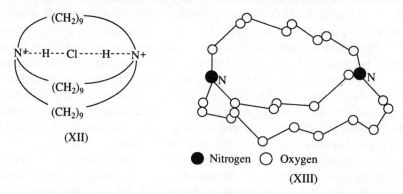

● Nitrogen ○ Oxygen

(XIII)

In this structure two calixarene units formed a partial cone conformation to accommodate tetrahedral coordination geometry around titanium atoms (XV). On same lines complexes of iron or cobalt with calixarene were studied by X-ray

[50]. For iron it showed distorted trigonal pyramidal structure and for cobalt where in all three of cobalt being irregular, cobalt and its symmetry partner had all of their ligands different from one another. A complex of titanium with calix (6) arene was also studied by X-ray [51] indicating that two calixarene units in a cone conformation coordinate with two titanium atoms connected to one another by Ti-O- Ti bond. The titanium atoms were penta-coordinated in slightly distorted trigonal bipyramid. Interestingly this solid state structure was same in solution phase. On same lines lanthanide complexes were also studied [52]. The best example was europium which was also studied [53]. Similar complex of aluminum was also studied [54-55]. The structure of calixarene metal complexes in solution was specially studied in non-aqueous solvents [56]. The study showed that these compounds possessed unusual transport ability. In basic media McKervey carried out extensive studies [57–58] to understand the complexation behaviour of calix (4, 6, or 8) on non-aqueous media. Keto ether gave better information on complexation of alkali metal ions [59]. They were good as the extracting compounds. Finally the calixarene-cation interaction with amides (R=CH_2 CO NH C_4 H_9) was also extensively investigated [60]. Some recent work on structure of such complexes with crown ether cryptands and calix(n) arenes is summarised in subsequent section of this chapter.

(XIV)   (XV)

The complexation of metals with crown ethers cryptands or calix(n) arenes provided an interesting area for further investigation. The complexes between B18C6 and some π-acceptors in methylene chloride when studied spectrophotometrically showed that the Ph-O-CH_2-CH_2- group was responsible for complexation [61]. So far number of reviews have appeared dealing with redox process in the supramolecular coordination compounds [62]. The functionalisation and recognition ability of calixarenes, a third host molecule was discussed, to cover their stereochemistry, complexation phenomena, and organic guest. The selective recognition of alkali metals with calix(n) arenes was most enchanting [63]. The photophysical properties of complexes of lanthanides encapsulated with supramolecular ligand and their application as fluoroscent labels in fluoroimmuno assay was also recently reviewed [64]. An excellent review has appeared on metal complexes on calixarenes, which covered broadly synthesis, rim modification by substitution; attachment of polymers and receptors for s, p and d-block metals with calixarenes [65]. The novel macrocycles were tailored for metal binding, in the process of metal ion recognisation by studying ring size, donor atom, substitution of ring and study of stability. The factors influencing metal ion discrimination were also identified [66]. A comprehensive review on calixarene and calix resorcinarene as ligands in transition metal chemistry was recently published [67] where the emphasis given on synthesis and interesting properties of calixarene such as ability to act as bulky ligand or behave as polyoxo surface, or act as catalytic center. Few applications in liquid crystal sensors are also indicated [67] in this review.

Much light on structure of metal complexes with these compounds was thrown by reports on structure elucidation by conventional method, X-ray crystallography, thermal methods and radiometric techniques.

4.3 Complexations and Structure Elucidation by Classical Methods

Although modern methods like UV, Visible IR, mass spectrometry furnished valuable information on characterisation of complexes, conventional methods provided a definite clue for understanding the geometry of the molecule. Such classical methods were generally resorted to the study of the complexes specially in solution [68-168]. Many methods were very effective not only to investigate complexes of crown ether as well as cryptands but also for study of complexes with calixarene compounds [116-168].

The stability of any complex was usually related to its conformation. It was also possible in such studies to ascertain the rate of the reaction [72]. In many separations it was observed that selectivity ratio was not dependent upon nature of anion used. The consideration of molecular mechanism for some crown complexes with alkali metal showed that difference in strain energy between uncomplexed and complexed entity decided remarkably the stability of the complex [78]. The asymmetrical crown ethers controlled cation selectivity. Further such stability was enhanced with number of donor atoms from 2-7 leading to demonstration of macrocyclic effect [80] in complexation of lanthanide series of elements. The kind of helical wrapping or folding around cation with 18C6 from lanthanide was worth further investigated [83]. The square planner or tetrahedral geometry was observed for cobalt complex with tetraozamacrocyclic tropocard and ligand [92]. Tetrahedrally bound complex of silver with four thiogroup of thia crown ether was confirmed [96]. In many places stereochemistry was controlled by methylene groups present in the structure [97] of the complex.

In comparison to crown ether the cryptand formed stable complexes with the metals. They were called 'cryptates'. An exhaustive review on complexation ion binding properties of cryptand was also published [104]. It was discovered that metallocryptands could be easily solubilised by binding them with alkali metal cation (107). The thermodynamic investigations were usually carried out in propylene carbonate and acetonitrile [108]. In complexation studies of zinc showed less stable complex formation on account of loss of enthalpy and also the domination of entropic part enhancing complexation [114] with no cryptate effect with distorted trigonal bipyramidal structure.

However very useful information was obtained for the complexes of metal with calixarenes. Several review articles have appeared in this area [116-120]. The carboxylic acid substituted calixarene of lower rim indicated wide possibility of variety of complexes with alkali metals [121]. They contained H-bond with nature of complex as 1:1. Further such carboxylation led to unusual stability of such complexes. The platinum metals reacted with calix(4) arene (mono or di) phosphinites to form stable complexes [123]. Amongst various series of calix(n) arene studied (where n = 4, 5, 6) the tetramer was most exhaustively studied. The generation of ϕ mesophase depended on stabilisation by endo-recognition and exo-recognisation [131]. The p-tert butyl calix(4) arene crown-5 complex with barium exhibited ping-pong mechanism commonly encountered in enzyme kinetics [135].

The luminescent compound were derived by complexation of yttrium (III) with calix(4) resorcinarene [140]. In fact calixresorcinarene unfortunately were still not extensively studied from the viewpoint of not only their structures but also from angle of analytical applications. In such cases solvating solvents influenced intensity of the luminescence phenomena [141]. Linear chain polymer of oxa group of calixarene was obtained with Vanadium. A cone conformation was in existance for calix(4) tetramide complexes with the number of transition metals belongings to group VIII and Group IB of periodic classification [151] and the common noble metals.

The p-tert butyl calixarene thioamide derivatives were very promising for complexation of copper and its separation from other interfering elements [153]. An ionic calixarene complexes as cation receptor were also synthesised [154]. The aluminum complex was formed by reaction of the substituted calix(4) arene in toluene [158] to give penta aluminum complex.

The 'CHEMFETS', i.e. chemically modified field effect transistors were prepared by interaction of calix(4) arene thioamide derivative [161]. Octopus calixarene were used as the catalyst for ether preparation. Chemical sensors were obtained from them [163]. Calixarene were used as polyfunctional core molecule to construct with several dendritic polyether macromolecules [165]. Water soluble calixarene were also tested in phase transfer catalyst in nucleophillic substitution reaction [167]. There were numerous practical applications of these supramolecular compounds.

A review on complexation of cryptand and their ion binding properties was published [104]. The caged cation could be obtained by coordination between photo cleavable ligands like cryptand-2, 2 2 formed with 2-6 dihydroxybenzoic acid in toluene [105]. A out-out form doubly protonated macrobicyclic compound with good stability due to presence of C-H—O hydrogen bonds [106]. Heteroditopic alkali metals and heterotritopic receptor sited cryptands were synthesised with alkali ions as templates, they formed inclusion complexes with d-block metals and behaved as amphiphiles and derivatisation with 9-methylanthracene exhibited fluoroscences [107].

Calix(6) arene ester bound arendiazonium ions and did not inhibit deazoniation [115]. A review was available on photochemistry of supramolecule cage complexes [116]. A calix(4) arene and similar compounds formed 'metacyclophanes' [117]. A review of novel methods for the synthesis of calix(4) arene was presented [118]. Thermodynamics of such compounds was listed [119]. The potential industrial applications of calixarenes as cyclic oligomers were also reviewed [120].

Various conventional methods used to determine the structure of metal complexes with macrocyclic and supramolecular compounds are summarised in Table 4.3.1.

Table 4.3.1   Complexation studies for crown ether complexes by conventional method

Element	Complexation Conditions
Alkali metals	When DC18C6 complexes were studied to know the steric factors influencing complexation, it was indicated that the stability was related to the conformation energies of ligands [68]. Plasmadesorphon Mass spectrometry was used to study alkali metal complexation of crown ethers; the ratio of molecular ion abundance reflected the complexation trend [69]. The study of complexes of s-block metals with crown ether carboxylic acids gave information on the stability constant [70]. A linear correlation between number of oxygen donor atoms and reaction enthalpies was observed in study of complexation of crown ether with alkaline earths in solvents like propylene carbonate [71]. In the reaction of crown ethers and metals methyl transfer reactions were accelerated by alkali metals [72]. Six alkides and electrodes containing sandwich complex of caesium was stoichiometric compound so also those of rubidium potassium complexes [73]. Chiral crownether incorporating D-glucose was synthesised to study complexing characteristics with alkali metals [74].
NH_4	Ferrioxamine-B with cis –Syn –cis anticis isomers of DC18C6 in chloroform and complexation with alkali metal and ammonium cation in picrate form when studied showed that the selectivity ratio was independent of anion used and the discrimination between syn- and anti-isomers of DC18C6 was dependent on cation size and it was dominated by preference for bulky complexes [75].
K (I)	Thermodynamic studies of crown ether complexes with alkali metals were studied using 18C6 and 12C4 [76]. The aggregation of crown ether substituted phthalocyanides with the metal gave facially ordered phthalocyanine stocks and network in case of barium [77]. Molecular mechanism calculations were made for structure of potassium ion with crown ether. It showed that the complex stability was correlated to difference in strain energy between uncomplexed and potassium complex and failure to recognise orientation of the ether C-O-C- mostly inter-relation to metal ions big flaw detected in size match selectivity [78].
Rb (I)	A use of asymetrical crown ether showed promise for controlling cation selectivity. Such crown was bis (benzocrown ether) [79].
Ln (III)	They were complexed with 12, 15, 18 and 21 membered crown ethers in propylene carbonate, lighter lanthanone formed stable complex with 15C5 A 1:1 and 1:2 complex was formed with 12C4 and 15C5 respectively; Stability rose with number of donor atoms, their number ranging from 2-7 to show macrocyclic effect [80]. The complexes of tribromides of

(Contd)

Element	Complexation Conditions
	lanthanides were formed with B15C5 in acetonitrile with stoichiometry of complex as 1:1 [81].
La (III)	B12C4 was used to separate it from cerium and yttrium by precipitation [82]. The complexation of lanthanum chloride and other rare earths with 18C6 and pentaethylene glycol showed crown ethers fold around cation while polyethylene glycol showed helical wrapping around lanthanum [83].
Sm (III)	The 1:1 complexes of samarium and thulium were formed with 18C6 by precipitation from tetrahydrofurane from their iodide [84] solutions.
Lu (III)	Bisphthalocyanines complexes with 15C5 and hexyl hexanate was formed [85].
U (VI)	The inclusion of uranyl ion in the cavity of 18C6 was confirmed by the complexation with 15C5, 18C6 21C7 20C10 or there substituted benzo or carboxyl group derivatives in propylene carbonate or acetonitrile. The substitution had no beneficial effect; Log K for 18C6 in propylene carbonate was 5.4 [86].
Zr (IV)	1, 14, 10, 13, tetraoxa 7, 16 diaza cyclo-octadecane when reacted with its phosphate showed in complete reaction with O-variety, but γ-variety was reactive [87].
Cr (III)	18C6 adduct of bis (μ-hydroxo) octa-aquodi chromium (III) mesitylene-2-sulphonate trihydrate showed four bonded coordinated to water molecule [88].
W (VI)	The nature of complex of tungsten carbonyl reacting with 21C7 as showed $H_5O_2^+$ it was coordinated through H-bonds to 21C7 as $[H_5O_2^+ 21C7]$ (WOCl$_5$) while one with DB30C10 complex was (2H$_3$O, DB30C10) [W (Co)$_4$Cl$_{3-}$]$_2$ (C$_6$H$_5$Me) leading on the exposure to formation of (2H$_3$O, DB30C10) (WOCl$_4$CH$_2$O$^-$)$_2$ [89].
Fe (III)	[1, 4, 7, 10, 13, 16 hexaoxa cyclo-octade-2 yl methylthio) methyl] ferrocene and bis (crown substituted) analogue showed uptake of potassium but not lithium or calcium [90]. The 4, 7 bis azacrownether phenanthroline complex of iron (II) was synthesised indicating reorganisation within barium crown ether subunits [91].
Co (III)	Tetraaza macrocyclic tropocorehand ligand formed four coordinate complex in form of square planner or tetrahedral geometry [92].
Ru (III)	New complexes of some polynucleating ligands incorporating terpyridyl and macrocyclic azacrown ether binding sites were formed. In such complex azacrown ether was linked directly to C-4 position of two-terpyridyl group via. N-atom [93] Ru-polypyridine complexes with crown ethers containing dipyridyl group were synthesised and cation binding property was studied by various methods [94].
Pt (IV)	H$_2$PtCl$_6$ with 18C6 formed complex as (H$_{13}$O$_6$) (Pt Cl$_5$(H$_4$O$_2$)$_2$ 18C6 where in (H$_{13}$O$_6$)$^+$ cation was embedded in three-crownether molecules [95].
Ag (I)	The crown thio ether silver (I) and copper (II) complexes were prepared, where in silver was tetrahedrally bound with 4-thiogroups from 4-thiacrownether molecules [96].
Cd (II)	Complexes of tetraaza macrocyclic tropo coronand were prepared and when studied showed stereochemistry was controlled by number of methylene groups [97].

(Contd)

Element	Complexation Conditions
Sb (IV)	Hydronium ion crownether antimonic salt as [H$_3$O18C6; SbCl$_6$], and similar compounds were characterised with 15C5, and 12C4 in acetonitrile [98].
Bi (III)	The trichlorides or bromides when reacted with 12C4, 15C5, B15C5 gave pyramidal complex (BiX$_3$) the first one with 12C4 was seven coordinated while the rest were octacoordinated. Ethylene glycol formed similar complexes with same geometry [99].
Sn (III)	With 12C4 and 15C5 in methanol complexes were formed the species being (SnCl$_2$12C4 MeOH) and (Sn15C5 5MeOH), where [Sn (15C5)$_2$]$^{2+}$ was cationic species [100].
Pb (II)	Napthochromene bound it to give complex which was photodissociated [101]. Tropylium ion complexes with DB24C8 [102] and electron density studies of 18C6 was studied [103].

Table 4.3.2 **Complexation studies of cryptand complexes by conventional methods**

Element	Complexation Conditions
K (I)	Metallocryptand could be solubilized (e.g. $Ga_2L_3^1$) by binding with potassium if the ligand is H_2L^1 when internal oxygen atom was used [108].
La (III)	With cryptand-222 and 221 in acetonitrile and propylene carbonate the complex formed were thermodynamically investigated with lanthanum (III) [109] cryptand-222 complex of lanthanum hydrolysed 4-nitrophenylphosphate to mono ester, best results were obtained with europium complex [110].
Eu (III)	Unsymmetric mononuclear europium cryptates were formed by template condensation and studied by the spectroscopy techniques to ascertain stereochemistry [111].
Cu (II)	Heteroditopic cryptand with 3-palmitoyl side groups when complexed with copper acted as amphiphile and formed vesicles [112]. The diprotonated cryptand complex of copper (I and II) were synthesised they gave homobinuclear complex of metal [113].
Zn (II)	4, 7, 10, 17, 23 pentamethyl –1, 4, 7, 10, 13, 17, 23 heptaazabicyclo (11.7.5) penta cosane (1, 1) and its monocyclic precursor 1, 4, 7, 13 tetramethyl-1, 4, 7, 10, 13, 16 hexaazacycloactadecane were complexed with zinc. Second ligand formed stable complex due to loss of enthalpy but entropic part promoted complexation. They showed no cryptate effect with distorted trigonal bipyramidal structure [114].

4.4 Metal Complexes Structure Studies by X-ray Methods

We discussed how X-crystallographic methods were used to asign structure and understand bonding of complexes with crown ethers cryptands as well as calixarenes. Now it will be our endeavour to consider application of such methods for elucidation of structure of complexes and quantitative analysis. The various

Table 4.3.3 Complexation studies of calixarene with conventional methods

Element	Complexation Conditions
Alkali metals	Calix(4) possessing carboxylic acid ligating group at lower rim formed complexes with s-block metals. When they were potentiometrically studied such studies showed that the presence of H-bonding in 1:1 complex, calixarene acids were strong bond former. The carboxylation led to enhanced stability at the same time decreasing selectivity for sodium [121].
	Bis (benzo crownethers) substituted calix(4) arene at lower rim formed complex, which was 1:1 in nature. B15C5 molecules were four or two attached to calixarene, latter readily complexed with barium, ammonium and potassium as sandwich structure [122].
	Calix(4) arene derived mono- and diphosphorites formed complexes with platinum, palladium and rhodium where they may behave as bridging or chelating ligand [123].
	Calix(4) arene coordination was contained by alkali metal interaction with phenoxide units [124].
	Calix(4) arene derivatives were used for encapsulation of metal ions, e.g. 1.3 dialkloxy calix(4) arene crown 5 fixed in 1-3 alternate position were best for potassium and sodium while calix(6) crown derivative was best for caesium [125].
	Selective bridging of p-tert butyl calix(4) arene with tetra and pentaethylene glycolditosylases gives p-tert butyl calix(4) arene crown ether provided a ligand for selective complexation of potassium [126]. Calix(4) arene like p-tert calix(4) arene was connector to oxygen with single hypervalent main group atom [127].
	Tetraester derivative of calix(4) arene with polymethylene chain spanning two opposite para positions showed sharp change in complexation ability with alkali cations however with short chain compound no extraction of picrates was possible [128].
Li (I)	Torands were host complex compounds for lithium and H-bond donor they were plannar and give 1:1 complex in the presence of picrates [129].
Na (I)	Calix(4) arene tetraethylester was hydrolysed by trifluoroacetic acid with loss of ester group to give monoacid triester but sodium inhibited this process [130]. The influence of complexation of these two metal triflate on self-assembly of tubular supramolecular architecture displaying a columnar mesospace based on tapeshaped monoesters of oligoethylene oxide with 3, 4, 5 tris (n-dodcane 1-glaxy) benzoic acid and polymethyl acrylate was intensively studied. It showed that the generation of a mesophase depended on the stabilisation by endo-recognition in core cylinder and also exo-recognition [131]. Sodium salt of complex p-tert butyl calix(4) arene tetraethyl acetate where suitable anion was present. It had undergone solvent dependent decomplexation if irradiated at low pressure [132] 1-3 dimethyl ether p-tert butyl calix(4) arene when reacted with sodium hydride in tetrahydrofurane gave dimeric metallocalixarene species [133].
K (I)	The mechanism of new reaction involving electron transfer from alkali

(Contd)

Element	Complexation Conditions
	metal supramolecular complex to oxacyclic compounds were observed [134].
Ba (II)	A complex of p-tert butyl calix(4) arene crown-5 exhibited what is called as ping-pong mechanism usually observed in enzyme kinetics [135].
Ln (II)	Complexes with macrocyclic devices as europium cryptate exhbited luminescence [136].
La (III)	5, 11, 17, 23, tetratert butyl 25, 26, 27, 28 tetrakis (diethylcarbomoylmethoxy calix(4) arene reacted with lanthanum-picrate to give 1:1 complex with partial cone structure [137].
Pr (III)	The ionisable calixarene carboxylates were complexed with it, and with europium, ytterbium and thorium. A complex of 1:1 type was formed the binuclear and methoxy forms of type [ML(OMe)] and (M (H$_2$L)] were encountered, p-tert butyl calix(8) arene was best for rare earths [138].
UO$_2$ (II)	Bimetallic complex with cyclic bis (hydroxymethyl) phenolic was obtained as oligomer [139].
Yb (III)	Two calix(4) resorcinarenes were complexed to give luminescent compounds solvating solvents influenced the luminescent phenomena [140]. Transition metals were used for shaping the cavity of calix(4) arene, by binding alkali metal cations inside the cavity specially in π-basic cavities [141].
Zr (IV)	4-tert butyl calix(4) arene activated the enatioselective allylation of the aldehydes catalysed [142]. A square array of zirconium from the full metallation of calix(4) resorcinarene was noted also for iron same reaction was fissible [143].
V (V)	p-methylhexabromotrioxa calix(3) arene reaction with oxovanadium (V) ion to give linear chain polymer with oxo group placed in cup of macrocycle [144].
Mn (II)	A diamondoid structure of rigid tetra functional H-bond donor network with 1, 2 diamino ethane was formed [145].
Co (II)	A upper rim cobaltocenium bridged calix(4) arene was synthesised which exhibited carboxylate anion selectivity [146].
Ru (III)	Electrochemical reduction of 2, 2' bipyridine and bis (2-pyridyl pyrazine) complex was studied in dimethylformamide [147]. Photoinduced energy transfer process for supramolecular species like tris (bipyridine) complexes of ruthenium and osmium was studied [148]. Mono-and bivalent ruthenium upper rim pyridyl complex were studied for selectivity [149].
Pd (II)	Upper rim functionalised calix(4) arene, at 1, 3 position with organopalladium showed first and second sphere coordination of substrate [150].
Fe (III)	Calix(4) arene tetramide was used for complexation of iron, nickel, copper, zinc and lead to give cone conformation, the metal atom was coordinated to 8-oxygen atoms of ligand [151].
Ni (II)	Bis (NN' dimethyl-1, 4, 17 triazo cyclononane) calix(4) arene formed a ferro magnetic dinuclear complex [152].
Cu (II)	p-tert butyl calix(4) arene in thioamide derivative formed complexes with it. They were extractable in dichloromethane and were separated from s-block metals [153]. Anionic calixarene complex of copper (I) as

(Contd)

Element	Complexation Conditions
	action receptor was known when tetraphosphonito calix resorcinarene was reacted [154].
Zn (II)	Dimerisation of mono zinc N-N' bis (4' (meso-triphenyl porphyrinyl) benzyl) 4-13 diaza 18C6 was considered by entropy and enthalpy measurement [155] t-butyl calix(4) arene with ethyl zinc in toluene gave deprotonated bis calixarene complex, calixarene adopted doubly flattened partial cone structure [156].
Hg (II)	p-tetrabutyl-1, 3 disulphanyl calix(4) arene reacted with its acetate to give mononuclear complex, where in mercury was linearly coordinated to thiolate group [157].
AI (III)	1, 3 dimethylether p-tert butyl calix(4) arene when reacted with a aluminum in toluene gave five coordinated metallocalixarene species [158]. The p-tert butyl calix(6) arene reacted with trimethylaluminum to give aluminum complex [159].
Ga (III)	p-tert butyl calix(4) arene reacted in toluene with trimethyl gallium or aluminum to form deprotonated alkyl metal rich complex having double flattened partial cone conformation with the cavity [160].
Pb (II)	The chemically modified field effect transitors (CHEMFETS) were obtained by action of calix(4) arene-thioamide derivative [161].
Organic Molecules	The octopus calixarenes were used as catalyst for ether preparation from phenols [162]. A chemical sensor was obtained by coating thin layer of calixarene to detect solvents [163]. Calix(4) resorcinarene was used for mass sensitive detection of solvent vapour rigid cone calix(4) arene was tested as π-donor systems in polar organic media toward ammonium ion [164]. Calixarene was used as poly functional core molecules to construct with several dendritic polyether macro molecules [165]. Molecular complexation of C-60 with p-benzyl calix(5) arene in toluene led to change in molar volume [166]. Water soluble calixarene were tested as inverse phase transfer catalyst in nucleophillic substitution reaction [167]. Calix(4) arene functionalised at 1, 3 positions of the upper rim with amido groups acted as neutral H-bonding receptors [168].

metal studied by the X-ray technique are summarised in Tables 4.4.1 to 4.4.3. Such work encompassed crown complexes of alkali metals and few transition metals. The crown [169-184] (Table 4.4.1) complexes and cryptates for transition metals were also studied [185-193] (Table 4.4.2). Only in recent year calixarene complexes were studied by X-ray [193-214] (Table 4.4.3).

The intramolecular cation-anion bonding of sodium with lariat ethers was studied [169]. The dimeric nature of sodium complex with DB15C4 was studied by X-ray [170]. The orthorhomic structure of Rb ozonide and 18C6 was ascertained [172]. Single crystal X-ray technique was used to know structure of yttrium with DB30C10 [174]. On same lines copper complex with substituted cyclophone [180] was also examined. The main group elements like gallium [182] thallium [183] germanium [184] were studied for their complexation behaviour with crown ethers. The distorted trigonal pyramidal structure of germanium then was assigned [184].

Table 4.4.1 Complexation studies with crown ethers by x-ray methods

Element	Complexation Conditions
Na (I)	Lariat ethers like system DB14C4 in acetate and propionate derivatives when complexed with sodium showed intramolecular cation-anion bonding [169]. DB15C4 complex with sodium iodide in ratio of 2:2 showed that each sodium atom was coordinated to four oxygen atoms of crown ether unit and one oxygen atom of 15 membered ring of the dimer [170].
K (I)	In [K(18C6) (Co)$_4$ Fe$_2$PPh$_2$Cs$_2$]$_2$ the phosphorus coordination of ligand showed weak contacts between sulphur atom and potassium ion of the crown ether [171].
Rb (I)	The 1:1 complex of rubidium ozonide and 18C6 in liquid ammonia when studied by X-ray showed it to be orthorhombic [172].
Cs (I)	The complex with α-cyanobenzothiazole α-carbaldydeoxime with 18C6 gave [Cs$_2$(18C6)$_3$] [H(Cbto)$_2$]$_2$ 4Hcbto·4H$_2$O, a club sandwich complex with caesium between two crown ligands leading to stability. H-bonding was in crystal packing [173].
Y (III)	A 3:2 complex with DB30C10 was studied by single crystal X-ray technique [174].
W (VI)	21C7 with [W(CO)$_6$] in aqueous solution containing hydrochloric acid and toluene gave complex with bowl like configuration, 21C7 was bound to $(H_5O_2^+)$ in cis form [175].
Re (VII)	1, 4, 7 trithia-cyclononone when reacted with ammonium rhenate in acidified methyl cyanide gave a complex containing thioether group [176].
Ru (III)	Bis (1, 4, 7 trithia cycolononone) complex of metal was used in cancer therapy [177].
Co (III)	Crown ether derivatives containing dipyridylino unit was obtained by direct [178] nucleophillic displacement by glycolates of ethyleneglycolon 6–6^1 bis(chloromethyl) 2, 2^1 dipyridine with cobalt, copper, zinc and palladium as 1:1 and 2:2 complex [179].
Cu (II)	1, 11, 21, trioxa-8, 14 dithias (2, 9, 2) para cyclophane was obtained and reacted with a copper iodide to form novel complex, single crystal X-ray studies showed 2:1 metal complex [180].
Hg (II)	bis (1-3 xylene-18 crown 5) 2-yl) mercury had intramolecular Hg-O coordination. Similar kind of complex was prepared for magnesium [181].
Ga (III)	18C6 was reacted in diethyl ether with metal iodide having structure as Ga$_2$I$_2$ (18C6) and Ga$_2$I$_4$.
In (III)	Also for indium (III) similar type of complex was identified [182].
Tl (I)	DB18C6 extracted it quantitatively in the presence of hexanitrodiphinylaminate ion. The extracted complex was dissociated and thallium was analysed by X-ray fluoroscene [183] method.
Ge (IV)	A penta coordinated 1–2 oxagermetanide was synthesised with [K (18C6)$^+$] X-ray studies showed that it had distorted trigonal bipyramidal structure with two-oxygen atoms at apical position [184].

The only noble metal complexes with cryptand [185-192] were investigated by this technique. The encapsulation of lanthanide in three dimensional network of cryptands was ascertained by this method. The cascade complex of dicopper with cryptate when studied showed that Cu-NNN geometry [187]. Dinuclear

Table 4.4.2 Complexation studies of cryptand complexes by X-ray methods

Element	Complexation Conditions
Gd (III)	The cryptate complex was obtained by reaction of gadolinium thiocyanate and potassium thiocyanate with sodium salt of cryptand in DMF solution; Gadolinium was encapsulated in cryptand [185].
Fe (III)	Metal complexes of ferrocene cryptand was obtained from diaza 15C5 and compound 1, 1, ferrocenediyl (bismethylene-bispyridinum) chloride tosylate, large cavity could accommodate two metal ions simultaneously [178] in the space.
Cu (III)	Hydrolytically sensitive hexaamino and hydrolytically inert octaamino cryptand complexes showed encapsulation of metal ion [186]. A cascade complex of pseudohalide by dicopper cryptate was studied. It showed existance of Cu-NNN geometry [187]. Thiophene linked azacryptand complexes of copper and silver were structurally characterised they showed dinuclear separation [188].
Ag (I)	Pentanuclear silver cluster of polyazocryptand was synthesised it contained pentanuclear cluster incorporated in a polazocryptand [189].
Hg (II)	The di iodide reacted with cryptand-2.2 2 crystal to give [Hg Cy-222]$^{2+}$ cations and [Hg$_3$ I$_8$]$^-$ anion as characterised by single crystal X-ray diffraction [190].
Te (IV)	A compound [K(222cryptand)]$_2$ (Cr(CO)$_5$Te$_3$]$_{1/2}$ C$_2$N$_2$H$_8$ was synthesised from K$_2$Te$_4$, and chromium carbonyl and cryptand −222 [191].
Ge (IV)	Closo- Ge$_5^{2-}$ anion was used to prepare (222 cry K$^+$)$_2$ Ge$_5^{2-}$ THF complex, with trigonal pyramidal geometry [192].

complex of copper and silver with cryptand were also characterised [188]. Pentanuclear cluster with polyazo cryptand was studied by this method. In most of these studies single crystal X-ray diffraction technique proved to be of great significance [179-181].

The structure elucidation in case of supramolecular compounds was very aptly carried out by this technique [193-214]. The bridging of calix(4) arene at upper rim with pyridine moieties was thus ascertained. Caesium was bound within cup formed by p-tert butyl calix(4) arene [194]. The crystal structure of 1-3 calix(4)bis crown diprotic receptor was also examined to know breaking of crown chain. The identical pattern for europium complexes with p-nitrocalix(8) arene as well as p-tert butyl calix(8) arene was discovered [198]. These were binuclear complexes generally unsuitable for the purpose of extraction. The encapsulation of uranylion in calix(4) tetramide complex was ascertained by X-ray technique [199]. The calixresorcinarene complexed with cobalt formed monodentate complex with square planner geometry [200-201]. X-ray methods were extensively used to study bond length as well as bond angles [202]. The dimer trapping of oxomolybdenum (VI) in calix(4) arene in nitrobenzene was also investigated [203]. The coordinating ability of calix(6) arene diphosphite with different platinum metals was ascertained [205]. A polynuclear band sandwich structure of mercury with trimeric perfluoro-phenylene was studied in depth [206]. The abnormal degradation of the calix(6) arene with arylazide [209] was also examined. The inclusion of anionic species like BF_4^-, I^-, ReO_4^- in bowl

Table 4.4.3 Complexation studies of calixarene by x-ray method

Element	Complexation Conditions
NH_4^+	Calix(4) arene bridged at upper rim with pyridine were complexed with ammonium ion the strength of such complexation was increased by electron donor ability of p-substituent on pyridine [193].
Cs (I)	The complex with macrocyclic tetraphenol p-tert butyl calix(4) arene showed that metal atom was bound within the cup formed by ligand, metal was hence close to aromatic carbon atoms than phenolic oxygen with polyhaptobonding [194] bis (synproximally) functionalised calix(4) arene was synthesised. Cation complexation in picrate extraction method gave change in regioselective functionalisation which in turn influenced the selectivity of alkali metals including sodium [195]. Crystal structure of 1-3 calix(4) bis crown diprotic receptor was examined, it showed that preorganisation of ligand to complex of ion second crown ether chain was ruptured [196].
Eu (III)	Reaction of p-tert butyl calix(8) arene in DMF gave 1:1 complex, with calixarene in bidentate coordination mode, with planner structure [197]. The binuclear complex of p-nitro calix(8) arene and p-tert butyl callix(8) arene in DMF when investigated showed similar geometry for both complexes and ligand conformation [198].
UO_2 (II)	Calix(4) tetramide complexes were studied indicating that cation was encapsulated in ligand, uranyl ion was less bound incomparison to alkali metals and solvent content depended on size [199].
Co (III)	Calixresorcinarene obtained from acetaldehyde and resorcinol with 3-ethylene bridge acted as ammonodentate ligand with complex (calix)$_2$ ML which was square plannar complex [200].
Nb (V)	p-butyl calix(4) arene with niobium pentachloride led to formation of mononuclear complex [201].
Cr (VI)	Water soluble calix(4) complexes were studied by X-ray for bond length and bond angles. There were three modes of metal incorporating in layered solid state structure [202].
Mo (VI)	Calix(4) arene was bound by four oxygen atoms of oxomolybdenum (VI) giving dimer trapping nitrobenzene in crystal structure [203].
Fe (III)	Ferrocene carbonyl chloride was condensed with p-t-butyl calix (4) arene and cobaltanium carbonyl chloride to give redox active bisferrocene calixarene [204].
Pd (II)	Calix(6) arene diphosphite was obtained from phosphorus trichloride in turn coordinated with platinum, palladium, rhenium [205].
Hg (II)	Cyclic trimeric perfluorophenylene mercury complex showed polynuclear bent sandwich structure [206].
Li (I)	It formed 1:1 complex as torand dimer threaded by hydrated chain of two lithium ions, with tetrahedral coordination [207].

shaped cavities of calixarene because of cooperative action of two metal ions was revealed at common binding site [211]. It is seen that H-bonding assisted self-assembly at upper rim of calixarene. A lot of interesting information as illustrated by X-ray method is summarised in Table 4.4.1.

A novel p-tert butyl calix(5) arene tris ferrocenoyle ester molecule was prepared showed that guest molecule was bound deep in the cavity of calix(5) arene

[208]. Calix(6) arene had undergone abnormal reaction with arylazide [209]. A H-bonded dimers of tetraurea calix(4)arene was obtained by self-assembly of calix(4) arene molecules [210]. Inclusion of anionic species, e.g. BF_4^-, I^-, ReO_4^- in electron rich bowl shaped cavities of calixarene due to cooperative effect of two or more metal ions around common binding sites was observed [211]. A 1:2:2 inclusion cascade complex with acetone or dichloromethane were isolated as heterocalix (9) arene molecules [212]. H-bonding helped the self-assembly at upper rim of calix(4) dimer [213]. The first acid dissociation constant of calix(4) arene with two p-nitrophenol units was determined in 2 methoxy ethanol water; calix(6) arene molecule was present in cone conformation. [214].

4.5 Structure of Complex by Thermal Methods

Apart from conventional methods as well as the X-ray crystallographic methods thermal methods were explicitly used to ascertain the composition of the metal complexes with macrocyclic, macrobicyclic and supramolecular compounds.

The most important thermal parameters to be considered were free energy change (ΔG), enthalpy change (ΔH), entropy change (ΔS) and heat capacity change (ΔC_p). Usually ΔG) was evaluated from K value and ΔH was measured by thermometric titractions [215-220]. A magnitude of ΔH from temperature dependence of K was also known [221].

The calorific value and entropy change during the complexation were important pertaining to stability constant evaluation. The complex was more stable if larger calorific value and entropy change in complexation reaction was encountered. ΔH reflected bond energy and solvation energy of reacting materials, while ΔS was result of electrostatic factor in relation to hydration and reactants [220]. The influence of ΔS on K was small unless conformation had drastically changed. ΔC_p value was related to conformational changes. For example in 30C10 complex of potassium it was large [222]. Amongst cations with smaller ionic dimeter gave small value for ΔH and ΔS due to large hydration energy. For multivalent ions solvation energy had increased inspite of strong dipole interaction. Transition metals formed stable complexes due to the presence of covalent bonding for them. Large K, ΔH, ΔS values were noted. These parameters were measured by Izatt [219] in methanol water system for example ΔH; ΔS increased with increasing size of crown for sodium ion for 18C6. The difference for 18C6 and 15C5 was small but large for 24C8, 21C7 etc. Such studies were carried out for DC18C6 in THF [221]. The enthalpy change, ΔH and entropy change ΔS in the complexation of cryptand 221, and cryptand 222 for s-block metals had been measured [222-224]. Here on account of cage type structure cryptates were more stable and ΔS and ΔH increased with cationic diameter. The rates of dissociation of cryptates were smaller in comparison with crown ether. Cryptate was dissociated when protonic acid was added as two of bridged nitrogen atom were protonated and converted to quaternary ammonium cation. As a matter of fact the anion cryptate chemistry will experience great growth in future years.

The various thermal methods used to arrive at structure of metal complexes are summarised in Table 4.5.1. Many alkali metals were thermally titrated [225]. In complexation of ammonium ion with calixarene ΔG, ΔS and ΔC_p^o values were evaluated [226]. The geometry of crown complex changed from oxygen-in

Table 4.5.1 Thermal studies of complexation

Element	Conditions of Thermal Studies
Alkali metals	B12C4, B13C4, B15C5, B17C5, B18C6 and B19C6 in methanol were titrated thermometrically with sodium chloride, potassium chloride in methanol to study phenomena of the complexation [225].
Alkyl and NH_4^+ ion	25, 26, 27, 28 tetrahydroxycalix(4) arene-5, 11, 17, 23 tetrasulphonate was titrated thermally at pH 7.1 with alkylammonium ions, viz. $H(CH_2)_n$ NH_3^+ to show presence of four negative charges due to sulphonato, one due to hydroxy group which were deprotonated favouring complexation $\Delta G°$, $\Delta S°$ and high ΔC_p^o were obtained [226].
Na (I)	Complexes with 18C6 and 15C6 occulded on zeolite cages when studied by TGA DTA showed that in oxidative atmosphere crown ethers degraded to the aldehydes and oxidised then to acetic acid and finally to oxides of carbon [227].
K (I)	The complexation of potassium with 18C6 in various solvents showed best results in propylene carbonate. When studied calorimetrically it showed that the geometry changed for crown ether from oxygen in and oxygen out conformation of macrocycles [228].
La (III)	The complex with DB24C8 in thiocyanate media in acetonitrile gave species as La $(SCN)_3$ DB24C8 $2H_2O$ CH_3CN as indicated by TGA DTA [229].
	The thermal lensing detection method was used to study complexation of lanthanide elements with 18C6, 15C5 by solvent extraction, such signals were large in the organic phase, the cavities in crown ethers limit complexation unless ionic size had matched with the cavity as it happened in europium and prosdymium with 28–42% extraction [230].
	The lanthanide isothiocyanate complexes with 13C4 in ethylacetate were $Ln(NCS)_3 (13C4)_2 2H_2O$, TGD DTA studies indicated metal ligand bonding [231].
Ln (III)	N-methoxyethyl aza 16C5 lariat in acetonitrile was studied calorimetrically. The complexation showed to be entropy driven while aza 16C5 lariat ether complexation with cation selectivity was entropic loss. Such loss was reduced when matching of cavity of lariat ether with ionic diameter of lanthanide took place [232] during complexation.
	Light lanthanides were complexed with 15C5, 16C5 and derivatives in acetonitrile and studied calorimetrically by titration, 1:1 complexes were formed, with cation selectivity which was related to loss of entropy [233]. The entropy loss was minimised as before when size match of cavity and ionic diameter was materialised [233].
	Calorimetric titration of aqueous 18C6 with Ba $(ClO_4)_2$ was carried out for their analysis, 18C6 in methanol. It was also thermally titrated with potassium chloride [234] synthesis and thermal regeneration of polymeric crown ethers was studied [235]. The decomposition of macroazonitriles containing crown ethers was studied by DSC [236]. A calorimetric titration technique was used to study 1:1 and 2:1 complexes of crown ethers in solution [236]. The p-tert butyl calix(n) arene (where n = 4, 6) was studied by TGA to study inclusion of organic solvent molecules in it. The ratio varied from 1:2 to 4:1 for host guest [237].

oxygen-out [228-230]. La $(SCN)_3$ DB24C8 species was thermally established [229] another complex with 13C4 for lanthanum was examined [231]. The cation selectivity was ascertained from loss of entrophy [232] but such loss was minimised when size match of cavity and ionic diameter was checked [233].

The complex of Ba $(ClO_4)_2$ with 18C6 was studied by thermal titration in potassium chloride solution [234]. The thermal regeneration of the polymeric crown ethers was also investigated [235]. Differential Scanning Calorimetry method was used to known for decomposition of macroazanitrites entaining crown ethers [236]. The nature of complex formed between metal and crown ether was evaluated by calorimetry by [236]. The inclusion of organic solvent on cavity of calix(n) arene was thermally studied.

4.6 Structure of the complex by radiochemical methods
In comparison to thermal or X-ray methods the radiochemical methods were rarely used to ascertain the structure of metal complexes with macrocyclic macrobicyclic and the supramolecular compounds. However the alkali metals and few lanthanides were studied by radiochemical methods [238].

For instance sodium, rubidium and strontium complexes with crown ether were studied by these methods [239-247]. Sodium was irradiated with thermal neutrons (flux 10^{-3}) for 20 minutes and was extracted in DC18C6. The stoichiometry of the complex was then ascertained. The analysis of sodium was done by radio analytical technique [239]. The conversion of sodium borohydrate ($NaBH_4$) to amine borane [241] was similarly studied.

The technique of neutron activation analysis was much useful to determine metal after its solvent extraction with crown ether, e.g. Rb was extracted with 18C6 with picric acid or tetraphenylborate as the counter ion in nitrobenzene and then analysed by activation analysis [242] Sr^{85} was extracted by crown ether to indicate that amido group influenced process of extraction [243]. Thorium was extracted by 16C5 oxyacetic acid at pH 4-6 and analysed by activation analysis [244]. The analysis of uranium from water was carried out by first extracting it with long chain crown ether followed by analysis by activation technique [245]. Radio active ^{59}Fe was extracted at 1-9M hydrochloric acid with crown ethers [246].

However, so far no attempt was made to study metalocalixarene complexes by radiometric methods. Various radiochemically determined metals with crown ethers are summarised in Table 4.6.1.

Conclusion
Thus we note that study of structure of complexes irrespective of whether they were present in solid state or in the solutions throws clear light on nature of bonding between ligands and metals. They offer important clue to the chemistry of complexation of elements so far not studied by the research workers. Very valuable information was obtained pertaining to cone and cavity in calix(n) arenes and lariat crown ethers and double armed crown ethers. X-ray crystallography had extensively used to ascertain structure of complexes and the nature and symmetry of bonding. A good amount of work was reported on

Tabel 4.6.1 Complexation studies by radiochemical method

Element	Conditions of Analysis or Characterisation by Radiometry
Na (I)	The samples containing sodium was irradiated with thermal neutrons (10^{-3} flux) for 20 min and latter it was mixed at pH 5–8.5 with sodium tetraphenyl borate and DC18C6 in chloroform and it was extracted and determined substoichiometrically. Caesium, calcium, barium did not interfere but [239] potassium showed strong interferences. Sodium with many other metals was extracted with ether diamide, ether carboxylic acid and N-alkylamino diacetamide with picrate as counter anion, diamide proved to be inefficient for extraction of sodium but amino podants were good for other [240]. Sodium borohydrate ($NaBH_4$) when catalysed by crown ether with R_3N-HCl gave amine borane [241].
Rb (I)	With 18C6 at pH 6–0 with tetraphenyl borate as counter anion or picric acid was extracted in nitrobenzene and analysed by neutron activation analysis technique [242].
Sr (II)	The extraction with open chain ether diamide, ether carboxylic acid amide and N-alkylamino diameters and analysis of Sr^{85} showed that amido group had influenced the extraction properties [243].
Th (IV)	At pH 4-6, α-sym dibenzo 16C5 oxyacetic acid quantitively extracted it but not lanthanium in chloroform in absence of any counter anion and analysed by activation technique [244].
U (VI)	At pH 6, it was extracted quantitatively in chloroform and analysed by activation analysis with (6, 7, 9, 10, 13, 19 hexahydrodibenzo) 1, 4, 7, 10, 13) penta oxacyclohexa decin-18yl oxyacetic or hexaonic acid derivative uranium (VI) from water was analysed [245].
Fe (III)	Radio active ^{59}Fe (III) was extracted with crown ethers and podands in chloroform thus separated from divalent species in 1–9M hydrochloric acid [246].

structure of alkali metal complexes with crown ethers. Some alkaline earths were studied. It was interesting to note that structure of uranyl ion with 18C6 was sandwich type structure. Cryptands were also studied by X-ray methods to discover their unusual stability. The elucidation of titanium calixarene complex structure was real challenge. Several conventional methods were used to unearth structure of crown complex of *s*, *f* and few *d*-block metals. X-ray technique was principally used to study complexes of lanthanides and coinage metals. The thermal methods had come to forefront in recent years. Alkali metals, lanthanides were best studied for their complexation behaviour with crown and cryptands. Unfortunately thermal methods were not used for calixarene complexes. The biochemical methods were still in infancy stage and only inner transition metals were studied by these method. With the knowledge of structure, stereochemistry, and stability of these complexes it would be quite simple task for chemist to develop novel methods for the separation of metals by the process of solvent extraction. Such methods are described extensively in next chapter of this monograph.

References

1. K. Maruyama, H. Sohmiya, H. Tsukube, Tetrahedron letter **26**, 3583 (1985).
2. K. Maruyama, H. Sohmiya, H. Tsukube, J. Chem. Soc. Perkin **1**, 2089 (1986).
3. S. Shinkai, H. Koreishi, K. Ueda, T. Arimura, O. Manabe, J. Am. Chem. Soc. **109**, 6371 (1987)
4. J.W. Huliteriwijk, C.J Vanstavern, D.N. Reinhoudt, H.J. der Hertog Jr. L. Kruise, S. Harkema, J. Org. Chem. **51**, 1575 (1986).
5. R.B. Davidson, R.M. Izatt, J.J. Christensen, R.A. Schultz, D.M. Dishong, G.W. Gokel, J. Org. Chem, **49**, 5080 (1984).
6. L. Echegoyen A. Kaifer, H. Durst, R.A. Schultz, D.M. Dishong, D.M. Goli, G.W. Gokel J. Am. Chem. Soc. **106**, 5100 (1984).
7. H. Tsukube, K. Yamashita, T. Iwanchido, M. Zenki, Tetrahedron Letter **29**, 569 (1988).
8. Y. Inoue, C. Fujiwara, K. Wada, A. Tai, T. Hakushi J. Chem. Soc. Chem. Comm. 393 (1987).
9. F.M. Menger, Topics in current chem. Vol. 136 Springer Verlag (1986).
10. H. Tsukube, K. Yamashita, T. Iwachido, M. Zenki, J. Chem. Res. 104 (1988).
11. I. Willner, Z. Goren, J. Chem. Soc. Chem. Comm. 1158 (1986).
12. A.M. Bond, G.A. Lawrance, P.A. Lay, A.M. Sargeson, Inorg. Chem. **22**, 2010 (1983).
13. J. Jazwinski, J.M. Lehn, D. Lilienbaum, R. Ziessel, J. Guilhem, C. Pascord, J. Chem. Soc. Chem. Comm. 1691 (1987).
14. D. Bright, M.R. Truter, J. Chem. Soc. B. 1544 (1970).
15. D. Bright, M.R. Truter, Nature **225**, 176 (1970).
16. M.R. Truter, C.J. Pedersen, Endeavour **30**, 142 (1971).
17. P.R. Mallinson, M.R. Truter, J. Chem. Soc. Perkin II 341 (1972).
18. M.A. Bush, M.R. Truter, Chem. Comm. 1493 (1970).
19. M. Mercer, M.R. Truter, J. Chem. Soc. Perkin II 2215 (1973).
20. D.E. Fenton, M. Mercer, M.R. Truter, Biochem. Bisphys. Res. Comm. **48**, 10 (1972).
21. D.E. Fenton, M. Mercer, N.S. Poonia, M.R. Truter, Chem. Communication 2469 (1973).
22. M. Mercer, M.R. Truter, J. Chem. Soc. Dalton Trans. 2469 (1973).
23. I.R. Hanson, D.L. Hughes, M.R. Truter, J. Chem. Soc. Perkin Trans. **2**, 972 (1976).
24. M.A. Bush, M.R. Truter, J. Chem. Soc. Perkin II **342** (1972).
25. M.R. Truter, Structure and bonding Vol. 16 p. 71 Springer Verlag (1973).
26. J.D. Dunitz, M. Dobler, P. Seiler, R.P. Phizackerley, Acta Crystallogr. B **30**, 2733 (1974).
27. J.D. Dunitz, M. Dobler, P. Seiler, Acta Crystallo **30**, 273a (1974).
28. N.K. Dalley, J.S. Smith, S.B. Larson, J.J. Christensen, R.M. Izatt, Chem. Comm. 43 (1975).
29. N.N. Dalley, D.E. Smith, R.M. Izatt, J.J. Christensen, Chem. Comm. 90 (1971).
30. R.B. King, P.R. Heckley, J. Am. Chem. Soc. **96**, 3118 (1974).
31. G. Bombieri, G. Devpaoli, A. Cassol, A. Immirzi, Inorg. Chem. Acta **18**, 223 (1976).
32. P.G. Eller, R.A. Penneman, Inorg. Chem. **15**, 2439 (1976).
33. D. Live, S.I. Chan, J. Am. Chem. Soc. **98**, 3769 (1976).
34. B. Dietrich, J.M. Lehn, J.P. Savuage, J. Blanzet Tetrahedron **29**, 1629 (1973).
35. B. Dietrich, J.M. Lehn, J.P. Savuage, Tetrahedron **29**, 1629 (1973).

36. F. Mathieu, R. Weiss, Chem. Comm. 816 (1973).
37. F. Dejehet, R. Debuyst, Inorg. Nucl. Chem. Letter **11**, 711 (1975).
38. R.M. Costes, G. Folcher, P. Plurien, R. Rigny, Inorg. Nucl. Chem. Letter **12**, 491 (1976).
39. J. Cheney, J.M. Lehn, Chem. Comm. 487 (1972).
40. R.A. Bell, G.G. Christoph, F.R. Fronzeck, R.E. Marsh, Science **190**, 151 (1975).
41. B. Metz, D. Moras, R. Weiss, Chem. Comm. 217 (1970).
42. J. Peisach, P. Aisen, W.E. Blumberg (Ed), Biochemistry of Copper Academic Press (1966).
43. R. Malkin, Inorganic Biochemistry (Ed. G.L. Eichhorn) Vol. 2. p. 689 Elsevier (1973).
44. Y. Sugiura, Y. Hirayama, Kagaku no. Ruoiki **31**, 874 (1977).
45. E.R. Dockla, T.E. Jones, W.F. Sokol, R.J. Engerer, D.B. Rorabacher, L.A. Ochrymowycz, J. Am. Chem. Soc. **98**, 4322 (1976).
46. C.D. Gutsche, B. Dhawan, K.H. No, R. Muthukrishana, J. Am. Chem. Soc. **103**, 3782 (1981).
47. M. Coruzzi, G.D. Andreetti, V. Bocchi, A. Pochini, R. Ungaro, J. Chem. Soc. Perkin Trans. **2**, 113 (1982).
48. G.D. Andreetti, R. Ungaro, A. Pochini, J. Chem. Soc. Chem. Comm. 1005 (1979), 533 (1981).
49. M.M. Olmstead, G. Sigel, H. Hope, X-Xu, P.P. Power, J. Am. Chem. Soc. **107**, 8087 (1985).
50. C.D. Gutsche, Calixarenes-Royal Chemical Society, London (1999).
51. G.D. Andreetti, G. Calestani, F. Ugozzoli, A. Ardunini, E. Ghidini, A. Pochini, R. Ungaro, J. Incl. Pheno. **5**, 123 (1987).
52. B.M. Furphy, J.M. Harrowfield, D.L. Kepert, B.W. Skelton, A.H. While, F.R. Wilner, Inorg. Chem. **26**, 4231 (1987).
53. B.M. Furphy, J.M. Harrowfield, F.R. Wilner (unpublished work) cited in ref. 52.
54. S.W. Bott, A.W. Coleman, J.L. Atwood, J. Incl. Pheno. **5**, 747 (1987).
55. S.G. Bott, A.W. Coleman, J.L. Atwood, J. Am. Chem. Soc. **108**, 1709 (1986), **110**, 610 (1988).
56. C. Rizzoli, G.D. Andreetti, R, Ungaro, A. Pochini, J. Mol. Structure **82**, 133 (1982).
57. M.A. Mckerrey, E.M. Seward, G. Ferguson, B.L. Ruhl, J. Org. Chem. **51**, 3581 (1986).
58. C.D. Gutsche, J.R. Rogers, G.F. Stanley, R.K. Sugeeth (unpublished work) cited in ref 50.
59. V. Bocchi, D. Foina, A. Pochini, R. Ungaro, Tetrahedron **38**, 373 (1982).
60. C.D Gutsche, L.G. Lin, Tetrahedron **42**, 1633 (1986).
61. El-Din-Nour, M. Ahmed, Spectrochemica Acta **42**, 637 (1986).
62. G. De Santisgiancarlo, Fabbrizzi Luigi, Licchelli Maurizio Pallauicini Piersandro, Cord. Chem. Review **120**, 237 (1992).
63. M. Takeshita S. Shinkai, Bull Chem. Soc. Japan **68**, 1088 (1995).
64. N. Sabbatini, M. Guardigli, J.M. Lehn, Coord. Chem. Review **123**, 201 (1998).
65. D.M. Roundhill, Prog. Inorg. Chem. **43**, 533 (1995).
66. L.F. Lindoy, Pure and applied Chem. **69**, 2179 (1997).
67. C. Wieser, C.B. Dieleman and D. Malt, Coordination chem. Review **165**, 93 (1997).
68. V.E. Kuzmin, L.P. Trigub, N.F. Aletkina, Y.A. Simonov, Izv Akad Nauk, Mold SSR Ser fiz Tekh Mat Nauk 23 (1989); CA **113**, 58195K (1990).
69. N. Malhotra, P. Roepstorff, T.K. Hansen, J. Becher, J. Am. Chem. Soc. **112**, 3709 (1990).

70. B. Heinz, E. Uhlemann, K. Gloe, P. Muchl, Polyhedron **10** 1591 (1991).
71. H.J. Buschmann, E. Schollmeyer, Thermochim Acta **211** 13 (1992).
72. R. Cacciapaglia, L. Mandolini, Pure and applied Chem. **65**, 533 (1993).
73. R.H. Huang, J.L. Eglin, S.Z. Huang, L.E.H. McMills, J.L. Dye, J. Am. Chem. Soc. **115**, 9542 (1993).
74. P.P. Kanakamma, N.S. Nani, U. Maitra, V. Nair, J. Chem. Soc. Perkin trans-I 2339 (1995).
75. I.B. Haberle, I. Spasjevic, R.A. Bartsch, A.L. Crumbliss, J. Chem. Soc. Dalton trans. 2503 (1995).
76. K. Ohtsu, Buneski Kagaku **46**, 167 (1997); AA **59**, 6D30 (1997).
77. O.E. Sielcken, W. Drenth, R.J.M. Nolte, Rec. Trav. Chim. **109**, 425 (1990).
78. B.P. Hay, I.R. Rustad, C.J. Hosteler, J. Am. Chem. Soc. **115**, 11158 (1993).
79. T. Ikeda, A. Abe, K. Kikukawa, T. Matsuda Chem. Letter 369 (1983), AA **45**, 3A15 (1983).
80. J.G. Bunzli, F. Pilloud, Inorg. Chem. **28**, 2638 (1989).
81. T. Minyu, H. Anquan, G. Xinmin, T. Ning, Lanzhou Daxue Xuebao Ziran Kexyeban **26**, 65 (1990), CA **118**, 159780f (1990).
82. N.I. Pedrurova, N.I. Snezhko, L.I. Martynenko, Z.M. Alibaeva, Vysokochist Veshehestova 207 (1989); CA **110**, 1999 87 V (1989).
83. R.D. Rogers, A.N. Rollis, R.D. Etzenhouser, E.J. Voss, C.B. Bauer, Inorg. Chem. **82**, 3451(1993).
84. A.N. Kamenskaya, N.B. Mikheev, V.I. Spitsyn, Dokl Akad Nauk SSSR **275**, 913 (1984); AA **46**, 12B 68 (1984).
85. D. Pernin, K. Haberroth, J. Simon, J. Chem. Soc. Perkin Trans-I 1265 (1997).
86. J. Lagrange, J.P. Metabanzoulou, P. Fux, P. Lagrange, Polyhedron **8**, 2251 (1989).
87. T. Kijima, K. Sakoh, M. Machida, J. Chem. Soc. Dalton Trans. 1245 (1996).
88. A. Drljaca, D.C.R. Hockless, B. Moubaraki, K.S. Murray and L. Spiccia, Inorg. Chem. **36**, 1988 (1997).
89. P.C. Junk, J.L. Atwood, J. Chem. Soc. Dalon 4393 (1997).
90. B. Czech, A. Czech, S.I. Kang, R.A. Bartsch, Chem, Letters 37 (1984) AA **46** 9B 14 (1984).
91. M. Schmittel, H. Amman, J. Chem. Soc. Chem. Commun, 687 (1995).
92. L.H. Doerrer, M.T. Bautista, S.J. Lippard, Inorg. Chem. **36**, 3578 (1997).
93. B. Whittle, S.R. Batten, J.C. Jeffery, L.H. Rees, M.D. Ward, J. Chem. Soc. Dalton trans. 4249 (1996).
94. V.W. Yam, V.W. ManLee, F.Ke. Kam wing Micheal Siu, Inorg. Chem. **36**, 2124 (1997).
95. D. Steinbarn, O. Gravenhorst, H. Hartung, U. Baumeister, Inorg. Chem. **36**, 2195 (1997).
96. M. Munakata, L.P. Wu, M. Yamamoto, T.K. Suwa, M. Maekawa, J. Chem. Soc. Dalton trans. 3215 (1995).
97. L.H. Doerrer, S.J. Lippard, Inorg. Chem. **36**, 2554 (1997).
98. G.R. Willey, M. Ravindran, Inorg. Chim. Acta **183**, 167 (1991)
99. R.D. Rogers, A.H. Bond, S. Aguinaya, A. Reyes, J. Am. Chem. Soc. **1141**, 2967 (1992).
100. B. Beagley, D.G. Nicholson, Acta. Chem. Scand **43**, 527 (1989).
101. M.T. Stauffer, D.B. Knowles, C. Brennan, L. Funderburk, F.T. Lin, S.G. Weber, Chem. Commu. 287 (1997).
102. M. Lamsa, T. Surosa, J. Pursiainen, J. Husskonen, K. Rissanen, J. Chem. Soc. Chem. Comm. 1443 (1996).
103. A. Knochel, A. Hannch, F. Lunger, T. Koristanszky, J. Buschmann D. Zobel, Pure and applied Chem. **65**, 503 (1993).

104. J. Smid, Ind. Eng. Chem. Prod. Res. Dev. **19**, 364 (1980) AA **41**, 4A11 (1981).
105. E. Grell, R. Warmuth, Pure and Apllied Chem. **65**, 373 (1993).
106. L.R. MacGillivray, J.L. Atwood, Chem Commun. 477 (1997).
107. D.K. Chand, P, Ghosh, R. Shukla, S. Sengupta, G. Das, P. Bandopadhyay, P.K. Bharadwaj, Proc. Ind. Acad. Sci. **108**, 229 (1996).
108. M. Albrecht, O. Blau, Chem. Comm. **4**, 345 (1997).
109. F. Angela, D. deNemor, M. Claude, R.M. Joreschling-Wiell F. Arnaud-Neu, J. Chem. Soc. Fardn trans. **86**, 89 (1990).
110. S.J. Oh, K.H. Song. J.W. Park, J. Chem. Soc. Chem. Comm. 575 (1995).
111. S.Y. Yu, Q. Huong, B. Wu, W.J. Zhang X.T. Wu, J. Chem. Soc. Dalton trans. 3883 (1996)
112. P. Ghosh, T.K. Khan, P.K. Bhardwaj, J. Chem. Soc. Chem. Comm. 189 (1996).
113. H. Adams, N.A. Bailey, D.E. Fenton, M.B. Hursthouse, W. Kanda, M. Kanesato, K.M. Abdul Malik, E.J. Sadler, J. Chem. Soc. Dalton Trans. **6**, 921 (1997).
114. A. Bencini, A. Bianchi, P. Dapporto, V. Fusi, P. Paoli, E.G. Espana, M. Micheloni, P. Paoletti, A. Rodrigues, B. Valtancoli, Inorg. Chem. **32**, 2753 (1993).
115. S. Shinkai, S. Edamitsu, T. Arimura, O. Manabe, J. Chem. Soc. Chem. Commun. 1622 (1988).
116. V. Balzani, Pure and applied chem. **62**, 1099 (1990).
117. I. Alam, C.D. Gutsche, J. Org. Chem. **55**, 4487 (1990).
118. L.C. Grocnen, D.N. Reinhoudt, Nat. Ass. Ser. 371 (1992); CA **118** (1993).
119. A.F.D. Namor, Pure and applied Chem. **65**, 193 (1993).
120. R. Perrin, R. Lamartine, M. Perrin, Pure and applied Chem. **65**, 1549 (1993).
121. F.A. Neu, G. Barett, S.J. Harris, M. Ownens, M.A. Mckervey, M.J.S. Weill, D. Schwinte, Inorg. Chem. **32**, 2644 (1993).
122. P.D. Beer, M.G.B. Drew, R.J. Knubley, M.I. Ogden, J. Chem. Soc. Dalton trans. 3117 (1995).
123. C. Locber, D. Matt, P. Briard, D. Grandjean, J. Chem. Soc. Dalton trans. 513 (1996).
124. V.C. Gibson, C. Redshaw, W. Clegg. M.R.J. Elsegood, Chem. Comm. 1605 (1997).
125. R. Ungaro, A. Aruduini, A. Casnati, A. Pochini, F. Ugozzoli, Pure and Applied Chem. **68** 1213 (1996).
126. E. Ghidini, F. Ugazzoli, R. Ungaro, S. Harkema, A.A. Fadl, D.N. Reinhoudt, J. Am. Chem. Soc. **112**, 6979 (1990).
127. D.V. Khasnis, M. Lattman, C.D. Gutsche, J. Am. Chem. Soc. **112**, 9422 (1990).
128. M.A. Mckervey, M. Owens, S. Cremm, E.M. Collins, V. Bohmer, W. Wogt, H. Goldman, J. Org. Chem. **55**, 2569 (1990).
129. T.W. Bell, P.J. Cragg, M.G.B. Drew, Angewante Chem. **31**, 348 (1992).
130. V. Bohmer, V. Gogt, S.J. Harns, R.G. Leonard, E.M. Collins, M. Deasy, M.A. Mckervey, M. Owens, J. Chem. Soc. Perkin Trans. I 431 (1990).
131. V. Percee, J.A. Heck, D. Tomazus, G. Ungar, J. Chem. Sco. Perkin trans. II 2381 (1993).
132. G. Barret, D Corry, B.S. Creaven, B. Johnson, M.A. Mckarvey, A. Rooney, J. Chem. Soc. Chem. Comm. 363 (1995).
133. S.R. Dubberley. A.J. Blake, P. Mountford. Chem. Comm. 1603 (1997).
134. Z. Jedlinski, Pure and Applied Chem. **65**, 483 (1993).
135. C. Roberta, J. Am. Chem. Soc. **114**, 10956 (1992).
136. N. Pietraszkiewics, J. Karpiuk, A.K. Rout, Pure and Applied Chem. **65**, 563 (1993).
137. P.D. Beer, M.G.B. Drew, M.I. Ogden, J. Chem. Soc. Dalton trans. **9**, 1489 (1997).

138. M.A. Mckervey, M.S. Weill, P. Schwinte, A. Walker, J. Chem. Soc. Dalton trans. **3**, 329 (1997).
139. P. Thurey, M. Nierlich, J. Chem. Soc. Dalton **9**, 1481 (1997).
140. S.V. Sherchuk, N.Y. Rusakova, A.M. Turianskaya, Y.V. Karovin, N.A. Nazarenko, A.I. Gren, Y.E. Shapiro, Anal. Comm. **34**, 203 (1993) AA **59**, 10D40 (1997).
141. A. Zanotti, Gerosa, E. Solari, L. Giannini, C. Floriani, A.C. Villa, C. Rizzoli, Chem. Comm. **183**, (1997).
142. S. Casolari, P.G. Cozzi, P. Orioli, E. Tagliavini A.U. Ranchi, Chem. Comm. 2123 (1997).
143. E. Salan, W. Lesuer, A. Klose, K. Schenk, C. Floriant, A.C. Villa, C. Rizzolic, J. Chem. Soc. Chem.Comm. 807 (1996).
144. P.D. Hampton, C.E. Daitch, T.M. Alam, E.A. Pruss, Inorg. Chem. **36**, 2879 (1997).
145. S.B. Copp, S. Subramaniam, M.J. Zaworotko, J. Am. Chem. Soc. **29**, 8719 (1992).
146. P.D. Beer, M.G.B. Drew, D. Hesek, K.C. Nam, Chem. Common. 107 (1997).
147. F. Poolucii, V. Balzani, G. Denti, S. Serroni, S. Campagna, Inorg. Chem. **32**, 3003 (1993).
148. L.D. Cola, V. Balzani, F. Bangelletti, L. Flamigni, P. Belser A.V. Zelewsky, M. Frank, F. Vogtle, Inorg. Chem. **32**, 5228 (1993).
149. P.D. Beer, M.G.B. Drew, D. Hesek, M. Shad, F. Szemes, J. Chem. Soc. Chem. Common 2161 (1996).
150. B.R. Cameron, S.J. Loeb, J. Chem. Soc. Chem. Common 2003 (1995).
151. P.D. Beer, M.G.B. Drew, P.B. Lesson, M.I. Ogden, J. Chem. Soc. Dalton Trans. 1273 (1995).
152. P.D. Beer, M.G.B. Drew, P.B. Lesson, K. Lyssenko M.I. Ogden, J. Chem. Soc. Chem. Comm. 929 (1995).
153. F.A. Neu, G. Barett, D. Corry, S. Cremin, G. Ferguson, J.F. Gallagher S.J. Harris, M.A. Mckierrvay, M.J.S. Weill, J. Chem. Soc. Perkin trans II **3** 575 (1997).
154. W. Xu, J.J. Vittal, R.J. Puddephatt, Inorg. Chem. **36**, 86 (1997).
155. J. Martensson, K. Sandros, O. Wennerstrom, Tetrahedron Letters **34**, 541 (1993).
156. M.G. Gardiner, S.M. Lawrence, C.L. Raston, B.W. Skelton, A.H. White, J. Chem. Soc. Chem. Comm. 2491 (1996).
157. X. Delaigue, M.W. Hossani, N. Kyritsakas, A.D. Cian, J. Fischer, J. Chem. Soc. Chem. Comm. 609 (1995).
158. M.G. Gardiner, G.A. Koulsantonis, S.M. Lawrence, P.J. Nichols, C.L. Raston, J. Chem. Soc. Chem. Comm. 2035 (1996).
159. J.M. Smith, S.G. Bott, J. Chem. Soc. Chem. Comm. 377 (1996).
160. J.L. Atwood, M.G. Gardiner, C. Jones, C.L. Raston, B.W. Skelton, A.H. White, J. Chem. Soc. Chem. Comm. 2487 (1996).
161. R.J. W. Lugtenberg, R.J.M. Egberink, J.F.G. Engbersen D.N. Reinhoudt, J. Chem. Soc. Perkin Trans. II **7**, 1353 (1997).
162. H. Taniguchi. Y. Otsuji, E. Nomura, Bull. Chem. Soc. Japan **68**, 3563 (1995).
163. F.L. Dickert, U.P.A. Baumler, H. Stathopulos. & O. Schuster, Anal. Chem. **69** 1000 (1997); Mikrochim Acta **119**, 55 (1995).
164. A. Arduini, W.M. McGregor D. Paganuzzi, A. Pochini, A. Secchi, F. Vgozzoli R. ungaro, J. Chem. Soc. Perkin trans II 839 (1996).
165. G. Fergusson, V.F. Gallagher, M.A. Mckervey, E. Madigan, J. Chem. Soc. Perkin Trans. I 599 (1996).
166. N.S. Isaacs, P.J. Nichols, C.L. Raston, C.A. Sandgra D.J. Young, Chem. Comm. 1839 (1997).
167. S. Shimizu, K. Kito, Y. Sasaki, C. Hirai, Chem. Comm. 1629 (1997).

168. B.R. Cameron, S.J. Loeb, Chem. Comm. 573 (1997).
169. J.H. Burns, R.A. Sachleben, Inorg. Chem. **29** 788 (1970).
170. G.W. Buchanan, V.M. Reynolds, K. Boarque, C. Benismon, J. Chem. Soc. Perkin Trans, II 707 (1995).
171. W. Petz, F. Weller, J. Chem. Soc. Chem. Comm. 1049 (1995).
172. N. Korber, M. Jansen, J. Chem. Soc. Chem. Comm. 1654 (1990).
173. K.V. Domasevitch, V.V. Ponomareva, E.B. Rusanov, J. Chem. Soc. Dalton Trans. **7**, 1177 (1997).
174. T. Lu, X. Gon, M. Tan. K. Yu, J. Cord. Chem. **29**, 215 (1993).
175. P.C. Junk, J.L. Atwood, J. Chem. Soc. Chem. Comm. 1551 (1995).
176. H.J. Küppers, B. Nuber, J. Weiss, S.R. Cooper, J. Chem. Soc. Chem. Comm. 979 (1990).
177. S.O.C. Matando, P. Mountford, D.J. Watkin, W.B. Jones, S.R. Copper, J. Chem. Soc. Chem. Comm. 161 (1995).
178. H. Plenio, R. Diodine, J. Org. Metallic Chem. **492**, 73 (1995).
179. G.R. Newkoma, G.E. Kiefer, D.K. Kohli, Y.J. Xia, F.R. Fronczek G.R. Baker, J. Org. Chem. **54**, 5105 (1989).
180. P.R. Ashton, A.L. Burns, C.G. Claessens, G.K.H. Shimizu, K. Small, J.F. Stoddart, A.J.P. White, D.J. Williams, J. Chem. Soc. Dalton Trans. **9**, 1493 (1997).
181. P.R. Markies, A. Villena, O.S. Akkerman, F. Bickelhaupt W.J.J. Smeets, A.L. Spek, J. Org. Metallic Chem. **463**, 7 (1993).
182. L.A. Kloo, M.J. Taylor, J. Chem. Soc. Dalton **15**, 2693 (1997).
183. M.K. Beklemisher, L.I. Gorodilova, N.I. Shevtsov, L.M. Kordivarenko, N.M. Kuzmin, Zhur anl. Khim **44**, 1058 (1989) CA **112**, 47894r (1990).
184. T. Kawashima, Y. Nishiwaki, R. Okazaki, J. Org. Metallic Chem. **499**, 143 (1995).
185. S.Y. Yu, Q.M. Wang, B. Wu, X.T. Wu, H.M. Hu, L.F. Wang, A.X. Wu, Polyhedron **16**, 321 (1997).
186. C.J. Harding, Q. Lu, J.F. Malone, D.J. Marrs. N. Martin, N.Mckee, J. Nelson, J. Chem. Soc. Dalton Trans. 1739 (1995).
187. C.J. Harding, F.E. Mabbs, E. J.L. MacInnes, J. Nelson, V. Mckee, J. Chem. Soc. Dalton 3227 (1996).
188. M.G.B. Drew, C.J. Harding, O.W. Howarth, Q. Lu, D.J. Marrs, G.C. Morgen V. Mckee, J. Nelson, J. Chem. Soc. Dalton 3021 (1996).
189. J. Wang, O.H. Luo, M.C. Shan, X.Y. Huang, Q.J. Wu J. Chem. Soc. Chem. Comm. 2373 (1995).
190. J. Pickardt, B. Kuhn, J. Chem. Soc. Chem. Comm. 451 (1995).
191. S. Stauf, C. Reisner, W. Tremel, J. Chem. Soc. Chem. Comm. 1749 (1996).
192. J. Campbell, G.J. Schrobilgen, Inorg. Chem. **36**, 4078 (1997).
193. A. Arduini, W.M. McGregor, A. Pochini, A. Secchi, F. Ugozzoli and, R. Ungara, Org. Chem. **61**, 6881 (1996).
194. J.M. Harrowfield, M.J. Ogden, W.R. Richmond, A.H. White, J. Chem. Soc. Chem. Comm. 1150 (1991).
195. J.A.J. Brunink, W. Verboom. J.F.J. Engbersen, S. Harkema D.N. Reinhault, Rec Trav Chem. **111**, 511 (1992).
196. P. Thuery, M. Nierlich, Z. Asfari, J. Vecens, J. Incl. Phen. and Mol. Recog. **27**, 169 (1997).
197. J.M. Harrowfield, M.I. Ogden, W.R. Richmond, A.H. White, J. Chem. Soc. Dalton Trans. 2153 (1991).
198. J.C.B. Bunzli, F. Ihringer, P. Dumy, C, Sager, R.D. Rogers, J. Chem. Soc. Dalton Trans. 497 (1998).
199. N. Muzet, G. Wipff, A. Casnati, L. Domiano, R. Ungaro, F. Ugozzoli, J. Chem. Soc. Perkin Trans II 1065 (1996).

200. T.N. Sorrel, F. Christopherpigge, P.S. White, J. Org. Chem. **33**, 632 (1994).
201. F. Corazza, C. Floriani, A.G. Villa, C. Guastini, J. Chem. Soc. Chem. Comm. 1083 (1990).
202. J.L. Atwood, G.W. Orr, N.C. Means, F. Hamada, H. Zhang, S.G. Bolt, K.D. Robinson, Inorg. Chem. **31**, 603 (1992).
203. F. Carazza, C. Floriani, A. Chiesivilla G. Guastini, J. Chem. Soc. Chem. Comm. 640 (1990).
204. P.D. Beer, A.D. Keele, M.G.B. Drew, J. Org. Met. Chem. **378**, 437 (1989).
205. F.J. Parlerliet, A. Oliver, W.G.J. Longe, P.C.J. Kamer H. Kooijaman A.L. Spek; W.N.M. Leeuwen, J. Chem. Soc. Chem. Comm. 583 (1996).
206. V.B. Shur, I.A. Tikhonova, A.I. Yanovsky, Y.T. Struchkov, P.V. Petrovskii S.Y. Panov, G.G. Furin, M.E. Volpin, J. Org metal. Chem. **418**, 139 (1991).
207. T.W. Bell, P.I. Cragg, M.G.B. Drew, A. Firestone, D.I.A. Kwok Angew wante Chem. **31**, 348 (1992).
208. P.D. Beer, Z. Chon, M.G.B. Drew, P.A. Gale, J. Chem. Soc. Chem. Comm. 1851 (1995).
209. N. Tokiton, T. Saiki, R. Okazaki, J. Chem. Soc. Chem. Comm. 1899 (1995).
210. O. Morick E.F. Paulus V. Bohmer, I. Thondorf, W. Vogt, J. Chem. Soc. Chem. Comm. 2533 (1996).
211. J.L. Atwood, K.T. Holman, J.W. Sleed, J. Chem. Soc. Chem. Comm. 401 (1996).
212. J. Trepte, M. Czugler, K. Gloe, E. Weber, J. Chem. Soc. Chem. Comm. 1461 (1997).
213. O. Struck, W. Verboom, W.J.J. Smeets, A.L. Spek, D.N. Reinhoudt, J. Chem. Soc. Perking Trans. II 223 (1997).
214. M. Backes, V. Bohmer, G. Ferguson, C. Gruttner, C. Schmidt, W. Vogt, K. Ziat, J. Chem. Soc. Perking trans II 1193 (1997).
215. R.M. Izatt, J. H. Rytting, D.P. Nelson, B.L. Haymore, J. J. Christensens–Science **164**, 443 (1969).
216. R.M. Izatt, D.P. Nelson, J.H. Rytting, B.L.Haymore J.J. Christensen, J. Am. Chem. Soc. **93**, 1619 (1971).
217. R.M. Izatt, D.E. Terry, B.L. Haymore, L.D. Hansen, N.K. Dalley, A.G. Avondet, J.J. Christensen, J. Am. Chem. Soc. **98**, 7620 (1976).
218. R.M. Izatt, D.E. Terry, D.P. Nelson, Y. Chan, D.J. Eatough J.S. Bradshaw L.D. Hansen, J.J. Christensen, J. Am. Chem. Soc. **98**, 7626 (1976).
219. J.J. Christensen, J.O. Hill. R.M. Izatt, Science **174**, 459 (1971).
220. J.J. Christensen, R.M. Izatt, Physical methods in advanced Inorganic Chemistry John Wiley (1968).
221. U. Takki, T.E. Hugen Esch, J. Smid, J. Am. Chem. Soc. **93**, 1955 (1970).
222. B. Dietrich, J.M. Lehn, J.P. Savage, Tetrahedron Letter 2889 (1969).
223. J.M. Lehn, J.P. Savage, Chem. Comm. 440 (1971).
224. V.M. Loyola R.G. Wilkins, J. Am. Chem. Soc. **97**, 7382 (1975).
225. Y. Liu. Y. Wang Guez, S. Yang Jin, Huazne Tongbia **17** (1985) AA **48** 3C52 (1980).
226. M. Stoderman, N. Dhar, J. Chem. Soc. Forady Trans. **94**, 899 (1998).
227. E.J.P. Feijen J.A. Martens, P.A. Jacob, J. Chem. Soc. Faraday Trans. **92**, 3281 (1996).
228. V.P. Solovev, N.N. Strakhova, O.A. Raevsky, V. Rudiger, H.J. Schneider, J. Org. Chem. **61**, 5221 (1996).
229. L. Tongbu, G. Xinmin, T. Ning, T. Minyu, Polyhedron **9**, 2371 (1990).
230. C.D. Tran, W. Zhang, Anal. Chem. **62**, 830 (1990).
231. T. Lu, M. Tan H. Su, Yu Liu Polyhedron **12**, 1055 (1993).

232. Yu. Liu, B.H. Han, Y.M. Li, R.T. Chen, M. Ouchi, Y. Inoue, J. Phy. Chem. **100**, 17361 (1996).
233. Yi Lu, T.B. Lu. M.Y. Ton, T. Hakushi, Y. Inoue, J. Phy. Chem. **97**, 4548 (1993).
234. J.D. Lamb, J.F. King, J.J. Christensen, R.M. Izatt, Anal. Chem. **53**, 2127 (1981).
235. W. Abraham, K. Nava. p. 227 Ed. A.R. Cooper in Polymer Sepn media Plenum Press 222 (1982) AA **44**, 3A3 (1983).
236. U. Tunca, Y, Yagei, Poly Science Part. I **28**, 1721 (1970).
237. H.J. Buschnan, E. Cleve, E. Schollmeyer, Therm. Chem. Acta. **207**, 329 (1992).
238. J. Schatz, F. Schildbach, A. Lentz, S Rastatter, J. Chem. Soc. Perkins Trans. II 75 (1998).
239. J.W. Mitchell, D.L. Shanks, Anal. Chem. **47**, 642 (1995).
240. H. Stephan, K. Gloe, J. Beger, P. Muehl, Solvent Extraction and Ion Exchange **9**, 459 (1991).
241. V. Kampel, A. Warshawsky, J. Orgmetallic Chem. **469**, 15 (1994).
242. W. Xiaolin, L. Yinong, F. Vibei, J. Radio and Nucl. Chem. **189**, 127 (1995).
243. H. Stephan, K. Gloe, J. Beger, P. Muehl, Solvent Extraction and Ion Exchange **9**, 435 (1991).
244. H.S. Du, D.J. Wood, S. Elshani, C.M. Wai, Talanta **40**, 173 (1993).
245. J. Jang, C.M. Wai, J. Radio and Nucl. Chem. **128**, 61 (1988).
246. V.V. Yakshin, O.M. Vilkova, L.T. Makarova, N.A. Tsarenko B.N. Laskorin, Doki Akad. Naak. **326**, 117 (1992) CA **118**, 133180k (1992).

5
Solvent Extraction Separations

5.1 Introduction

Solvent extraction is poweful method of the separation in analytical chemistry. It was used to concentrate and separate metal cations and get stoichiometry and stabilities of complexes which were extracted in to an organic phase. The use of macrocyclic compounds for this purpose was of recent origin [1, 2]. The cost of the macrocyclic compounds was prohibitive for large scale applications. However, for analytical separations one could exploit advantage of their selectivities with low reagent concentration. They had ability to induce increased solubility of metal salts in the nonpolar solvents, there by increasing the reactivity of anions in nucleophillic substitution reactions which in turn had great relevance in phase transfer catalyst reactions.

The selectivity of crown ethers for metal ions mainly depended on the relation between the size of the metal and the hole or cavity of the crown ether. Those metal ions which fitted best the crown ether cavity were most favourably extracted. In fact Pedersen was the first to demonstrate the use of the crown ether, viz. DB18C6 in solvent extraction of alkali metal like potassium. Frensdorrff [3] elaborated extraction of sodium with DC18C6 with picrate as the counter anion. Crown ethers were initially not very popular for liquid-liquid extraction because the stability of complexes formed between metal ion and crown ethers were too low. However, such stability was easily enhanced by use of organic solvents with effective substitution.

A very limited review articles on solvent extraction with macrocyclic and macrobicyclic polyethers are available. Two such reviews [2, 4] had appeared in japanese journals. The analytical chemistry was covered by Kolthoff [1]. The recent review by Gandhi and Khopkar however, is quite comprehensive [5, 6].

These macrocyclic compounds containing nitrogen, or oxygen or sulphur as donor atoms were commonly used in solvent extraction. Those containing oxygen as donor atom were usually termed as "Macrocyclic polyethers". The group of all these compounds was collectively termed as "Crown ethers". The specialty of these compounds was special relationship between the crystal radii of the cation and the size of cavity of crown ethers. One could get clear insight in solvent extraction procedure by considering extraction equilibria with adequate understanding of complex formation in aqueous phase and distribution of the uncomplexed and complexed crown ether in the aqueous-organic phases.

5.2 Extraction Equilibria With Crown Ethers [5]

Let us consider an extraction equilibria of metal ion M having valency 'n$^+$' with crown ether R in the presence of counter anion as X^- then we have:

$$M^{n+} + R_{org} + n X^- = (MRX_n)_{org}$$

where subscript 'org' represent the organic phase.
The overall extraction equilibrium constant K_{ex} shall then be given as

$$K_{ex} = \frac{[MRX_n]_{org}}{[M^{n+}][R]_{org}[X^-]^n}$$

The crown ether R is dissolved in organic phase it gets distributed as

$$[R] \rightleftharpoons [R]_{org}$$

Hence $\qquad K_{D_R} = [R]_{org}/[R]$

The formation constant K_f of metal complex is obtained as

$$M^{n+} + R \rightarrow MR^{n+}$$

$$K_f = [MR^{n+}]/[M^{n+}][R]$$

When the complex get distributed between organic and aqueous phase then

$$MR^{n+} + nX^- \rightleftharpoons [MRX_n]_{org} \quad \text{i.e. ion pair formation phenomena}$$

$$K_{D_x} = [MRX_n]_{org}/[MR^{n+}][X^-]^n$$

where K_{D_x} represents distribution ratio of the complex
Now overall distribution ratio D of K_{ex} is given as

$$D = \frac{|M|_o}{|M|} = \frac{[MRX_n]\,org}{[M^{n+}]+[MR^{n+}]}$$

Since $[M^{n+}] > [MR^{n+}]$ in aqueous phase.

By substituting values of $[MRX_n]$ and $[M^{n+}]$ in terms of K_{D_R}, K_{D_x} and K_f; since $K_{ex} = K_{D_R} K_f K_{D_x}$ from above equations, we have finally

$$D = \frac{[MRX_n]_{org}}{[M^{n+}]} = \frac{[MRX_n]_{org}}{[M^{n+}][MR^{n+}]} = K_{D_x}[X^{-n}][R]_{org}$$

In actual practise when concentration of crown ether $[R]_{org}$ is high in the organic phase, better extraction is achieved. Further when concentration of counter anion $[X^-]$ is high in aqueous phase or when complex has high extraction coefficient $[K_{D_x}]$ in organic phase we have best extraction, i.e. D or K_{ex} be large. For all practical purposes K_{DR} should be small to get satisfactory extraction with maximum magnitude for distribution ratio (K_{ex} or D).

5.3 Interpretation of Equilibria

Now it is possible to interpret various equations in extraction equilibria in absolute terms of the process of extraction, with suitable illustrations.

The total extraction equilibria could be summarised in three simple equations viz.

(i) Partitioning of the ligand K_{DR} in the organic phase

$$K_{D_R} = [R]_{org}/[R] \quad \text{(i)}$$

(ii) Partitioning of complexed salt in organic and aqueous phases

$$K_{D_x} = [MRX]_{org}/[MR^+][X^-] \quad \text{(ii)}$$

(iii) Complexation of M^+ in the aqueous phase was represented by formation constant,

$$K_f = [MR^+]/[M^+][R] \quad \text{(iii)}$$

The dissociation of extracted ion pair $(MRX)_{org}$ into $(MR^+)_{org}$ and $(X)_{org}$ was neglected in nonpolar media. It was experimentally shown that ion-pair formation constant (K_{D_x}) of crown ether complex in dichloroethane was 10^{-5} so dissociation was noticeable below 10^{-4} concentration. The ion pair formation constant depended on the polarity of medium or solvent. In polar solvents complexed salt behaved as strong electrolytes. The macrocyclic ligands had the advantage in polar organic solvents because of their ability to inhibit ion pair formation through complexation of cation. In above terms the distribution ratio D of metal ion was given by

$$D = [MRX]_{org}/[M^+]_{aq} + [MR^{n+}] \text{ since } [M^{n+}] > [MR^{n+}]$$

in the aqueous phase which is simplified to

$$D = K_{ex}[R]_{org}[X^-]_{aq} \quad \text{(iv)}$$

In most of extractions of alkali metals with crown ethers, picrate was used as the counter anion. The reasons for preference of picrate as counter anion during the extraction was because of the large molar volume of the picrate better extraction was favoured. The extractibility of the crown ether complex with univalent metal picrates depended on the interaction of metal ion in crown ether cavity with molecule [7, 8] 2, 4, 6 trinitrophenolate i.e. picric acid was best counter anion as it had absorbance in the visible region and extraction in the organic solvent was thermodynamically favoured [9].

Another factor to be accounted during such extraction was the presence of water molecule. Amongst alkali metal ion included in crown cavities, the interaction of ligand with water seemed to be largest hindrance for hydration of B15C5 [10. 11]. The complex by benzo group attached to 15C5 was more effective for lithium than caesium. The Table 5.3.1 summarises extraction behaviour of alkali metals with different crown ethers.

It was worth examining ionic radii and cavity size (Table 5.3.2 and 5.3.3) as it had definite relationship for formation of complex and hence ensuring quantitative extraction [12].

This showed that 15C5, B15C5, 18C6 and DB18C6 would show good cation selectivity for alkali metals, e.g. Na (15C5) K (18C6) while 12C4 and 14C8 due to too small or too big cavity size were useless for extraction of alkali metals

Table 5.3.1 Extraction behaviour of alkali alkaline earths in terms of log K_{ex} [5]

Crown	Na	K	Rb	Cs	Ca	Sr	Ba
15C5	3.90	2.58	2.14	1.90	3.72	5.70	5.20
18C6	3.39	5.97	5.43	4.38	7.03	9.72	9.72
DB24C8	2.05	2.79	3.0	3.15	3.07	3.99	6.86

Table 5.3.2 Ionic size of metals [12]

Ion	Li	Na	K	Rb	Cs	Ca	Sr	Ba	Pb	Cd	Mn
Ionic Radii Å	0.60	0.95	1.33	1.48	1.69	0.99	1.13	1.35	1.20	0.97	0.81
Ionic diameter Å	1.48	1.99	2.66	2.96	3.44	1.98	2.32	2.68	2.40	1.95	1.62

Table 5.3.3 Cavity size of crown ethers [12]

Crown ether	12C4	15C5	18C6	24C8	DB18C6	DC18C6	DB24C8
Cavity size radii Å	0.60–75	0.85–1.1	1.3–1.6	> 2.0	1.3–1.6	1.3–1.6	> 2.0
Cavity size diameter Å	1.2–1.5	1.7–2.2	2.6–3.2	> 4.0	2.6–3.2	2.6–3.2	> 4.0

respectively. For alkaline earth metals selectivity tendencies coincided with size fit idea, e.g. DB24C8 showed best selectivity for barium among alkaline earths. The lanthanides and actinides, formed stable complexes with crown ethers [13]. It would be interesting to consider various factors influencing extraction process. The Table 5.3.4 summarises the distribution coefficient of crown ethers in few selected organic solvents.

All these tables contain log K_{DR}, log K_f, and log K_{ex} or D used in extraction equilibria. Thus some important inferences from above tables are worth noting.

The size of cavity and ionic radius of metal ion decided the distribution ratio

Table 5.3.4 Distribution ratio of crown ethers (log K_{DR}) in selected solvents [14, 15]

15C5	18C6	DB18C6		DC18C6		DB24C8
Benzene	Benzene	Methylene chloride	Chloroform	Methylene chloride	Ethylene chloride	Methylene chloride	Benzene
−0.81	−1.2	0.05	3.9	3.5	2.8	3.6	3.0

Table 5.3.5 Stability constant of alkali metals with 18C6 (log K_f) [16]

Crown	Na	K	Rb	Cs	Ca	Sr	Ba
18C6	0.80	2.03	1.56	0.99	< 0.50	2.72	3.87

Table 5.3.6 Extraction constant for metal in crown ethers [17–20] (log K_{ex} or D)

Metals	Li	K	Cs	Ag	Ca	Ba	Pb
15C5	1.29	2.58	1.90	4.45	3.71	5.20	6.46
18C6	1.92	5.97	4.38	4.44	7.03	9.72	11.7
DB18C6	–	4.65	3.07	3.51	5.77	–	7.16
DB24C8	0.67	2.79	3.15	3.09	3.07	6.86	–

(D) and percentage extraction of metals [21-23]. Further the extractability was maximum when crystal radius of cation fitted well with the holes or cavity size of the crown ether. Using smaller size of cation and flexible crown ether did change the extent of dehydration and solvation energy of cation crown complexes and deformation energy of crown ring. This was further discussed along with effect of temperature on extraction.

Table 5.3.7 Distribution coefficient of crown ether 18C6 in some solvents [24]

Solvent	Benzene	Nitrobenzene	Chloroform	Methylene chloride	Octanol	Hexadecane
K_{DR}	0.063	0.10	6.10	5.0	0.21	0.00039

5.4 Factors Influencing Extraction

There were various parameters which generally influenced the process of extraction. Such factors influencing extraction are considered in following sections. An endevour is made to consider in details each and every factor influencing the process of extraction by either crown ether or cryptands, with no specific reference for calixarenes.

5.4.1 Distribution of Crown Ether in Two Phases

If crown ether was more soluble in water it exhibited higher extractability for metal ion. If thus crown ether itself was very soluble in water then crown ether complex with metal ions was also soluble in water. This in turn was reflected in the extractability of the complex. Usually if the molar volume of chemical species was large then complex was more extractable. When a substituent groups like cyclohexyl in DC18C6 or benzo in DB18C6 was attached to a crown ring the D, i.e. distribution ratio value was considerably increased. Thus distribution of ligand between water and organic solvent was of great importance in deciding extractability, i.e. (K_{DR}) the solubilities in water for three series of crown ethers increased as the number of oxy ethylene units ($-OCH_2 CH_2^-)_n$ increased. A regular variation in K_{DR}, i.e. distribution of ligand size and number of benzo groups attached had encouraged to find out quantitative fitment for empirical equations. As rule the distribution coefficient were solvent dependent, e.g. chloroform and methylene chloride had greater affinity for 18C6, than water which in turn was seen in the dependence of the distribution coefficient on the solvent polarity. Such dependence was explained for 18C6 for various solvents in Table 5.3.7 [24]. The solubility studies of cryptand in polar solvents such as

DMSO, DMF acetonitrile was carried out from view point of solute solvent interaction. It was seen that cryptand–2, 2, 2 was more soluble in water than in all of the dipolar aprotic solvents. However, cryptand 2, 2, 2, B was not soluble in water. Another important factor influencing partitioning of ligand and the extraction equilibria was related to the conformational mobility of the ligands. High flexibility gave better adoption to conformation in which donor atoms were rotated outwards to allow H-bond formation with water molecule. Thus the hydrophobic group was exposed to the solvent enhancing extraction by transfering the ligand and complex. The high solubility was not always favourable for extraction; but high degree of substitution in polar solvents was favourable for extraction.

5.4.2 Stability of Crown Ether Complex in Aqueous Phase

The stability of the crown ether complex with many metal ions in the aqueous solution governed the value of K_{DX}. The more stable the crown ether complex in the aqueous phase the more extractable it was in the organic phase. But the extractions of the 15C5, 18C6, DB18C6, between benzene and water revealed that the selectivity tendencies of a crown ether for metal ion was necessarily in agreement with that in the aqueous solution, e.g. log K_{Dx} was varying for different crown ethers for different alkali metal as shown in Table 5.4.1 [25, 26].

Table 5.4.1 Variation of log K_{DX} for different metals in various crown ethers

Metals	Na	K	Rb	Cs	Ca	Sr	Ba
15C5	0.70	0.74	0.62	0.80	–	1.95	1.71
18C6	0.80	2.03	1.56	0.99	< 0.5	2.72	3.87
DB18C6	1.16	1.67	1.08	0.83	< 0	1.0	1.95

The above table showed values of log K_{Dx} indicating the extent of distribution of complex in the aqueous solution. The value of log K_{Dx} for 1:1 complex showed no direct relation with the extractability of such complexes in the organic solutions.

5.4.3 Extraction of Complexed Salt

The Table 5.4.2 give extraction of complexed salt (picrate) in to benzene for alkali metals [27].

Table 5.4.2 log D or log K_{ex} for alkali metals with different crown ethers [27] from picrate solution

Metals	Na	K	Rb	Cs
15C5	2.40	1.04	0.72	0.30
18C5	1.39	2.74	2.67	2.19

It showed log D values were not always related closely to the ability of a ligand to extract cation; but it predicted the nature of extraction process and characteristics of the complex. Such value was not dependent upon cation and it increased directly with hydrophobicity of the ligand. The complex of potassium with cryptand – 2 2 2 i.e. $K[2, 2, 2]^+$ was completely extracted into dichloromethane incomparison with DB18C6 or DC18C6 in spite of the fact that the distribution coefficient of crown ether (K_{DR}) was 1000 times larger than that of cryptand–2 2 2 [28]. This was because cryptand-2 2 2 adopted conformation in free state in water in such manner that solvation took place through O and N-atoms while the complex had hydrophobic exterior and the cation and donor atoms were very effectively shielded from direct interaction of the solvent.

When one or two benzo groups were attached to a crown ether ring to render it more hydrophobic, the crown ether complex with metal ion was readily extractable (B15C5 or 18C6 and DB18C6). The size sequence of these crown ethers was in order of extractability : 15C5 < B15C5 < 18C6 < DB18C6. Further as discussed earlier the more closely the metal ion fitted into the crown ether cavity, more extractable was the crown ether metal complex. This was because the charge of the metal ion held in the crown ether cavity was effectively shielded by the crown ether and thus the complexed cations were equal in chemical nature. The extractability of metal ion complex with one and the same crown ether was determined by both the ratio of ion to cavity size and the interaction of the metal ion trapped in the cavity with the water molecule [29]. Thus some metal ion complexes of 18C6 and DB18C6 were relatively more extractable in methylene chloride and chloroform than in benzene. This was because of the presence of stronger interaction of hydrogen atoms of methylene chloride or chloroform with the crown ether oxygen atoms compared to those interactions of benzene [30, 31]. Generally coextraction of water was caused by the hydration of the cations. The complexation between these cations and crown ethers caused a remarkable decrease of the hydration number. However the number and type of donor atoms and the presence of fused benzo rings in the crown ether had no influence on the coextraction of the water.

5.4.4 Effect of Counter Anion

The extractability of the ion pair into the solvent was decided by the nature of the counter anion. Usually counter anion with large molar volume such as picrate, dinitrophenolate, dipicrylamine or tetraphenylborate rendered the ion pair extraction very good. This is examplified in Table 5.4.3.

Table 5.4.3 Variation in extraction of potassium with DC18C6 in methylene chloride with different counter anions

Counter ion	Picrate	Dipicrylamine	2–5 dinitro phenolate	2, 6 dinitro phenolate
log K_{ex}	6.36	8.16	4.20	3.98
Molar Volume	Large	V. large	Medium	Small

The extractability increased with the hydrophobicity of the counter anion, but the selectivity showed decrease. The extractability sequence of the counter anions with DB18C6 for alkali metals was of the order as: $I^- > NO_3^- > Br^- > Cl^- > F^-$. The incorporation of carboxylic acid group into crown ether molecule had enhanced the selectivity as well as extractability [32–35]. The extractability was not dependent on change in concentration of an anion and kind of anion explored. The anion was not necessarily extracted in to the organic phase. The extent of extraction of complex depended upon the nature of counter anion used, due to free energies of transfer of the anions from water to organic phase. Picrate was 70 times more effective than 2.4 dinitrophenolate in extraction of potassium [36]. The variation in extent of extraction was attributed to the presence of water molecules in the organic phase or interaction with anion in the organic phase. Ion pair attained higher stability due to higher charge density of anions. Fluoride was not very good counter anion [37] as it was stabilised by hydrogen bond formation. For anions the smaller was the hydration energy the larger was the ionic radius and smaller was the number of charges. The hydration was due to electrostatic force. Usually univalent bulky and low electron density ions were easily extracted. Amongst organic anions plannar anions were preferred, e.g. 2, 5 dinitrophenol, 2, 6 dinitrophenol, 2, 4, 6 trinitrophenol (i.e. picric acid) bromocresol, a dipicrylamine were all plannar hence gave good extraction. Methylene blue was plannar cation. Interestingly plannar species, viz. picrate anion extracted well plannar potassium crown ether cation (Fig. 5.4.1).

5.4.5 Extraction of Metal Complex Anion:

The extraction of the metal complex anion was also possible, e.g. $KMnO_4$ with DC18C6 or DB18C6 in the aromatic hydrocarbons [38]. Also $[Co(SCN)_3^-]$ or $[Fe(SCN)_3^-]$ complex [39–40] were extracted with 18C6 as an extractant in chloroform or dichloroethane respectively. Anionic copper or zinc complex was

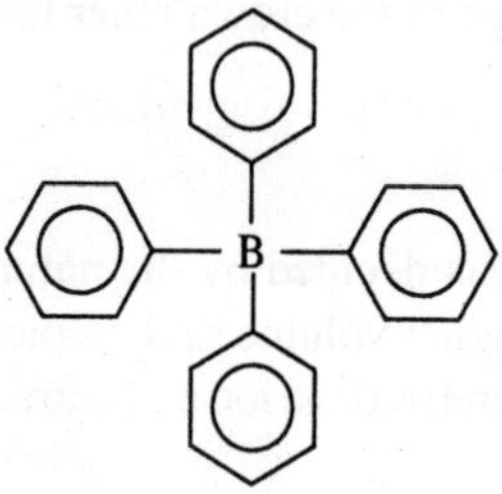

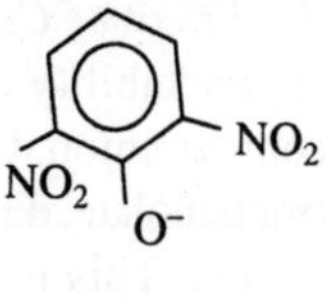

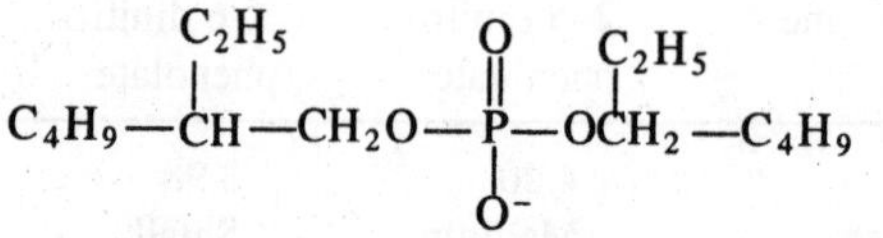

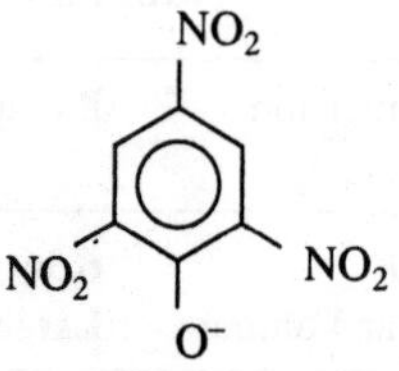

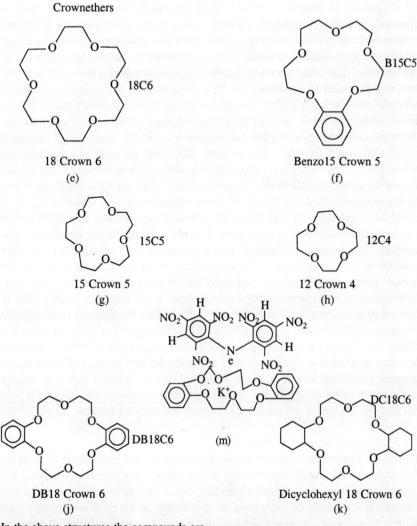

Fig. 5.4.1 Structure of typical counter ions and crown ethers

In the above structures the compounds are—
- (a) Tetraphenyl borate
- (b) 2-6 dinitrophenolate
- (c) bis (2 ethyl hexyl phosphoric acid)
- (d) Picric acid
- (e) 18C6
- (f) B15C5
- (g) 15C5
- (h) 12C4
- (j) DB18C6
- (k) DC18C6
- (m) Structure K^+ (DB18C6) picrate complex

paired with [K18C6$^+$] in chloroform to complete extraction [41]. The bulky and low mass compounds were good extractant but were not selective reagents.

5.4.6 Effect of Diluent
The extractability and selectivity of crown metal complex was greatly influenced

by the nature of diluent or solvent used for extraction with crown ether or cryptand. The extractability varied with change in dielectric constant of diluent. In conventional solvent extraction procedures the nonpolar solvents were best diluents, e.g. benzene, toluene, xylene, chloroform carbon tetrachloride or cyclohexane. They had very small dielectric constant hence association of ion pair was favoured in complexation. The extracted ion pair was dissociated in solvents with high dielectric constant. A protic solvents which could solvate anions effectively were ideal for the extraction. A protic solvent should have low dielectric constant to favour ion pairing, the low Gibb's free energy facilitates transfer of anion to water. The oxygen containing solvents did not improve extractability. The main decision on the selection of diluent should be that solvent should be promoted effectively to central cation. Best solvents were one where in electrolytes or ion pair could be easily dissociated. The value of K_{ex} or D could be increased with high dielectric constant solvents and one having good solubility of crown ether in the solvent. Usually solvents having larger dielectric constant offer better extraction in solvent extraction with crown ethers. When a hard anion such as chloride was counter anion and if it was to be transferred to organic phase, then it required high dehydration energy. The macrocylic ligands have ability to increase solubility of electrolytes in the organic solvents. This in turn could substantially enhance the partitioning of the electrolytes from aqueous phase to organic phase of low polarity [42]. In extraction with cryptand the change in solubility product due to cryptate formation showed either increase or decrease in solubility. The solubilities of more soluble salts were reduced due to cryptate formation and vice versa. The solubility of cryptate salt was usually increased relative to solubility of corresponding uncomplexed salt. This was because the ability of the medium to solvate electrolytes became poorer.

5.4.7 *Effect of Nature of Crown Ether*

The ligand concentration in the organic phase was of vital importance in deciding the course of the extraction. Usually crown ether extractions were carried out with low reagent concentration. Cram [43] developed picrate extraction scale to compare free energies of complexation of different crown ether which were least soluble in aqueous solution. He found out that there was a definite relationship between ligand binding power and the number of ligand binding sites for maximum complex formation. The hydrophobic nature of ligand did not have much significance in complexation reaction as the lack of sensitivity of the hydrophobic nature of the ligands was influenced by K_{ex}. It was function of the difference in solvation energy of free ligands and complexed salt in the organic phase.

Secondly the structure of crown ether was also very important. An introduction of t-butyl group in DB18C6 ring enhanced its solubility in the organic solvent [44]. The extraction of sodium was less effective to the substituent effect on extractant than in extraction of potassium. The crown ethers with large electron withdrawing substituent exhibited a reversal of cation extraction selectivity. The orientation of the substituted amide group also was of the importance to decide extraction behaviour. The metal extraction properties of photoresponsive crown ethers showed that those containing photo isomerisable functional groups like –

– N = N – showed conformational changes of the crown ring in response to photo radiation during extraction. The photo controlled selective metal ion extraction could be carried out with such crown ethers, where the donor atom was equally important.

The thia crown ethers had better affinity for metals belonging to soft acids [45–47]. The increase in the number of the sulpur atoms in extractants in turn increased the extractability [48]. This was specially valid in solvent extraction of metals like zinc, copper, cadmium.

The modification of structure of crown ether by effective substitution also influenced the process of extraction. The introduction of chromophoric group in crown ether ring altered the nature of extraction. The introduction of an ionic anchor made complex more stable. The introduction of substituent group in benzene ring changed extraction pattern. The optically active attachment helped to separate enantiomers. A good separation was possible with polymer attached to crown ether.

5.4.8 Selectivity of Extraction

As described earlier the Crams picrate extraction scale [49] provided a basis for a correlation between ligand structure and the free energies of complexation of ligands of different structures. The stability constant threw much light on the ligand selectivity. Depending upon the kind of ligand the metal selectivity changed. Polythioethers were best for silver and mercury [50]. However many ligands with more complex structures had been synthesised for their role as specific complexing and extracting agents like lariat ethers having side arms to enhance complexation [51]. The bis-poly crown ethers were also effective. The metal ion extractability and selectivity of poly and bis (crown) ethers containing pendant crown ether moieties were different from those monomeric analogues [52]. Further various kinds of polymers as polystyrenes, polymethylacrylates, polypeptides and polyethylene eneimnes containing crown ethers in their side chain had been used [50, 53–58]. The poly and bis (crown ethers) exhibited better extraction and selectivity for metal ions forming 2:1 crown ether unit cation complexes. This was due to cooperative effects of the adjacent crown ether unit on the metals.

5.4.9 Effect other Species in Aqueous Phase on Extraction

The effect of ionic strength and salting-out on extraction by crown ethers was studied [59–62]. The salting out agent enhanced the extraction. They favoured ion pair formation due to high anionic concentration, removal of water molecule from the coordination sphere of metal and decreasing dielectric constant of the aqueous phase favouring extensive ion pair formation to promote extraction.

5.4.10 Effect of Temperature on Extraction

Various thermodynamic factors like ΔG, ΔH and ΔS with usual notations were studied for crown ether extractions [63–64]. The extractability was controlled by entropy, e.g. potassium metal as shown in Table 5.4.4.

Table 5.4.4 Extraction of potassium with DB18C6 in m-cresol from chloride solution

Thermodynamic parameter	ΔG	ΔH	ΔS
Complex	−3.06	−9.69	−22.2
Ligand	−4.0	+2.0	20.0

The enthalpy changes for the cations was nearly constant while complexing DC18C6 and potassium, entropy term did not play significant part. A large entropy could be occurring for K_{Dx} or K_{ex}. When complex was extracted in the organic solvent, dehydration from MRX was important. Such dehydration process contributed to the difference of entropy term. The difference in ΔS might be due the preorganisation of the structure [65] in sandwich structure in preference to 1:1 complex.

5.4.11 Synergic Solvent Extraction

Along with crown ethers other extractants like trioctyl phosphine oxide (TOPO) [66], tributyl phosphate (TBP) [67] and tributyl phosphine oxide (TBPO) [68] were studied as synergic agents. The extractability was substantially enhanced during such extractions. This could be explained in terms of extraction equilibria. When complex had under gone further complexation with another ligand (L) in the organic phase then equilibrium could be described as

$$[MRX_0] + [L_0] = [MRLX_0]$$

where L is the synergic agent, R the crown ether and X the counter anion

Therefore β-the stabilisation factor of mixed complex is

$$\beta = \frac{[MRLX_0]_0}{[MRX]_0 + [L]_0} \text{ if } K_{ex} = D \text{ then}$$

We have

$$\beta = \frac{D_{MRLX}}{D_{MRX}} \text{ Since } D_{MRLX} = \frac{[MRLX]_0}{[M^+][L]_0[R]_0[X]} \text{ and } D_{MRX} = \frac{[MRLX]_0}{[M^+][R]_0[X^-]}$$

β is the formation constant for $MRLX$ complex in the organic solvent. It was clear from table 5.4.5 that β was more for TBP and B15C5 extraction

Table 5.4.5 Synergic extraction of rubidium with TBP as synergist

Crown	Synergic agent	Log β
15C5	TBP	0.81
B15C5	TBP	1.32

That for same alkali metal like say rubidium [69] larger the number of oxygen donor atoms of another ligand and more closely the metal ion crown ether cavity the greater was the complex formation constant for the synergic reaction, e.g. with TBP as synergist the extraction was more. Similarly extraction of silver with

DB18C6 was enhanced by TOPO. This was due to adduct formation of the complex [Ag (DB18C6) (TOPO)$_2^+$] in chloroform phase. The synergic effect was size selective it was more for metal ions that best fitted the cavity of crown ether. The effect was less for alkaline earths in comparison with alkali metals [70].

In summary of all the factors so far considered, we see that (i) partition coefficient of ligands depended upon balance between the donor atom (like O, S) which tended to increase their solubility in water and hydrophobic character which was decreased (ii) The complexed salts showed trends in extraction coefficient similar to the free ligands (iii) K_{ex} or D values were not sensitive to organic nature of the ligand (iv) The selectivity depended upon cation binding ability of the ligand as it was reflected in magnitude of stability constant (v) The anion dependence of electrolyte extraction followed the trends in free energies of transfer of the anions from water to organic phase for instance picrate was favoured. Such difference between anions was attributed to ion pair formation in the organic phase. The effective system was one involving ligand insoluble in water and one which formed stable complex with metal and anion with high degree of charge dispersion (vi) Finally the coordination of the cation by macrocyclic ligand tended to increase the solubilities of poorly soluble salts in weakly solvating media [71, 72].

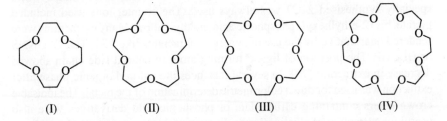

(I)   (II)   (III)   (IV)

5.5 Solvent Extraction Separations with Crown Ethers

Most commonly used crown ethers for extraction were 15C5 (II), 18C6 (III), 24C8 (IV) and their benzo or cyclohexyl derivatives. The small 12C4 (I) is less commonly used.

Table 5.5.1 **Physical properties of crown ethers**

S.No.	Name	Formula	Mol. wt	Cavity size °A	m.p. °C	% moisture	State
1	12C4	$C_8H_{16}O_4$	176	1.2–1.5	16°	0	Semisolid
2	15C5	$C_{10}H_{20}O_5$	220.27	1.7–2.2	–32.4°	7.38	Liquid
3	18C6	$C_{12}H_{24}O_6$	264.32	2.6–3.2	39–42°	2.29	Solid
4	24C8	$C_{24}H_{48}O_8$	464.0	4.5–5.0	>100°	—	Solid
5	B15C5	$C_{16}H_2O_5$	356.0	1.7–2.2	38°	—	Semisolid
6	DB18C6	$C_{20}H_{24}O_6$	360.41	2.6–3.2	163°	0.40	Solid
7	DC18C6	$C_{20}H_{36}O_6$	372.47	2.6–3.2	38–54°	0.72	Solid
8	DB24C8	$C_{24}H_{32}O_8$	448.52	4.5–5.0	103–104	0.60	Solid
9	DC24C8	$C_{24}H_{44}O_8$	460.61	4.5–50	< 26°C	1.30	Liquid

There were several metals apart from alkali and alkaline earths which were extracted by crown ethers in recent years.

Amongst alkali metals, most extensively studied metals were sodium and potassium and caesium. In comparison, lithium and rubidium were not so exhaustively investigated. Within alkaline earths, most commonly studied metal. was strontium, while little attendation was paid to extraction chemistry of radium.

Amongst lanthanide and actinide series of elements apart from lanthanum, europium was very exhaustively studied while in actinide series of elements most of the work pertain to solvent extraction of uranium and some on thorium.

Within transition metals, molybdenum, manganese, rhenium iron, cobalt palladium, copper, silver, zinc, mercury were best studied. Within p-block elements, only gallium, indium, thallium and lead were studied in greater details.

Most of the extractions of alkali metals were carried out at pH 2–4 but lipophilic phosphoric acid monoalkyl ester derivatives were used at pH 7.0–9.0. Generally the pH of extraction fluctuated between pH 4–7 depending upon the kind of the crown ether used. Variety of diluents were used but the diluent most commonly used was invariably chloroform for extraction of alkali metals. In few instances dichloromethane, nitrobenzene were also used as the diluents. 18C6 was best for sodium and potassium while 15C5 was good for sodium, but 12C4 was bad for lithium inspite of small cavity. For caesium apart from 24C8, a specially synthesised 21C7 was always used. The counter ions used included picrate, bis (2-ethylhexyl) phosphoric acid, methylyellow. Many extractions were rendered quantitative by the use of synergic extractants like TBP, TOPO, DNNS or HDEHP. The position of ligand forming atom in neutral side chain showed decrease in extraction [78]. In several instances the colored organic phase after extraction was used for direct colorimetric determination of the metal. The ionisable crown ethers containing carboxylate or phosponic acid derivatives were also useful for extraction of alkaline earths. In most of the cases complex formed was of type 1:1 for metal to ligand.

Lanthanides were invariably extracted around pH 2–4 in benzene with variety of counter anions. Usually high concentration of the extractant was employed. During extraction of higher member of series, no correlation to size of metal and the cavity of crown ether was ever noticed [156]. Good extractions were possible from media containing trichloroacetate when 2-thenoyltrifluoroacetone (TTA) was used as the synergic extractant along with crown ethers. The basicity sequence as well as steric effect determined the bonding between crown ether and metal-diketonate complex [167]. The effect of antisynergism was encountered with TOPO and DNS as extractants.

The large number of papers have appeared on solvent extraction of uranium (VI) with crown ethers like DC18C6, DB24C8, DC30C10 [172]. The extractions were carried out from highly acidic solutions (viz. 7–8M hydrochloric acid) methylene chloride was best diluent but benzene nitrobenzene were also used as the effective diluents. In several extractions involving mineral acid media, metal from the organic phase was preferably stripped with dilute acid and then quantitatively determined. Among transuranic elements the element which was best studied was Americium [181] but such extractions were fisible only in the presence of synergic extractants like TTA or 3 methyl-4 benzyl pyrazolone [PMBP]. Some work on extraction of thorium was also worthy of consideration.

In transition metals, group VII, VIII and Group I, II of the periodic classification of elements found favoured position. For extraction of technetium, ascorbic acid proved to best counter anion, with 18C6 or DB18C6 was best extractant. The anionic species of rhenium, viz. [ReO_4^-] was ion paired with [K^+ B15C5] to form neutral species in dichloroethane to form 1:2 complex. Iron (III) was extracted from highly acidic solution and it was stripped with very dilute mineral acid. DC18C6 proved to be best extractant for iron. Silver was extracted with aza or thia crown ethers due to strong bonding through nitrogen or sulphur respectively. This was also true for extraction of mercury. In main group elements, four metals like gallium, indium, thallium and lead were extracted selectively by crown ethers. They also permitted sequential group separations. For extraction of gallium linear polyether were effective extractants [215]. The separation of thallium in two oxidation state was possible with crown ethers [217]. The complex was 1:1. Lead was also extracted quantitatively but methods involving cryptands are more useful.

The various methods for solvent extraction separation of metals with crown ethers are best summarised in Table 5.5.1.

5.6 Solvent Extractions with Cryptands

The discovery of the cryptands by Lehn made a spectacular progress in the analytical chemistry of heavy metals. It provided a powerful tool for the isolation of metal pollutants from aquatic environment. Gandhi and Khopkar developed excellent methods for the separation of lead [222] thallium [223] nickel [227] manganese [224] cadmium [225] copper [226] and mercury [227]. These methods were also extended for the separation of uranium [228] zirconium [229] and calcium [230] from fission products by same workers. First attempt to use cryptand to extract alkali metals such as sodium or potassium was made with picrate as counter anion [231] and nitrobenzene as diluent. Toluene or ethylene chloride were also used as diluents for extraction of cryptates of alkali metals [232]. The substituted cryptands such as cryptand 222B (VII) and cryptand 221 (V) were respectively used for the extraction of potassium and sodium [233]. The several alkali metal cryptates with 222 (VI), 322 and 222B (VII) or 333 were extracted in transport membrane studies [234]. The most commonly used cryptands are shown (V) to (VII).

One interesting difference in relation to extraction with cryptands was that they had three dimensional structure with better stability for the ion pair complex

Table 5.5.1 Solvent extraction separations with crown ethers

Elements	Extraction Conditions
Alkali metals	At pH 2–4, DB18C6 in chloroform extracted them with bromophenol as the counter anion and subsequently determined spectrophotometrically. When potassium and thallium were co-extracted and separated from lithium [73] from 1–2M nitric acid they were extracted with DC18C6 in dichloroethane to investigate order of the extraction. It was noted that lithium was least extracted irrespective of the acid concentration used while there was maximum extraction observed for potassium [74] perhaps due to the size-cavity fixment. Cyclohexyl 15C5, extracted sodium while DC18C6 extracted potassium in benzene in the presence of bis (2-ethylhexyl phosphoric acid) due to synergistic extraction of latter ligand [75]. Alkaline earths were also coextracted. Poly and bis 12C4 in chloroform extracted alkali metal picrates [76] as 2:1 complex. S-dibenzo 14C4 oxyacetic acid, i.e. crown ether carboxylate derivative extracted them from pH 4.0–7.5. At high salt concentration extraction was better. The position of ligand forming atom in neutral side chain decreased the extraction [77]. At pH 7–9, lipophillic crown phosphoric acid mono alkyl ester derivatives were used for the extraction of sodium [78] and potassium with 15C5, 18C6, 21C7 and 24C8 as the parent compound [79]. 2, 4 dinitro 5–tri-fluoromethylphenylaminomethyl 12C4 and (picrylaminomethyl) 15C5 in chloroform was used for extraction and extractability was then compared with other crown ethers [80] B15C5, bis 18C6 were used for extraction of alkali metals when large size metals like rubidium exhibited bis crown effect with caesium [81]. Benzo crown ethers were synthesised and extraction and association constants were calculated [82]. Secondly amine type bis crown ethers like benzol15C5 extracted them form picrate solution as 2:1 complexes [83]. Alkyl derivatives of DB18C6 in chloroform extracted all alkali metals with the exception of lithium from picrate solution as the 1:2:1 complexes [84]. DB18C6 in chloroform extracted them from picrate media and distribution ratios were evaluated [85] but alkaline earths were also coextracted. Schiffs bases and secondary amine type bis crown ethers of B18C6 in chloroform formed 2:1 complex with caesium [86]. Lipophillic 14C4 carboxylic acid derivative showed best selectivity for sodium, potassium, and caesium specially for crown ethers with small cavity [87]. Secondary amine type B18C6 from picrate solution extracted sodium or potassium as 1:1 complexes while rubidium and caesium formed 2:1 sandwich complexes [88]. Lariat 16C5 crown ether extracted alkali metals with picrate as counter anion indicating large extractability, however during conductometric studies reduced stability of the complex was noted [89]. Triodophenol, nitrophenol and picrates were explored to study structural effects of the counter anions during extraction [90]. The proton ionisable dibenzo crown carboxylic and phosphoric acid esters in chloroform were successfully used for the extraction purpose [91]. The novel cage functionalised 17C5, 17C6 and analogues compounds were also used and a comparison was made with 15C5 and B15C5 as the extractants [92]. Tröger base crown ethers in chloroform from picrate media [93]

(Contd)

Elements	Extraction Conditions
	and lipophillic derivatives of 24C8 in dichloroethane were also explored [94, 95] for the purpose of extraction and to investigate the effect of the substitution on the process of extraction.
Li (I)	From 0.1M picric acid it was selectively separated from other alkali metals by 1, 4, 7, 12, 15, 18 hexaoxocyclodocaso 9–20 diene, 8, 11, 19, 22 tetrane [94] spheroid azophenol in chloroform with pipredine extracted it as the colored complex [96]. Didecalin 14C4 is another alternative extractant for lithium [97]. Chromogenic 14C4 with azonitrophenol group on ring in dichloroethane was very effective with substituent in sterically sensitive position [98] 18C6, 24C8, DB24C8 were used for its extraction when they gave dicatonic complexes with first two extractants but other alkali metals were also coextracted while last compound formed dicationic complex [99]. 14C4 was best for silver on account of matching of ionic size with cavity of crown ether however in both cases sodium was coextracted [100] 4-t-butyl B15C5 in nitrobenzene was used for multistage extraction [101]. Didecaline 14C4 and tetradecaline 14C4 both were used effectively in ion selective electrode, specific for lithium [102].
Na (I)	At pH 5–10 it was quantitatively extracted with 2 (S-dibenzo 16C5) oxybutaric acid in chloroform [103]. With sodium other alkali metals were extracted with water soluble ligands in polar solvents [104] 12C4 from picrate media extracted it as 2:2 complex but potassium and rubidium showed strong interference [105]. Lipophillic dibenzo crown ether carboxylate derivative in chloroform extracted it with potassium but such extraction varied with varying size of crown ether cavity [106]. Lariat crown ethers with pendant chain of N-alkyl and N-aryl sulphonyl derivative in chloroform extracted it in preference to potassium [107].
K (I)	So far this was the only alkali metal most extensively studied with the crown ether as extractant. From hydrochloric acid 4-picrylaminobenzo 18C6 extracted it and analysed photometrically at 490 nm-potassium from cement was thus analysed [108] DB18C6 in different diluents extracted it with bromocresol blue or green as the counter anion when chloroform was rated as the best diluent [109, 110]. Crown ether azo dyes of 15C5, DB15C6 were synthesised and extracted it with sodium and caesium (111) with p-tertiary butylphenol it was also extracted with caesium with the formations of phenolates with simultances process of dehydration [112] DB18C6 or 18C6 from the picrate solution extracted potassium with halogenated hydrocarbons as the diluent. Carbon tetrachloride was not very effective as the diluent [113]. The addition of trioctylphosphine oxide enhanced the extraction due to synergic effect. Bis crown ethers containing triazine ring in chloroform with lipophillic substituents extracted it from picric acid [114] but rubidium was also coextracted B15C5, B18C6 hydrozones in ethylene chloride extracted it quantitatively and determined colorimetrically at 494 nm [115].
Rb (I)	With tributylphosphate or trioctylphosphine oxide as the synergic agent crown ethers extracted it from benzene with picrate as counter anion in preference to caesium [116].
Cs (I)	4, 4 (5') bis (1-hydroxyheptyl) cyclohexane or (1-hydroxyl-2 ethylhexyl) B18C6 quantitatively extracted it from nuclear waste but latter extractant

(Contd)

Elements	Extraction Conditions
	also extracted strontium, however, better results were obtained with tributyl phosphate or dinonyl sulphonates as the synergistic agent [117]. DB21C7 in nitrobenzene extracted it from chloride solution from nuclear waste with pentachloro-antimonate as the counter anion to form adduct compound [118]. Radioactive caesium was also extracted from 5M nitric acid with dicarbolide crown ether but process was most unsuitable for large scale pilot plant operations as strontium was coextracted [119]. Numerous crown ethers in dichloroethane extracted it from solution containing trichloroacetate as the counter anion when it was observed that the extraction increased with the increase in counter anion concentration [120]. From 0.1M nitric acid it was extracted with bis 14 (5) tert butyl benzo 21C7 in toluene with 0.025 dinonylnaphthalene sulphonic acid as the counter anion [121] however rubidium, sodium and potassium were coextracted with high reagent concentration good quantitative extraction was possible. DB18C6 in nitrobenzene with bis (1–2 dicarbolytylcobaltate) quantitatively extracted it [122] and was determined radiometrically.
Alkaline earths	At pH 4–10, symm dibenzo 16C5, 13C4, and 19C6 oxyacetate derivative in chloroform extracted them specially barium with good selectivity, but magnesium was least extracted [123]. 0.25 M lipophillic cyclic polyether dicarboxylate derivative in the chloroform extracted them as 1:1 complex and were measured colorimetrically at 273 nm [124]. Several crown ethers including aza compounds were used for their extraction from picrate solution and composition of extracted species was ascertained [125]. With trioctylphosphine oxide as the synergist 18C6 was extracted at pH 4-7 with phenyl 3 methyl 4 benzyl pyrozolone (PMBP) [126] with cyclohexane or benzene as the diluent. 18C6 with trioctylphosphine oxide extracted it with bis (ethyl hexyl) phosphoric acid as the counter anion but PMBP was certainly better synergistic extractant [127] for alkaline earths.
Mg (II)	From 3.5M hydrochloric acid it was extracted and enriched with DC18C6 for various isotopes like Mg^{24}, Mg^{25} and Mg^{26}. One could observe isotope effect during extraction [128].
Ca (II)	At pH 8.5 symm DB14C4 oxycarboxylate or symmdibenzo 16C5 oxy analogue extracted it along with several alkalimetals to form 1:1 complex [129] analogues to sandwich type complexes.
Sr (II)	Dibenzo or dicyclo substituted 18C6 and 24C8 extracted it from 0.1M perchloric acid in chloroform [130]. At pH 2–6 crown ethers in benzene chloroform extracted it with either picric acid or trialkyl phenol as the counteranions from nuclear waste however lead, silver and mercury were coextracted [131] 18C6 in the presence of surface active agents like polyamine, ECA-4360 or SPAN-80 (i.e. sorbitol monooleate) extracted it from picrate solution [132]. DC18C6 in chloroform extracted it from milk and consequently separated from calcium while barium was extracted with 21C7 [133]. Polymer bound 18C6 polyether extracted it from nitrate media and stripped with water at 60°C [134] 12C4, 15C5, 18C6 and DB18C6 were also used but only 18C6 from nitric acid gave best separations [135]. Latter it was stripped with water.
Ba (II)	0.45M of DB24C8 in nitrobenzene extracted it from 0.2M picric acid solution containing nitric acid [136]

(Contd)

Elements	Extraction Conditions
Ra (II)	Symm DB18C6 oxyacetic acid extracted it in chloroform from alkaline solution in absence of any anion [137].
Alkali + alkaline earth mixture	B15C5, DB18C6 extracted best from solution containing p-tertiary butyl phenol or octanic acid and analysed radiometrically [138] 0.5M of symm DB18C6 acetic acid in chloroform extracted them at pH~6.0–13.0 a 1:1 complexes for alkali metals and 1:2 complex for alkaline earths [139]. About 0.1M DC18C6 in chloroform extracted them in the presences of saltingout agents like lithium nitrate [140] but magnesium was not extracted. Ionisable crown ethers including 21C9, 30C10 in the presence of salicyclic acid extracted them [141]. The ionisable crown ethers as carboxylic or phosphoric ester derivatives in chloroform extracted them, stripped with 0.1M hydrochloric acid and determined by the technique of ion chromatography [142]. Sym DB14C4 or DB18C6 oxycarboxylate were also used [143]. 16C5 in benzene extracted from picric acid [144]. At pH 8.5, DB14C4, DB16C5 tyloxyhexanonic acid were also used for extraction [145]. 15C5 B15C5 analogues were used for thermodynamic investigations [146] B16C5, B15C5 in dichloroethane were investigated from picric acid solution [147–149]. Several thermodynamic parameters were evaluated. With polyethylene glycol-200 were also used as the saltingout agents with 15C5 and 18C6 [150]. Biscrown ether containing metallocene like iron or ruthenium were used to form the complexes. Such complexes were 1:1 if ionic size cavity used matched or 2:1 when sandwich type complexes were formed where in metal was embedded between two crown ether moiety [151].
Y (III)	About 0.01M DC18C6 in chloroform extracted an isotope of Y^{90} from 0.01M picric acid and thus separated it from an isotope of strontium viz. Sr^{90} [152].
Ln (III)	At pH 3.0, 18C6 and DC18C6 in benzene extracted them but DC24C8 and 15C5 had proved selective for europium, dysposium, erbium and lucetium [153] (4-teriary butyl cyclohexyl) 15C5 with dinonylnaphthionic acid (HDDNS) as the synergistic extractant extracted them at pH 2.0 better extraction was possible with high concentration of crown ether [154]. 15C5, 18C6, or DB18C6 in chloroform extracted them from picrate solution as 1:2:3: complex where for metal to ligand to counter anion [155]. At low pH, 1×10^{-4}M of DC18C6, DB18C6, DC24C8 with dinonylnaphthionic acid sulphonates (HDDNS) extracted them at high crown ether concentration. Light lanthanides were extracted with no particular relationship between size of ion and cavity of the crown ether [156]. 15C5 and 18C6 extracted them from picrate solutions to form 1:2 complex of sandwich type specially with 15C5 [157]. Symm DB16C5 oxyacetic acid with increasing lipophillicity and its analogues were used for the extraction of lanthanides from solutions with high ionic strength [158]. Some crown ethers and cryptand were explored when hydrolysis phenomena was encountered for lanthanide ions [159–160]. DC18C6 or DB18C6 in chloroform extracted them from trichloroacetate as the counter anion [161]. Crown ethers with iso-oxazoyl-lariat homologue of the iso-xazolealdehyde linkage were synthesised when presence of the carboxylated form promoted extraction [162]. At pH 3.0 1×10^{-3}M of 18C6 extracted it from 0.9–1M of trichloroacetic acid and EDTA

(Contd)

Elements	Extraction Conditions
	complexion when light lanthanides were quantitatively extracted on account of the extra stability of EDTA complexes [163]. 18C6 in various polar solvent like dichloroethane or dichloromethane were extracted from the trichloroacetic acid when lipophillicity of counter anion and the steric hindrance helped the extraction [164]. 1–8 dioxaoctanemethylene bis (4' B15C5)$_4$ extracted them from picrate media. The formation of bimolecular sandwich complex increased the extractibility but cation selectivity showed decrease due to absence of intramolecular sandwich type formation [165]. DB18C6 or DC18C6 in benzene or chloroform extracted them with 2-thenoyl trifluoroacetone (TTA)
	As the synergistic agent [166] for several metals including uranium (VI), thorium (IV) which showed the basicity sequence as well as the steric effect decided the kind of bonding between crown ether and crown ether metal complex and complex with TTA [166].
La (III)	18C6 in different solvent extracted it with picrate or perfluoro-octanate acid trichloroacetate as the counter anions, it was then observed that the best extraction was noted with penta deca fluoro octanate [167].
Eu (III)	Crown ether with bis (2 ethyl hexyl) phosphoric acid i.e. HDEHP as the counter anion, with TTA as synergist or TBP, TOPO as complexing ligands gave excellent extraction when luminescence studies were carried out [168]. Hydration of crown ether complex was related to the steric hindrance by attached group to the crown ether [168].
	From nitric acid 18C6 or 15C5 extracted it with polyethylene glycol 3, 6 a trioxa undecane-1, 1, 1 diol with cobalt dicarbolide as hydrophobic counter anion [169] DB24C8 or DC18C6 with HDDNS or TOPO (trioctyl phosphoric oxide) extracted it to exhibit antagonistic effect [170] instead of synergistic extraction showed by other process [171-172]. At pH 6 – 6.5 a compound, 6.7, 9, 10, 18, 19 hexahydro 17-H dibenzol (1, 4, 7, 10, 13) pentaoxacyclohexadean-18 yloxy acetic acid extracted it in chloroform heptanol mixture [172].
U (VI)	From 8M hydrochloric acid various crown ethers like DC27C9, DC24C8, DC18C6, DC30C10 and DC21C7 extracted it from chloride solutions [173]. DC28C6, DC24C8, DC27C9 & DC30C10 extracted as 1:2 complex [174]. From 3–8M hydrochloric acid it was extracted with 0.1M of DC18C6 in dichloromethane and thus separated from thorium [175]. From the thiocyanate solutions, 15C5 in nitrobenzene or from perchlorate media DC18C6 in benzene extracted it with TTA as the synergistic agent but neptunium americium were extracted in the presence of hexanone [176]. From chloride solution it was extracted in dichloroethane indicating that variation in extractibilty with hydrochloric acid concentration was quite significant [177]. From nitric acid it was extracted with DC18C6 or DC24C8 with benzonitrile was stripped with oxalic, sulphuric or perchloric acid [178]. Also extracted from nitric acid solution, stripped with oxalic or sulphuric acid [179]. DC18C6 in toluene extracted it from mixed nonaqueous solvents, stripped with perchloric acid was separated from plutonium. It was seen that propylenecarbonte, acetone and acetonitrile or dioxane enhanced the extent of extraction whereas alcohols decreased the extractibility perhaps due to the antagonistic effect, however at low nitric acid concentration better extraction was observed [180].

(Contd)

Elements	Extraction Conditions
Am (III)	At pH 3.5, 18C6, DC18C6, DB18C6 extracted it in benzene with TTA as synergic extractant. It was noted that here the extraction showed increase in value due to increasing basicity of the crown ether [181]. DC18C6, 15C5 also extracted it [182]. From 0.1M perchlorate solution was extracted with 12C4, 15C5, 18C6 and their analogues in toluene with 1-phenyl 3-methyl 4-benzyl pyrazolone (PMBP) where in ether basicity did not indicate no specific order of extraction for amercium, thorium or uranium [183].
Mo (VI)	DB18C6 extracted it from hydrochloric acid [184].
Mn (II)	Tertiarybutyl B15C5 and tertiarybutyl cyclohexano 15C5 extracted it with dinonylnaphthalene sulphonic acid (HDDNS) as the synergistic extractant where as benzosubstituted derivatives showed better extraction [185]. At pH 9.0 macrocyclic Schifs base containing thiophene phenol in nitrobenzene extracted it with tetraphenylborate as counter anion however copper showed strong interference by coextracting it under similar conditions for the extraction of manganese [186].
Tc (VII)	DB18C6 in dichloromethane extracted it from sulphuric acid solution and separated from uranium (VI) [187]. The extraction from alkaline solution with DB18C6 or 18C6 in benzene or nitrobenzene in the presence of ascorbic acid providing counter anion extraction was successfully carried out [188–189]. DB18C6 or 18C6 in benzene or nitrobenzene in the presence of acetylacetone (HAA) extracted it quantitatively from 0.1–1M absorbic acid solution and was separated from molybdenum (VII) [190, 191].
Re (VII)	B15C5 in nitrobenzene extracted it with coextraction of molybdenum [192] but when DC18C6 was used with brilliant green as counter ion extraction was good [193] B15C5, DB18C6 in dichloroethane extracted it as anionic species with $[K^+B15C5]$ as cationic species as ion pair complex with $[ReO_4^-]$ as anion. A hydrophobic 1:2 complex was usually formed [194].
Fe (III)	From 8–10M hydrochloric acid it was extracted with DC18C6, DB18C6 or 18C6 in chloroform, stripped with 1M hydrochloric acid. Use of the titanous chloride enhanced the extraction. Out of ligands used DC18C6 proved to be best [195] DC18C6, DB18C6. Or 18C6 in chloroform had extracted it from hydrochloric acid solution as the anionic chlorocomplex when gallium, thallium, antimony and gold were coextracted with various crown ethers [196]. From 7–9M hydrochloric acid on solid base containing styrene-divinyl benzene polymer coated with 18C6, it was extracted quantitatively. Proton ionisable crown ether gave best results while aluminum was selectively extracted from 3–4M hydrochloric acid leaving iron (III), in the aqueous phase as unextracted [197] species.
Co (II)	2-(symm DB14C4) oxydecanoic acid in chloroform extracted it at pH > 10.0 [198]. Various crown ethers like 12C4, 15C5, 18C6 DB18C6, or DC18C6 extracted it from perchlorate solution as $[Co(OH)^+ ClO_4^-]$ 2HR [199]. With 1-phenyl 3-methyl 4-benzoyl pyrazol-5 one and B15C5, 18C6, DB18C6 it was extracted from 0.1M chloride solution but group IIB metals showed strong interference [200]. With same counter ions in dichloromethane it was extracted with DB24C8 and was consequently separated from caesium when synergistic extraction was possible [201].

(Contd)

Elements	Extraction Conditions
Pd (II)	At pH 6–9, DC18C6 in dichloroethane extracted it with 4 (2 pyridylazo)-resorcinol (PAR) as the counter anion. The change in entropy during the process of extraction was investigated [202].
Cu (II)	15C4 in chloroform or benzene extracted it from picrate solution along with copper [203].
Ag (I)	15C5, 18C6 in chloroform extracted if from picrate media. There was decrease in extractibility in the presence of lithium nitrate or perchlorate solution [204]. The structural investigations were made. Dibenzo (d, m) 1, 4, 8, 12 dioxadizacyclo-pentadeca-5, 10 diene and pentadecane in chloroform extracted it with dipicrylamine as the counter anion as 1:1:1 complex. The first extractant was good for copper and mercury [205]. Macrocyclic polyethers containing nitrogen or oxygen as donor atom extracted it along with mercury, palladium copper as coextracted species [206]. Thirteen DB18C6 nitrogen containing derivatives were used for its extraction when copper, mercury and gold were coextracted [207]. 0.05M of macrocyclehexacyclone (N-containing) compound in methylisobutylketone (MIBK) extracted it from 0.5M perchloric acid or DNBS (sodium salt), but mercury was coextracted, for the extraction of other metals warming of solution before extraction was essential [208]. Nitrogen and sulphur containing 18C6 or DB18C6 extracted silver along with copper cadmium and palladium as impurities [209].
Zn (II)	15C5 or 18C6, DC18C6 extracted it from picric acid solution [210] when along with zinc cadmium, calcium and strontium were coextracted [210]. Isotopic effect facilitated separation of Zn^{64} and Zn^{70} by selective extraction with crown ether [211].
Hg (II)	About 0.2M DC18C6 in dichloroethane extracted it from 2M nitric acid with water as strippant. It was separated from zinc and cadmium [212]. Crown ethers containing nitrogen or sulphur as donor atoms extracted it from picrate solution. Radioactive mercury was also extracted at pH 4–5 as 1:1:1 complex [213]. DB18C6 or nitrogen containing analogue was used for extraction from the alkaline solutions [214]. Proton ionisable crown ethers or tert butyl substituted dibenzodistrazalo crown ether in alcohol were also extracted it when supercritical fluid chromatography technique (SFC) was used to accomplish separation [215].
Ga (III)	From 0.50–9M hydrochloric acid it was extracted with 18C6 or analogue in different diluents [216]. Linear polyethers (n = 4, 5) were most effective.
Tl (I)	At pH 3.3, DB18C6 in benzene or chloroform extracted it from the picrate solution with 0.1M perchloric acid as the stripping agent. It was possible to separate thallium (I) from thallium (III) [217]. 15C5 or B15C5 in the presence of tributyl phosphate (TBP) as the synergistic extractant extracted it from picrate solution [218] as 1:1:1 complexes. Numerous crown ethers in dichloroethane extracted it from picric acid solution [219]. Such studies indicated that log D for metal picrate was inversely proportional to the dielectric constant of the diluents [219].
Pb (II)	At pH 5–6, 0.5M of DB18C6 extracted lead when cobalt nickel and copper were coextracted [220]. In addition during real analysis the thiourea, carbathione were also coextracted.
At (VII)	The radioactive isotope At^{85}_{211} was extracted from alkaline solutions in the absence of reducing agents. It was noted that species $MAtX_2R$ was extractable were R is crown ether, M metal ion, X is counter anion [221].

formation. The stability may be partly attributed to bonding through nitrogen instead of oxygen as donor atom. Secondly these complexes were more soluble in the nonpolar solvents like toluene, xylene. Many workers had used colored counter anions like eosine, erythrosine during extraction which in turn facilitated direct spectrophotometric determination.

Although cryptands in several ways were better extracting agents than crown ether so far they have not been fully exploited for the said purpose by the analytical chemists. They were exploited only for the solvent extraction separation of metal pollutants from the aquatic environment [222–227] and heavy metals [228–231]. Amongst them most effective was cryptand 222 [232–233]. The cryptand 222B proved to be most facinating for the extraction of lead [222, 234]. Lithium was quantitively extracted at pH 7–8 with chromogenic cryptand in chloroform, when other metals like sodium or potassium exhibited negligible extraction [237]. Similarly sodium was quantitatively extracted with cryptand 222 in chloroform [239]. The use of salting-out agent did not enhance the extent of the extraction. Calcium was extracted at pH 7.2–8.6 with cryptand 222 in cyclohexane with PMBP as the counter ion [240] such work facilitated its separation from alkaline earths like barium and strontium. The method was further modified with use of TOPO as the synergistic extractant. The method for calcium was modified [230] to get better selectivity during extractive separation. Further this method had permitted extractive photometric determination of calcium in single step, the details were worked out [230] for the purpose.

The various elements extracted by cryptand (without using simultaneous spectrophotometry) are best summarised in Table 5.6.1.

Table 5.6.1 Solvent extraction separations with cryptands

Elements	Extraction Conditions
Alkali metals	Cryptand 2 2 1, 2 2 2 and 2 2 2B were used for the extraction of alkali metals with various counter anions. The extraction behaviour of lead, thallium, manganese, cadmium and nickel was also investigated [235].
Li (I)	At pH 7–8 chromogenic phenolic cryptands in chloroform extracted it when other s-block metals were partially extracted [236; 237]. The chromogenic phenolic compounds containing azonaphthol chromophore were also used [238].
Na (I)	Cryptand-222 in chloroform extracted it but elements like lithium, potassium and rubidium showed strong interference [239]. The effect of saltingout was also investigated to understand mechanism of the extraction [239].
Ca (II)	At pH 7.2–8.6, 0.1M cryptand-222 in cyclohexane extracted it with PMBP as the counter anion. The method had facilitated its separation from strontium and barium [240]. With trioctylphosphine oxide as the synergistic agent the extractions were also carried out with cryptand-222 in cyclohexane, or in mixture of solvents (230).

5.7 Solvent Extractions with Aza and Thia Crown Ethers

The crown ethers formed stable complexes with many transition elements [241–243]. On the contrary the thia crown ethers formed stable complexes with alkali metals. Thia crown ethers could extract soft metals. Very Limited studies were available on thia crown ethers. Takagi [244] and his group made extensive studies in this area, e.g. silver was extracted with thia 12C4 compound (VIII) from 1M hydrochloric and perchloric acid but mercury, tin, cadmium and cobalt were co-extracted. The extracted species was 1:1:1 complex for metal: ligand: anion [245–246].

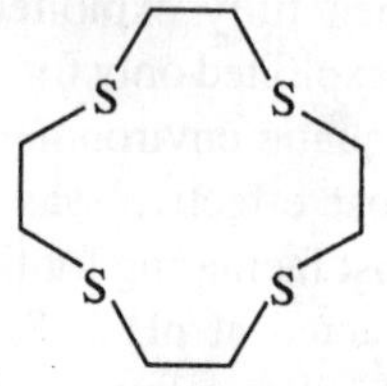

(VIII)

The metals like copper nickel were also extracted in thia crown ether [247]. Shono [248] carried out extraction of metals with thia crown ethers: but such extraction was not quantitative even with thia crown ether with aromatic substituent (IX).

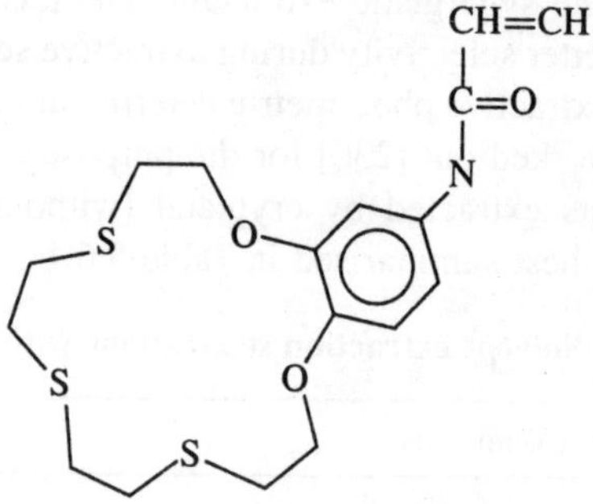

(IX)

As opposed to thia crown ethers, the aza crown ethers were extensively used for the solvent extraction of metals. The extraction with mixed donor type crown ethers such as aza 18C6 was tried as an extractant.

The substituted chromophoric group for amino proton of aza 18C6 was used for extraction of calcium [249]. The alkali metal picrates were extracted with substituted aza 18C6 [250]. In such extractions* group which in turn linked nitrogen atom to phthalmide [251, 252].

Aza crown ethers proved to be promising extractants for inner transition elements. For instance 4f transition series metals like lanthanides were quantitatively extracted between pH 4–8 in the presence of 2-thenoyltrifluoroacetone (TTA) as the synergic extractant [253]. Most favoured metal for extraction with sulphur containing ligand was copper. It was extracted with 14 membered ring aza compounds with thymolblue as the counter ion [255]. For extraction of gold, between aza and thia crown ethers as the extractant former proved to be best [256].

---

*The extractability and the selectivity depended largely upon the number of methylene.

Different chromogenic and fluorogenic aza crown ethers [259] were used for several extraction. In comparison to the aza crown ethers, the thia crown ethers have been more extensively [261-278] used. The important metals extracted by them were rhodium, palladium, copper, silver and mercury. Of these metals exhaustive work was carried out on copper [263-269] and silver [270-276]. In all such extractions common diluent like dichloroethane, MIBK nitrobenzene were used but ethylene chloride was most favoured diluent. The average pH of extraction for silver was around 4.0 the commonly employed counter ions were picrate, perchlorate and in rare eases tetraphenylborate. In several of the extraction the quantitative extraction was accomplished by use of synergistic extractants like tributyl phosphate [264] or dinonylnaphthalene sulphonate [266]. Solvent polarity played an important role during extraction of metals like mercury, silver which showed strong interferences by inhibiting some extractions [274]. The soft metals as silver copper mercury were easily extracted while hard metals such as indium, manganese were not at all extracted. The various methods used for the solvent extraction of elements with thia and aza crown ethers are summarised in Table 5.7.1 and Table 5.7.2 respectively.

Table 5.7.1 Solvent extraction separations with aza crown ethers

Elements	Extraction Conditions
NH_4^+	Chiral diazacrown ethers were used for complexation enantioselectivity was then investigated [251].
Na (I)	Macrocyclic compounds containing pyridine moiety in carbon tetrachloride was extracted when number of isotopes of elements like iron, manganese, cobalt, nickel, copper zinc, mercury and silver were investigated [252].
Ln (III)	At pH 4-8, 4.13 didecyl 1, 4, 10, 16 tetraoxa 4, 13 diazacyclo-octadecane in chloroform along with 2-thenoyltrifluoroacetone as synergic agent with cryptand 222D was used for extraction when larger organic molecule showed preference for better extraction. The stability constant of the complexes were also evaluated [253].
Cu (II)	About 0.01M of 5, 7, 7, 12, 14, 14 hexamethyl 1, 4, 8, 11 tetraza-macrocyclic compounds extracted it with bromothymolblue as the counter anion [254].
Au (II)	Aza crown ether in benzene extracted it. Such extraction was however not comparable with extraction with thia crown ether [256] but tera aza form extracted it [255].
Hg (II)	Polymeric azacrown 8, 9, 17, 18 dibenzo 1, 7, 13 trioxa, 4, 10, 16 triazacyclodecane extracted it with extraction constant indicating wide variation due to the cooperative effect of the ligands with varying molecular weight [257, 258]. Different derivatives of chromogenic and fluorogenic crown ethers were also used for the extraction. NN bis (2-hydroxy 5-nitrobenzyl) 4, 13 diaza dibenzo 18C6 and analogues compounds were used while NN' bis (7-hydroxy 4-methyl coumarin 8-methylene) 4, 13 diazadibenzo 18C6 was used as the fluorogenic ligand [259].

5.8 Solvent Extractions with Calixarenes

Calixarenes were macrocyclic phenol-formaldehyde condensation oligomers that possessed the capability for cationic, anionic and molecular inclusion. They were represented as (X). Here (n) represented the numbers of phenyl rings

Table 5.7.2 Solvent extraction wih thia crown ethers

Elements	Extraction Conditions
Rh (III)	1M dibutyldioctyl or dibenzyl sulphide in hexanol extracted it from 1–3.5 M nitric acid as [Rh (NO$_2$)$_6$]$^{3-}$ ion [260] while [Rh (H$_2$O)$_6$]$^{3+}$ was not extracted [261].
Pd (II)	Lipophillic polythio macrocycle in chloroform extracted it from nitric acid solution, the lipophillicity was caused by alkyl chain; macrocyclic cavity size; and number of sulphur as donor atoms which in turn influenced the extraction behaviour [262].
Cu (II)	Maximum work was carried out on extraction of copper and silver. At pH 4.0, 1, 4, 8, 11, tetrathiacyclotetradecane in dichloromethane or dichloroethane extracted it with picrate, perchlorate or tetraphenyl borate as the counter ion [263] 7, 10 dithiohexadecane and 2, 5, 9, 12 benzotetrathiocyclopentadecane in toluene with triphenyl phosphine or tributyl phosphate as auxiliary complexing ligand were quantitatively extracted [264] along with silver and gold as coextracted cations. Sulphuric acid with 1, 4, 7, 10, 13, 16 hexathiocyclo-octadecane in the presence of picrate solution extracted it but silver was coextracted along with copper [265]. Tetrathia 14C4 and tetrathia 16C4 in toluene with dinonylsulphonic acid was extracted along with partial extraction of silver [266]. At pH 4.8 bisthia crownether in chloroform extracted it from picrate solution accomplishing separations from several metals [267]. At pH 5.9, 0.5M of 1, 4, 8, 11 tetrathiacyclotetradecane in dichloromethane extracted it with tetraphenylborate as counter ion [268]. At pH 7.0, 3, 6, 10, 13 tetrathiacyclotetradecane-1-oxyacetic acid of hexanoic acid derivative with pendant carboxylic acid group extracted it with picrate as the counter anion [269]. The extraction was very much dependent on polarity of the solvent.
Ag (I)	1, 4, 8, 11 tetrathiacyclotetradecane in nitrobenzene extracted it and separated it from mercury in 2M perchloric acid [270]. Same extractant in dichloromethane extracted it from picrate media [271]. However few elements like mercury and palladium were coextracted. At pH 4.7, sulphide podand was used for the extraction in methylisobutylketone or dichloroethane with perchlorate as counter ion. It was stripped with 4–6M mineral acid [272] 15 (4-hydroxyphenylazo) benzo 2, 5, 9, 12-tetrathiacyclopentadecane and 15C5 chloro 2-hydroxyphenylazo derivative in ethylenechloride extracted it [273]. Benzothiocrownether in chloroform extracted it from picric acid solution and was thus separated from several elements but mercury inhibited the process of extraction [274] 1, 4, 8, 11 tetrathiocyclopentane and 1, 3, 4, 7, 8, 10, 11, 13 octahydro 2, 5, 9, 12 benzotetrathiocyclopenin benzene extracted it from picric acid with copper [275]. Number of benzodithiacrown ethers in picrate media were explored. The sulphur atom on crown was favourable for extraction [276]. At pH 5.5 cyclic monoazotetrathiaether in dichloroethane extraction it but soft metals like silver, copper, mercury were best extracted while hard metals like indium, manganese were not extracted [277].
Hg (II)	Polythiaether and cyclicpolyseleno ether also extracted the mercury ion quantitively [278].

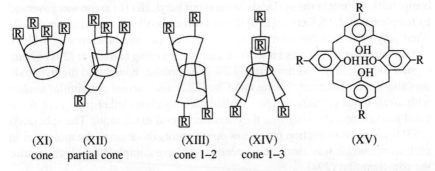

R = stands for group like tertiary butyl etc. while,
R′ = stands for anions like OH⁻, CH$_3$COO⁻ etc. The important members were calix(4) arene calix (6) arene and calix (8) arene. However, most interesting part was the conformation. For instance calix(4) arene had four conformation as shown in (XI), (XII), (XIII) and (XIV).

Calixarenes were first reported by Zinke and Ziegler [279] but Gutsche [280] furnished systematic procedure for the synthesis of calix(4) arene and calix (6) arene compounds. In calixarenes the rotation about the methylene bridges gave rise to numerous conformation that could be frozen out by substitution of larger groups. The water insolubility of p-tert butyl calixarene (4) was useful for cation transport [282] p-tert butyl calixarene (4) methylether exhibited slight selectivity for sodium. Ungaro [283] had functionalised the base of calixarene. Such structures showed both hydrophobic cone into which neutral molecule could bound and hydrophillic region was useful for cation binding. The better derivatives have been used for the extraction of alkali metal picrates in chloroform. Calixarenes were several times more efficient than crownethers like methyl B15C5 or methyl B18C6 for extraction of sodium [284].

The synthesis of water soluble calixarene opened up new era in chemistry. The sulphato derivative of calixarene was most popular [285]. Several studies of calix(4) arene for cation complexation centred around cone conformation. Atwood [286] found that 1, 3 alternate conformation presented bidentate coordination through methoxy groups in calixarene with R = SO$_3$ Na and R′ = CH$_3$. In comparison the studies on complexation by calix (8) arene were few in number. The stoichiometry was consistent with pinched conformation for the calixarene, the chemical shifts could be explained better by pleated loop conformation where in two guest were face to face in the cavity [287].

The future prospects for calixarene will revolve round following points (a) synthesis of new calixarene for complexation of specific metals ions, e.g. a calixarene specific for uranium (VI) was developed by Shinikai [288] (b) improved synthesis of odd number of calixarenes-so far no work has been done (c) structural characterisation of bonding modes of metal ions to calixarene in solid state. There was lack of information on transition metal complexes in solid state (d). Extraction experiment carried out in competitive manner. It was important to determine selectivities on the basis of separate extraction experiments for various cations. The field of complexation of cationic species by calixarenes was new one and has great future.

Izatt [289] first discovered the resemblance between calixarenes, cyclodextrins and crown ethers and had explored the possibility of utilising them for complexation of metal separations. They were considered in effective cation carriers in neutral solution. They demonstrated a transport ability of alkali metal complexes in basic solution. They proved that macrocyclic ring played an effective role during complexation. Further the synthesis of the p-tert butylcalix (6) arene was governed by template effect. McKervey [290] showed that carboalkoxymethyl ethers (XVII) p-tert buty calix(n) arene (if n = 4, 6, 8) were best cation extractants for s-block metals ketoether (XVI) was better for cation transporting extractant [291] giving highest extraction of sodium, viz. 99.2%. The another compound diethylamide was also good extractant. Chang and Cho [292] also carried out similar studies with alkali metal extraction. They studied carboxyethoxylether instead of p-tert butyl calixarenes, showing that they were not good extractants. The spherands [293] had similar extraction properties due to preorganisation of the molecule. In such compound it was absolutely necessary to have complete desolvation prior the complexation [294].

(XVI)   (XVII)

The calixarene with substituent in upper rim (XIX) or one with substituent in lower rim (XX) like acetyl derivative are very useful for the extraction of transition element. For instance, the water soluble calixarenes such as sulphato calixarene (XVIII) the higher homologues where n = 5, 6 were better extractant for uranyl ion [295, 296]. A 1:1 complex was formed in basic solution. It was found that uranyl ion had cavity of the larger calixarenes useful for fitting, i.e. hole size selectivity and the larger calixarenes provided (XVIII) better arrangement, i.e. hexacoordination. This was called coordination geometry selectivity.

The calixarenes were metacyclophane [297] which were composed of phenol and methylene bridge units. They were prepared from para substituted phenol and formaldehyde. They provided desired ring size selectivity. They possessed cyclic phenolic hydroxyl groups useful for constructing functionalised host molecules. They had several sized π-rich cavity for incorporating guest molecule. An interesting comparison is given in Table 5.8.1 for crown ether cryptand and calixarene.

(XVIII) (XIX) (XX)

Table 5.8.1 Comparison of extractants

S.No.	Property	Crownether	Cryptand	Calixarene
1.	Was change in ring size possible?	Yes	Yes	Yes
2.	Was mass scale preparation viable?	Yes	No	Yes
3.	How was spectral transparency?	Good	Fair	Not good
4.	How was neutrality?	Yes	Yes	Possibility of modification
5.	Did they have optical property?	Need modification	No	Modification
6.	Had synthetic derivative?	Yes	Yes	Yes
7.	Could function as ionophore?	Yes	Yes	Yes
8.	Was cavity shaped to suit host change?	Yes	Yes	Poor cannot change

The important quality of calixarene was for metal cation recognisation. They showed highest selectivity for rubidium [298]. On the whole the calixarene could form complexes with both organic and inorganic guest and as well as the anions in chemistry.

Most commonly extracted metals [299-333] included not only alkali and alkaline earth metals but also lanthanide series, the transition metals like iron, palladium, platinum, silver, mercury. Gallium and lead were also extracted. Ammonium ion was strongly bound by lariat type calixarene from picrate media [298].

Invariably the alkali metals were extracted with picrate as counter anion with chloroform or dichloroethane as the diluent. The ester and amido derivative of calixarene proved to be beneficial for quantitative extraction of s-block metals. Many of the extractions were possible due to change in conformation from cone to partial cone favouring phenomena of the metal complexation.

The extraction of lanthanide and actinide series of elements had some similarity, viz. tetra-alkyl ethers of calix (4 or 5) were efficient extractants for both series of elements in methylenechloride or ethylacetate or tributylphosphate as the solvent. Diethyl carbonyl methoxy derivative of tetramer was good for lanthanides.

This was because lower rim oxygen containing coordination sphere was good and lipophillicity improved selectivity of ligand extractant. Phosphoryl derivatives were good for extraction of few lanthanides. Carboxylate derivative were best for extraction of thorium.

Amongst group VIII of periodic classification important metals which were extracted by calixarenes were iron, palladium, cobalt and platinum. The hexa-acetato derivative of calix (6) arene proved to be best extractant for these metals [334-340]. Toluene was used as the diluent. The extractant concentration used was as small as 10^{-4}M of the reagent. The extraction of silver was best accomplished by the ketonic derivative of calixarene in chloroform from nitric acid media. Calixarene with hydroxamic acid as functionalised group readily extracted mercury and other trivalent metals. Alkylamide substituted calixarene were being used for extraction of transition metals.

The interesting information is summarised in Table 5.8.2. All the important factors influencing extraction such as pH, reagent concentration, diluent, strippant, interference and important applications are included in these Tables.

Table 5.8.2 Solvent extraction separations with calixarenes

Elements	Extraction Conditions
NH_4^+	Calixarenes, crown ethers analogues to lariat ethers were used from the picrate media for extraction [298]. To the hydrophobic side arm the ammonia was firmly bound in comparison with other alkali metals [299].
Alkali metals	2 pyridylmethyl(n) calixarene (n = 6, 8) was synthesised and used for the extraction, when the selectivity was improved with the introduction of the pyridyl group [300]. Noncycle anologue was also used for the extraction [301]. It exhibited cooperative effect between the crown ether and calixarene [301]p-tertiarybutyl calix(n) arene ester was used for the extraction from thiocyanate or picrate media [302]. Calix(4) arene tetramides in dichloroethane extracted it from picrate solution. Strontium was coextracted with other metals [303]. Ethyl p-tertiarybutyl calix(4) arene tetraethanol in benzonitrile extracted alkali metals with picrate as the counter anion [304].
Na (I)	25, 21 dicarboxy methyl 26, 28 dimethoxy, 5, 11, 17, 23 tetratertbutyl calix(4) arene in chloroform or dichloroethane extracted it along with potassium [305]. There was conformation change from cone to the partial cone conformation [306].
Cs (I)	1:3 alternate mono and bis crown calix(4) arene was used to separate it [306]. 1, 3 calix(4) bis-o-benzo crown 6 and 1, 3 calix(4) bisnaphthyl crown-6 was also used for the selective separation by membrane technique [307].
Ln (II)	Calix(4) arene tetra-alkyl ether and calix(5) arenepentalkylether in methylene chloride was used to separate from actinides [308]. Calixarene phosphates in ethylenechloride or ethylacetate or tributylphosphate in hexane extracted all lanthanides from the picrate solution [309].
La (III)	5, 11, 17, 23 tetrabutyl 25, 27 bis (diethyl carbonyl) methoxy 26, 28 dihydroxy calix(4) arene in dichloromethane was extracted due to lower

(Contd)

Elements	Extraction Conditions
	rim oxygen containing coordination sphere and lipophillicity gave selectivity to calix(4) arene. The encapsulation in 8-coordinate oxygen environment was beneficial [310] for the purpose of complexation.
Nd (III)	At pH 2–3.5 p-tertiarybutyl calix (6) arene hexacarboxylate in chloroform extracted it as the 1:2 complex when ytterbium was least extracted. Generally cavity size determined the order of extractibility [311].
Eu (III)	1, 3-acid diethylamide of calix(4) arene was used, samarium and lucetium were coextracted [312] calix(n) arene phosphine oxide extracted it [313]. Calixresorcinarene cavitand based cation ligand with carbonyl phosphonate extracted it from the picrate media [314] quantitatively.
Ac (III)	Carboxylate derivative of calix(n) arene (when n = 4, 5, 6) in chloroform extracted radioactive Ac^{225} [315] in the presense of alkali metals and zinc with very good selectivity.
Th (IV)	4-tertiary butyl calix(4) arene in tetrahydroxamate extracted it quantitatively but iron affected the selectivity due to coextraction [316].
U (VI)	Hexacarboxylate derivative of p-t-butyl calix (6) arene and p-hexa calix (6) arene was used for its extraction [317]. At pH 3.7–6.2, p-t-butyl calix (6) arene derivative bearing six hydroxame group in chloroform was used for extraction [318]. At pH 2.5 calixarene uranophiles having hydroxamate group in chloroform was explored for the extraction [319]. The dicarboxylated calix(4) arene in chloroform or 1, 2 dichloroethane was used when a mixed complex with sodium was formed [320]. Phosphate ester of C-undecyl calix (4) resorcinarene was explored for the analysis of uranium (VI) from sea water [321]. Calix (6) arene with carboxy groups in dichloromethane and benzene were also tested [322]. 1, 2 and 1, 3 disubstituted butyl calix(4) arene amide with oxy ions in toluene or iso-octane were used but molybdenum (VI) and chromium (VI) and selenium (IV) were coextracted with uranium (VI) [323].
Fe (III)	At pH 2.2 calix(n) arene based cyclic ligand in chloroform extracted it but copper, zinc palladium showed strong interference [324]. Calix(4) arene based ligands like tetramethyl p-tert butyl calixarene tetra ketone with dioximic group on lower rim was also used for the extraction [325]. Ligand groups circularly arranged on such rims offered cyclic metal receptor for selective extraction [325].
Pd (II)	Tertiary butyl calix (6) arene in methylene or ethylene dichloride extracted it [326]. The derivative of p-tertiary butyl calix(4) arene in chloroform extracted it but other platinum metals were also coextracted due to cation exchange reaction and adduct formation [327]. Methylthiomethyl and NN^1 dimethyl carbomoyl methyl calix(4) arene when used metals like gold were coextracted [328].
Pt (IV)	25, 26, 27, 28 (2, 2 pyridyl thio N-oxide) ethoxyl calix(4) arnene in chloroform extracted it quantitatively (329).
Ag (I)	Ketonic derivative of calixarene in chloroform extracted it from nitric acid with ammonium thiocyanate as the auxiliary complexing ligand. It was successfully separated from palladium [330].
Hg (II)	Calixarene based cyclic ligand having oxime group had extracted it but metals like copper, chromium showed strong interference. The polymeric compound was more selective [331].

(Contd)

Elements	Extraction Conditions
Ga (III)	Hydroxame and carboxylate derivative of calix(4) arene and calix (6) arene when used as the extractants indium was coextracted. Such extraction was not dependent on the ring size of cavity of calixarene but also upon the nature of hydroxamate group used. By and large hydroxamate derivative of tetramer as well as hexamer were best for the extraction of trivalent metal ions from the main groups of periodic classification [332].
Pb (II)	Calix(4) arene bis (N-X) sulphonyl carboxamides extracted it from nitric acid, it was separated from alkali and alkaline earths and some selected transition elements [333].

Conclusion

We can see that though the crown ethers have made great revolution in the chemistry of alkali and alkaline earth metals, the crown ethers containing donor atom other than oxygen namely sulphur and nitrogen have provided a handy tool for the solvent extraction of mainly divalent transition metals like silver, cobalt, cadmium etc. However the alkali metals also formed useful complexes with thia and aza crown ethers but with lesser stability.

The discovery of cryptands was epoc making. It provided relatively stable complexes, such complexes were having three dimensional structure; and due to stretching of bridge in macrobicyclic compound provided good cavity to be filled up by metals like manganese, copper, nickel, thallium, lead, calcium, thorium and zirconium which were normally encountered as metal pollutants in the aquatic environment. They thus proved selective extractants.

Within supramolecular compounds, calix(n) arene finds covated position [334–342]. It had several advantage over crown ethers and cryptands namely their cavity size could be expanded by synthesising higher homologues say hexamer, octamer instead of tetramer. Further the possibility of substitution of lower rim (with keto acetato group) or upper rim. (with sulphato nitrito or groups alkyl) it was possible to design a host molecule which could accomodate any guest molecule in order to satisfy its coordination number [334–342].

In relation to crownether, cryptand, calixarene in supramolecular compounds the calix(n) resorcinarene is yet to find applications in analytical chemistry. Since many extractions also involved colored complexation such discussion is taken in subsequent chapters depending upon method of quantitative analysis.

References

1. I.M. Kolthoff, Anal. Chem. **51**, IR (1979).
2. T. Sekine, Y. Hasegawa, kagaku No Ryoiki **33**, 464 (1979).
3. H.K. Frensdroff, J. Am. Chem. Soc. **93**, 600 (1971)
4. J. Hasegawa, T. Sekine, bunseki **55**, (1982).
5. S.M. Khopkar and M.N. Gandhi, J. Sci. and Ind. Res. **55**, 139 (1996).
6. M.N. Gandhi and S.M. Khopkar, Crown ethers and Cryptands in the Solvent Extraction (Unpublished monograph)

7. B.G. Cox and H. Schneider, Coordination and transport properties of macrocyclic compounds insolution, Elsevier (1992).
8. M. Yoshio, H. Noguchi, Analytical Letters **15**, A15 (1982).
9. Y. Takeda, F. Takahashi, Bull. Chem. Soc. Japan **53**, 1167 (1960).
10. Y. Takeda, F. Goto Bull, Chem. Soc. Japan **52**, 1920 (1979).
11. Y. Takeda, H. Kato, Bull. Chem. Soc. Japan **52**, 1027 (1979).
12. Y. Takeda, Bull Chem. Soc. Japan **53**, 2393 (1980).
13. V.V. Yakshin, Dokl Akad Nauk SSSR **241**, 159 (1978).
14. J. Raise, Proc. Int. Sol. Extn. Conf. **2**, 1705 (1974).
15. R.M. Izatt, J.J. Christensen, Synthetic multidentate macrocyclic compounds Academic Press N.Y. (1978).
16. S. Shinkai, Chem. Letter 283 (1980).
17. K. Kimura, Chem. Letter 611 (1979).
18. T. Kojima, Bull. Res. Lab, Nucl. React **6**, 23 (1981).
19. B.S. Mohite, Ph. D thesis entitled: Solvent extraction separation of alkali and alkaline earths with macrocyclic polyethers I.I.T, Bombay (1986).
20. Y. Marcus, Hydrometallurgy **7**, 27 (1981).
21. Y. Takeda, H. Goto, Bull. Chem. Soc. Japan **52**, 1920 (1979).
22. Y. Hasegawa, J. Inorg. and Nucl. Chem. **43**, 633 (1981).
23. L. Pauling, The nature of the chemical bond Cornell University Press (1960).
24. T. Iwachido, M. Minami A. Sadakare, K. Toei, Chem. Letters **12**, 1511 (1977).
25. R.M. Izatt, J. Am. Chem. Soc. **98**, 7620 (1976).
26. E. Shchori, J. Chem. Soc. Dalton Trans. 2381 (1975).
27. P.R. Danesi, H. Meider Goricon, R. Chiarizia, G. Scibona, J. Inorg. and Nucl. Chem. **37**, 1479 (1975).
28. E. Buned, H.S. Shin, R.A.B. Bannard, J.G. Purdon, Cand J. Chem. **62**, 926 (1984).
29. Y. Takeda, Bull. Chem. Soc. Japan. **55**, 3438 (1982).
30. Y. Hasegawa, Bull Chem. Soc. Japan. **54**, 2427 (1981).
31. T. Sekine, Bull. Chem. Soc. Japan. **41**, 571 (1979).
32. J. Strzelbicki, R.A. Bartsch, Anal Chem. **53**, 1894 (1981).
33. R.G. Vibhute, Ph. D. thesis entitled: Solvent extraction separation of main group element with macrocyclic polyethers–I.I.T, Bombay (1989)
34. N.G. Deorkar, Ph. D. thesis entitled: Solvent extraction separation of transition metals with crown ethers–I.I.T, Bombay (1991).
35. W. Charewicz, Anal. Chem. **54**, 2094 (1984).
36. G. Eisenman, S. Ciani, G. Szebo, J. Memb. Biol. **1**, 294 (1969).
37. M.N. Gandhi, Ph. D. thesis entitled: Separation of toxic metal pollutants from environment by extractions with cryptands–I.I.T, Bombay (1992).
38. C.J. Pedersen, M.K. Frensdorff, Angew want chemi. **11**, 16 (1972).
39. M. Yoshio, M. Ugamura, H. Noguchi, M. Nagamatsu, Anal. Letters A11 281 (1978).
40. Y. Takedo, Topic Current Chem. **121**, 1 (1984).
41. M. Yoshio, M. Ugamura, H. Naguchi, M. Nagamatsu, Anal. Letters **13**, 1431 (1980).
42. B.G. Cox, H. Schneider, Pure Appl. Chem. **61**, 171 (1981).
43. D.J. Cram, K.N. Trueblood, Topic Current Chem. **93**, 43 (1981).
44. L.J. Tusek, J. Inorg. and nucl. Chem. **37**, 1538 1975.
45. D. Sevdic, H. Meider Gorican, J. Inorg. and Nucl. Chem. **39**, 1403 (1977).
46. D. Sevdic, H. Meider Gorican, J. Inorg and Nucl Chem. **39**, 1409 (1977).
47. D. Sevdic, L. Fekete, H. Meider Gorican, J. Inorg Nucl. Chem. **42**, 885 (1980).

48. J. Rebek, R.V. Wattley, J. Heterocycl Chem. **17**, 749 (1980).
49. K.E. Gkoenig, G.M. Lein, P. Stukler, T. Kaneda, D.J. Cram, J. Am. Chem. Soc. **101**, 3553 (1979).
50. K. Kimura, T. Maeda, T. Shono, Talanta **26** 945 (1979).
51. G.W. Gokel, D.M. Dishong, C.J. Diamond, J. Chem. Soc. Chem. Comm. 1053 (1980).
52. S. Kopolow, Macromolecules **6**, 133 (1973).
53. N.G. Deorkar & S.M. Khopkar, Analyst **116**, 961 (1991).
54. M. Oue, Chem. Letters 275 (1982).
55. K. Kimura, Z. anal. Chem. **313**, 132 (1982).
56. T. Maeda, Z. anal. Chem. **313**, 407 (1982).
57. T. Maeda, Z. anal Chem. **298**, 363 (1979).
58. K. Kimura, Poly Bull **1**, 403 (1979).
59. I.M. Kolthoff, Candian J. Chem. **59**, 1548 (1981).
60. W.J. McDowell, R.R. Shoun, Proc. Int. Cont. Sol. Extn. Conf. (1977) Toronto **21**, 95 (1979).
61. K. Gloe, Z. Chem, **19**, 382 (1979).
62. V.M. Abashkin, Dokl. Akad Nasik SSSR **257**, 1374 (1981).
63. A. Sadakane, T. Iwachido, K. Toei, Bull. Chem. Soc. Japan **48**, 60 (1975).
64. Y. Marcus, L.F. Asher J. Phy. Chem. **82**, 1246 (1978).
65. K. Kimura, T. Maeda, T. Shono, Talanta **27**, 801 (1980).
66. Y. Hasegawa, K. Suzuki, T. Sekine, Chem. Letters 1075 (1981).
67. Y. Takeda, Bull. Chem. Soc. Japan **54**, 526 (1981).
68. I.H. Gerow, M.W. Davis, Jr, Sepn Sc. and Tech. **14**, 395 (1979).
69. B.S. Mohite & S.M. Khopkar, Talanta **32**, 565 (1985).
70. Y. Takeda, Y. Wada, S. Fujiwara, Bull. Chem. Soc. Japan **54**, 3727 (1981).
71. K. Gloe, P. Muhl, J. Beger, Z. Chem **28** 1 (1988), AA **50**, 6B1 (1988)
72. T.B. Stolwijk, L.E. Vos, C Sudholter. J.R. Ernst, D.N. Reinhaualt, Rec. Trav. Chem. 108 (1989) CA. **111**, 29074 Y (1989).
73. I.V. Pyatnitskii, A. Yu. Nazarenko, Zh. Nerg Khim **25**, 1064 (1980) AA **41**, 1B10 (1981).
74. V.M. Abashkin, V.V. Yakshin, B.N. Laskorin, Dokl Akad Nauk SSSR **257**, 1374 (1981) [AA **41**, 6B30 1981].
75. W.F. Kinard, W.J. McDowell, J. Inorg. nucl. Chem. **43**, 2947 (1981).
76. T. Maeda, K. Kimura, T. Shono, Bull Chem. Soc. Japan **55**, 3506 (1982).
77. J. Strzelbicki, H. Gwisuk, R.A. Bartsch, Sepn. Sci. and tech. **17**, 635 (1982).
78. H. Otsuka, H. Nakamura, M. Takagi, K. Ueno, Anal. Chim. Acta 147227 (1983).
79. J.F. Koszuk, B.P. Czech, W. Walkowiak, D.A. Babb, R.A. Bartsch, J. Chem. Soc. Chem. Commun. 1504 (1984).
80. B.P. Bubnis, G.E. Pacey, Tetrahetron Lett. **25**, 1107 (1984).
81. K.K. Kikukawa, X. HeGong, A. Abe, T. Goto, R. Arata. T. Ikeda, F. Wada, T. Matsuda, J. Chem. Soc. Perkin Trans-2135 (1987).
82. B.P. Czech, A. Czech, B.E. Knudsen, R.A. Bartsch, Grass Chim. Ital. **117**, 717 (1987) CA **109**, 170401 t (1987).
83. Goodeng Xuexia, D. Wang H. Du, X. Su, H. Hu Huaxue Xuebao **9**, 317 (1988) CA **109**, 1573834 (1988).
84. Y.G. Mamedova, A.M. Babazade, A.I. Shabanove, Zhur neorg Khim **34**, 173 (1989) AA **51**, 7B17 (1989).
85. Y. Takeda, M. Nishida, Bull. Chem. Soc. Japan **62**, 1468 (1989).
86. D. Wang, D. Wang, X. Xun, H. Hu, Goodeng Xuexia, Huaxue Xuebao **11**, 672 (1990) CA **114**, 309225 (1991).

87. W. Walkowiak, S.I. Kang, L.E. Stewart, G. Ndip, R.A. Bartsch, Anal. Chem. **62**, 2022 (1990).
88. S. Akabori, M. Takeda, H. Kawakami, Bull. Chem. Soc. Japan. **64**, 1413 (1991).
89. Y. Inoue, M. Ouchi, K. Husosyama, T. Hakushi, Y. Liu, Y. Takeda, J. Chem. Soc. Dalton trans 1291 (1991).
90. H. Bukowsky, F. Dietrich, E. Uhlemann, K. Gloe, P. Muchl, H. Mosler, Z. Chem. **30**, 33 (1990) CA **113**, 139800f (1991).
91. R.A. Bartsch, I.W. Yang, E.G. Jeon, W. Walkowiak, W. Charewicz, J. Cord. Chem. **27**, 75 (1992).
92. A.P. Marchand, K.A. Kumar, A.S. Mckim, K. Milinaric Majerski G. Kragol, Tetrahedron **53**, 3467 (1997).
93. A. Manjula, M. Nagarajan, Tetrahedron **53**, 11859 (1997).
94. A.V. Bogatskii, N.G. Lukyaneko, M.U. Mamina, V.A. Shapkin, D. Taubert, Dokla Akad Nauk SSSR **250**, 1389 (1980).
95. R.A. Sachleben, Y. Deng, B.A. Moyer, Sepn. Sci. and Techn. **32**, 275 (1997).
96. T. Kaneda, S. Umeda, H. Tanigawa, S. Misumi, Y. Kai, H. Morii, K. Mikik, N. Kasa, J. Am. Chem. Soc. **107**, 4802 (1985).
97. K. Kobiro, T. Matsuoka, S. Takada, K. Kakiuchi, Y. Tobe, Y. Odaira Chem. Letters 713 (1986) CA. **106**, 119864g (1987).
98. K. Kimura, M. Tanaka, S. Iketani, T. Shono, J. Org. Chem, **52**, 836 (1987).
99. Y. Inoue, Y. Liu, F. Amano, M. Ouchi; A. Tai, T. Hakushi, J. Chem. Soc. Dalton trans 2735 (1988).
100. Y. Liu, Y. Inoue, T. Hakushi, Bull Chem. Soc. Japan **63**, 3044 (1990).
101. S.Q. Fang, L.A. Fu, J. Radioanal and nucl. Chem. **187**, 25 (1994)
102. K. Kazuya, Cord. Chem. Revi. **148**, 135 (1996).
103. J. Strzeibicki, R.A. Bartsch-Anal. Chem. **53**, 2251 (1981).
104. M. Ates, O. Erk, Chim. Acta. Ture **14**, 65 (1986) A A. **107** 103774m (1987).
105. Y. Inoue, Y. Liu, L.H. Tong, A. Tai, T. Hakushi, J. Chem. Soc. Chem. Comm. 1556 (1989).
106. W. Walkowiak, W.A. Charewicz, S. I. Kang, I.W. Yang, M.J. Pugia, R.A. Bartsch, Anal. Chem. **62**, 2018 (1990).
107. V.J. Huber, N. I. Sheryl, J. Lu, R.A. Bartsch, Chem. Commun. 1499 (1997).
108. H. Nakamura, M. Takagi, K. Ueno Anal. Chem. **52**, 1668 (1980).
109. I.M. Kolthoff, Pharma weekly Sci. Ed. **5**, 184 (1983).
110. X. Aristotelis, S. Claude, T. Christain, Talanta **34**, 509 (1987).
111. S.W. Kang, C.M. Park, W.R. Koo, K.J. Kim, S.M. Lee C.H. Chang, Taihan Hwahadhoe Chi **32**, 443 (1988).
112. A.I. Kholkin, K. Gloe, N.D.Kuznetsova, P. Mühl, L.M. Gindin, L.Y. Marochkina, Izu. Sib. Otd. Akad. Nauk SSSR Ser Khim Nauk 17 (1988) CA **108**, 174480v (1988).
113. T. Makoto, N. Satoru, T. Shigeru, M. Mitsuo, Chem. Eng. J. **39**, 157 (1988) CA **110**, 64602u (1989).
114. Lu. Guoyuan, Zhu, Chusheng, W. Defen, H. Hongwen, Huaxue shiji **12**, 263 (1990) CA **114**, 185467g (1991).
115. H. Sakamoto, H. Goto, M. Yokoshima, M. Dobashi, J. Ishikawa K. Doi, M. Otomo Bull. Chem. Soc. Japan **66**, 2907 (1993).
116. Y. Takeda, Bull. Chem. Soc. Japan **56**, 2589 (1983).
117. D. Milton, W C.B. Bowers, Eur. Pat. Appl. Ep 216 473 (1987) CA **107**, 185702z (1987).
118. E. Blasius, Nilles, K. Heinz, US 4647440 (1987) CA **106**, 2226862 (1987).
119. W.N. Schulz, L.A. Bray, Sepn. Sci. and Technology **22**, 191 (1987).

120. E. Peimli, J. Radio anal. and Nucl. Chem. **144**, 1 (1990) [AA **53**, 5023 (1991)].
121. W.J. McDowell, G.N. Case, J.A. McDonough, R.A. Bartsch, Anal. Chem. **64**, 3013 (1992).
122. P. Novy, P. Vanura, E. Makrlik, J. Radio anal. Nucl. Chem. **207**, 237 (1996).
123. J. Strzelbicki, R.A. Bartsch, Anal. Chem. **53**, 2247 (1981).
124. W. Walkowiak, L.E. Stewart, H.K. Lee, B.P. Czech, R.A. Bartsch, Anal. Chem. **58**, 188 (1986).
125. A. Yu. Tsivadze: V.V. Basmanov: A.V. Levkin. Zh. Neorg. Khim. **35**, 2425 (1990) CA **114** 12922n (1991).
126. S. Umetani: M. Matsui: S. Tsurubou, J. Chem. Soc: Chem. Commun. 914 (1993).
127. S. Tsurubou; M. Mizutani; Y. Kadota: T. Yamamoto: S. Umetani: T. Saski: Q.T.H. Le M. Matsui. Anal. Chem. **67**, 1465 (1995).
128. K.N. Shizawa: T. Nishida: T. Miki: T. Yamamoto: M. Hosoe. Sepn. Sci. and Technol. **31**, 643 (1996).
129. H. Bukowsky: F. Dietrich: E. Uhlemann: K. Gloe: P. Muehl: H. Mosler Z. Chem. **30**, 33 (1990) CA **113**, 38900f (1990).
130. L. Matel T. Bilbao, J. Radioanal. Nucl. Chem. **137**, 138 (1989).
131. L.E. Juergen: S. Herbert: L. Helmut: P. Brigitta: R. Ekkart. German Patent CA. **111**, 122634r (1989).
132. P. Rajec: V. Mikulaj: J. Mackova, J. Radioanal Nucl. Chem. **150**, 315 (1991).
133. D. Tait: A. Wiechen, J. Radioanal and Nucl. Chem. **159**, 239 (1992).
134. G. Zirnhelt: M.J.F. Leroy: J.P. Brunette: Y. Freve: P. Gramain, Sepn. Sci. Technol. **28** 2419 (1993).
135. E.A. Shehata, J. Radioanal and Nucl. Chem. **185**, 411 (1994).
136. Y. Takeda: K. Oshio: Y. Segawa, Chem. Lett 601 (1979) A.A **38**, 1B37 (1980).
137. M.K. Beklemishev: S. Eishani: C.M. Wai, Anal. Chem. **66**, 3521 (1994).
138. K. Gloe: P. Mühl A.I. Kholkin: M. Meerbote: J. Beger, Isotopenpraxis **18**, 170 (1982) A.A **44**, 3B4 (1983).
139. E. Uhlemann: H. Geyer: K. Gloe: P. Muhl. Anal. Chim. Acta **185**, 279 (1986).
140. V.V. Yakshin: A.T. Fedorova: B.N. Laskorin, Izv. Akad. Nauk. SSSR Serkhim 509 (1986) A.A **48**, 11B19 (1986).
141. R.A. Bartsch, Energy Res. Abstr **11**, 3370 (1986) C.A. **106**, 23959r (1987).
142. R.A. Bartsch, Sol. Extn. and ion exchange **7**, 829 (1989).
143. E. Uhlelmann: H. Bukowsky: F. Dietrich: K. Gloe: P. Muchl: H. Mosler, Anal Chim. Acta **224**, 47 (1989).
144. Y. Takeda: T. Kimura: Y. Kudo: H. Matsuda: Y. Inoue: T. Hakushi, Bull. Chem. Soc. Japan **62**, 2885 (1989).
145. H. Bukowsky: F. Dietrich: E. Uhlemann: K. Gloe: P. Muehl: H. Mosler, Z. Chem. **30**, 33 (1990) A.A. **53**, 5d19 (1991).
146. Y. Inoue: T. Hakushi: Yu. Liu: L.H. Tong, J. Chem. Soc. Perkin Trans. 2. 1239. (1990).
147. Y. Inoue: Y. Liu: L.H. Tong: T. Hakushi, J. Chem. Soc. Perkin Trans. 2. 1247 (1990).
148. E. Weber: K. Skobridis: M. Ouchi: T. Hakushi: Y. Inoue, Bull. Chem. Soc. Japan. **63**, 3670 (1990).
149. Y. Takeda. N. Ikeo: N. Sakata, Talanta **38**, 1325 (1991).
150. R.D. Rogers: A.H. Bond: C.B. Bauer, Pure and Applied Chem. **65**, 567 (1993).
151. Yasuhiro Tate: Taeko Izumi: Satoshi Ohashi, J. Heterocyclic Chem. **30**, 967 (1993).
152. J.T. Chuang: J.G. Lo, J. Radioanal and Nucl. Chem. **189**, 307 (1995).
153. V.V. Yakshin: A.T. Fedorova: A.V. Val Kov: B.N. Laskorin, Dokl. Akad. Nauk. SSSR **271**, 1417 (1984) [A.A. **47**, 5B 55 (1985)].

154. D.D. Ensor: G.R. McDonald: C.G. Pippin, Anal. Chem. **58**, 1814 (1986).
155. Y. Hasegawa: M. Masuda: K. Hirose: Y. Fukuhara, Solvent Extr. and Ion Exch. **5**, 255 (1987).
156. D.D. Ensor: P.S. Reynolds, J. Less Common Metals **149**, 287 (1988).
157. K. Nakagawa: S. Okada: Y. Inoue: A. Tai: T. Hakushi, Anal. Chem. **60**, 2527 (1988).
158. J. Tang: C.M. Wai, Analyst **114**, 451 (1989).
159. K. Nakagawa: T. Hakushi: Y. Inoue, Kidorui **14**, 156 (1989) C.A. **111**, 241292n (1989).
160. K. Nakagawa: Y. Inoue: A. Tai: Hakushi, Chim. Express **4**, 429 (1989) C.A. **112**, 166233j (1990).
161. V.V. Sukhon: A. Yu. Nazarenko: O.I. Kronikovskii, Zhur. Neorg. Khim. **35**, 1221 (1990) A.A **53**, 7D 39 (1991).
162. X.B. Xia: M.S. Munsey: H. Du: C.M. Wai: N.R. Natale, Heterocycles **32**, 711 (1991) C.A. **115**, 92244b (1991).
163. R. Frazier: C.M. Wai, Talanta **39**, 211 (1992).
164. R. Frazier: C.M. Wai, J. Radioanal and Nucl. Chem. **159**, 63 (1992).
165. Y. Inoue: K. Nakagawa: T. Hakushi, J. Chem. Soc. Dalton Trans. 2229 (1993).
166. J.N. Mathur: G.R. Choppin, Sol. Extr. And Ion Exch **11**, 1 (1993).
167. H. Imura: H. Mito, J. Radioanal. and Nucl. Chem: **189**, 229 (1995).
168. S. Lis: J.N. Mathur: G.R. Choppin, Solvent Extr. And Ion Exch **9**, 637 (1991).
169. V.V. Proyaer: V.V. Romanovskii: V.N. Romanovskii, Radio Khimya **33**, 46 (1991) C.A. **115**, 36528k (1991).
170. A. Ramdan S.M. Khalifa: M. Mahomoud: N. Souka, J. Radioanal and Nucl. Chem. **176**, 457 (1993).
171. F.A. Shehata: S.M. Khalifa: H.F. Aly, J. Radioanal and Nucl Chem. **159**, 353 (1992).
172. J. Tang: C.M. Wai, Anal. Chem. **58**, 3233 (1986).
173. W. Zhang: S. Xu: Y. Han, Yuanzineng Kexue Jishu **20**, 420 (1986) C.A. **106**, 126793m (1987).
174. S. Xu: W. Zhang: Z. Gu, Yuanzineng Kexue Jishu **20**, 425 (1986) C.A. **106**, 126794n (1987).
175. W. Wang: Q.C. Sun: B. Chen, J.Radioanal and Nucl. Chem. **110**, 227 (1987).
176. A.G. Godbole N.V. Thakur: R. Swarup: S.K. Patil, J. Radioanal and Nucl. Chem. **108** 89 (1987).
177. W. Cao: W. Luo, Yanzineng Kixue Jishu **21**, 456 (1987) C.A. **108**. 211275e (1988).
178. J.P. Shukla: K.V. Lohithakshan, Chem. Sci. **29**, 341 (1989) C.A. **113**, 199908 (1990).
179. J.P. Shukla: R.K. Singh: A. Kumar, Radiochim Acta, **54**, 73 (1990) C.A. **115**, 36525h (1991).
180. J.P. Shukla: R.K. Singh: A. Kumar. Talanta **40**, 1261 (1993).
181. G.M. Nair: D.R. Prabhu, J. Radioanal and Nucl. Chem. **121**, 83 (1988).
182. J.N. Mathur: P.K. Khopkar, Solvent Extr. And Ion. Exch. **6**, 111 (1988).
183. C. Yonezawa: G.R. Choppin, J. Radioanal and Nucl. Chem. **134**, 233 (1989).
184. B.S. Mohite: J.M. Patil, J. Radioanal and Nucl. Chem. **150**, 207 (1991).
185. R.B. Chadwick: W.J. McDowell: C.F. Baes. Jr, Sep. Sci. Technol. **23**, 1311 (1988).
186. A. Shigeki: S. Tyo: F. Kazuhito: E. Masatoshi, Anal. Chim. Acta, **274**, 141 (1993).
187. M.G. Jalhoom, Radiochim Acta, **39**, 195 (1986) C.A. **106**, 92060c (1987).
188. M.G. Jalhoom, J. Radioanal and Nucl. Chem. **104**, 131 (1986).

189. T.M. Le: T. Lengyel, J. Radional and Nucl. Chem. **128**, 417 (1988).
190. T.M. Le: T. Lengyel, J. Radioanal and Nucl. Chem. **135**, 403 (1989).
191. T.M. Le: T. Lengyel, J. Radioanal and Nucl. Chem. **136**, 363 (1989).
192. Z. Zhou: J. Yang, Huaxue Shijii **9**, 50 (1987) A.A. **49**, 11B235 (1987).
193. H. Koshima: H. Onishi, Anal. Chem. Acta, **232**, 287 (1990).
194. X. Zhang: Z. Zhou: S. Ma: C. Shu, Solvent. Extr and Ion. Exch **11**, 585 (1993).
195. H. Koshima: H. Onishi, Anal. Sci. **1**, 389 (1985).
196. H. Koshima: H. Onishi, Analyst **111**, 1261 (1986).
197. V.V. Yakshin: O.M. Vilkova: N.A. Tsarenko, Zavod. Lab. **61**, 9 (1995) A.A, **57**, 10D33 (1955).
198. J. Strzelbicki: A. Charewicz. R.A. Bartsch, J. Inclusion phenom **6**, 57 (1988) C.A. **109**, 28321y (1988).
199. S.M. Khalifa: H.F. Aly: J.D. Navratil, Talanta, **36**, 406 (1989).
200. A.M. Sastre: A. Sahmoune: J.P. Brunette: M.J.F. Leroy, Solvent Extr. and Ion. Exch. **7**, 395 (1989).
201. B. Rusdiarso: A. Messaoudi: J.P. Brunette, Talanta **40** 805 (1993).
202. A.G. Gaikwad: H. Noguchi: M. Yoshida, Analyt. Letters **24**, 1625 (1991).
20.3 Y. Hasegawa: H. Tanobe: S. Yoshida, Bull. Chem. Soc. Japan. **58**, 3649 (1985).
204 Y. Hasegawa: T Nakano: Y. Odori: Y. Ishikawa, Bull. Chem. Soc. Japan. **57**, 8 (1984).
205. E.I. Morosanova: Yu. A Zolotov: V.A. Bodnya: A.A. Formanovsky, Mikrochim Acta III, 389 (1984).
206. E.I. Marosanova, E.D. Mateeva, Yu. G. Bundel, V.A. Bodnya, Yu. A. Zolotov, Dokl. Akad. Nauk. SSSR, **227**, 1151 (1984). A.A. **45**, 5A15 (1985).
207. M.K. Beklemishev, A.A. Formanovskii, N.M. Kuzmin, Yu, A. Zolotov, Zh. Neorg. Khim, **31**, 2617 (1986) C.A. **107**, 88660B (1987).
208. S. Arpadyan, M. Mileva, P.R. Bancher, Talanta, **34**, 953 (1987).
209. N.G. Yanifatova, N.V. Isakova, O.M. Petrukhim, Yu. A. Zolotov, Zh. Neorg. Khim, **36**, 792 (1991) C.A, **115**, 16390f (1991).
210. H. Date, K. Nozawa, M. Nakamura, Y. Hasegawa, Proc. Symp. Solvent Extr. Conference 59 (1986) C.A. **107**, 142361V (1987).
211. K. Nishizawa, T. Yamamoto, M. Nomura, Sepn. Sci. and Technol, **31**, 2831 (1996).
212. V.V. Yakshin, V.M. Abashkin, M.B. Korshunov, Zh. Anal Khim, **37**, 938 (1982) A.A. **44**, 3b49 (1983).
213. N.A. Pasekova, Yu. A. Zolotov, E.V. Malkhasyan, Vestn. Mosk. Univ. Ser 2 Khim, **29**, 603 (1988). C.A. **110**, 122498d (1989).
214. N.G. Vanifatova, N.V. Isakova, O.M. Petrukhin, Zh. Neorg. Khim, **36**, 804 (1991) C.A. **115**, 16391g (1991).
215. S. Wang, S. Eishani, C.M. Wai, Anal. Chem. **67**, 919, (1995).
216. V.V. Yakshin, O.M. Vilkova, N.A. Tsarenko, B.N. Laskorin, Dokl. Akad. Nauk. SSSR, **316**, 419 (1991) C.A. **115**, 419 (1991) C.A. **115**, 121268V (1991).
217. T. Sekine, H. Wakabayashi, Y. Hasegawa, Bull. Chem. Soc. Japan, **51**, 645 (1978).
218. Y. Takeda, Bull Chem. Soc. Japan, **62**, 2379 (1989).
219. Y.K. Kim, B.Z. Iofa, Vest Mosk Uni. ser Khim **32**, 258 (1991) CA **115**, 264470R (1991).
220. L.G. Shaidarova, I.L. Pedorova, N.A. Ulakharich, G.K. Budnikov, Zhr. Prikl Khim **69**, 778 (1996), AA **59**, ID24 (1997).
221. J. Jin, S. Jin, S. Zhang, D. Xu, M. Zhou, Sinchuan Daxue Xuebao Ziran Kexucban **24**, 444, (1987), CA **108**, 211273C (1988).
222. M.N. Gandhi and S.M. Khopkar, Ind. J. Chem. **30A**, 706 (1991).

223. M.N. Gandhi and S.M. Khopkar, Anal. Chim. Acta. **270**, 87 (1992).
224. M.N. Gandhi, S.M. Khopkar, Chemia Analytiza (warsaw) **37** 437 (1992).
225. M.N. Gandhi, S.M. Khopkar, Analytical Science **8**, 65 (1992).
226. M.N. Gandhi, S.M. Khopkar, Mikro chim. Acta. **111**, 93 (1993).
227. M.N. Gandhi, S.M. Khopkar, Chemical and Env. Res. **1**, 389 (1993).
228. V.J. Mathew, S.M. Khopkar, J. Radioanal and nucl Chem. **201**, 281 (1995).
229. V.J. Mathew, S.M. Khopkar, Chemia analytiza (warsaw), **42**, 651 (1997).
230. H.S. Ajgaonkar, S.M. Khopkar, Chemia analytiza (warsaw) **44**, 61 (1999).
231. M.J. Reyes, A.G. Maddock, G. Duplatre, J.J. Schelarfler, J. Inorg. and Nucl. Chem. **41**, 1365 (1979).
232. M. Takagi, H. Nakamura, Y. Sanui, K. Ueno, Anal. Chim. Acta. **126**, 185 (1981).
233. M. Kirch, J.M. Lehn, Angewant chem **14**, 555 (1975).
234. S.M. Khopkar, and M.N. Gandhi, J. Sci. and Ind. Res. **53**, 630 (1994).
235. M.N. Gandhi and S.M. Khopkar, Ind. J. Chemical Sciences **8**, 111 (1990), CA **118**, 72596 h (1991).
236. V.P. Ionov, Zhuv Neorg. Khim **32**, 825 (1987), CA **106**, 2027459 (1987).
237. A.F. Sholl, I.O. Sutherland, J. Chem. Soc. Chem. Comm. 1716 (1992).
238. A.F. Sholl, M. Dolman, A.J. Mason, K. Saman, K.R.A. Sandenoyke, A. Shendri Sutherland, Analyst **121**, 1775 (1996).
239. V.V. Yakshin, A.T. Fedorova, B.N. Laskorin, Dokl. Akad. Nauk. SSSR **276** 169 (1984), AA **47**, 1B26 (1985).
240. T. Saski, S. Umetani, M. Matsui, T. Tsurubou, Chem. Letter 1195 (1994), AA **57** 4D 57 (1995).
241. A. Kumar, S.M. Khopkar, Chem. and Env. Res. **4**, 145 (1995).
242. S.S. Rane, S.M. Khopkar, Ind. J. Chem. Techn. **3**, 363 (1996).
243. R.J. Namdeo, S.M. Khopkar, Ind. J. Chem. **34A**, 540 (1995).
244. M. Takagi, K. Ueno, Topics in current chem **121**, 39 (1984).
245. D. Serdic, L. Jovanovac, H. Meider-Gorican, Mikro. Chem. Acta **11**, 235 (1975).
246. D. Serdic, H. Meider-Grican, J. Inorg Nud. Chem. 43, 153 (1981).
247. T. Takekawa, Y. Masuda, E. Sekido, 30th conf. of Japan Soc. of analchem. p. 168 (1981).
248. Y. Matsui, T. Maeda, K. Kimura, T-Shono, 30th conf of Japan Soc. Of Anal. Chem. p. 170 (1981).
249. H. Nishida, M. Takagi, K. Ueno, MiKrochim Acta. **1**, 281 (1981).
250. I. Cho, S. Chang, Chem. Letters 515 (1981).
251. M. Pielraskiewicz, N. Spencer, J. Coordi. Chem. **27**, 115 (1992).
252. S.G. Dimtrienko, G. Pasekova, E.D. Slyusareva, A. Formanorskii, Y.A. Zolotov V. Mikhural, Vest Mosk Unisevr Khim **31**, 378 (1990), AA **53**, 12D4 (1991).
253. D.D. Ensor, M. Nicks, D.J. Pruett, Sepn. Sci. and Techno. **23**, 1345 (1988).
254. A. Yu. Nazarenko, T.A. Titarenko, Ukr. Khim. Zh, **49**, 630 (1983) A.A. **45**, 5B147 (1983).
255. V.V. Ageeva, A. Yu. Nazarenko, Visn. Khiv. Uni. Ser, khim, 27 (1987) A.A, **50**, 7B37 (1988).
256. K.B. Yaminskaya, N.V. Pertsov, V.L. Lishchinskii, V.S. Pshezhetskii, A.A. Rakhnyanskaya, A.A. Formanovskii, Kolloidn Zh. **53**, 191 (1991) C.A. **115**, 36287f (1991).
257. V.L. Lishchinskii, L.S. Nikolaeva, A.A. Rakhnyanskara, Yu. A. Kiryanov, M.K. Beklemishev, A.M. Evseer, V.S. Pshezhetskii, Koord. Khim. **17**, 1025 (1991) C.A. **115**, 190911 m (1991).
258. V.L. Lishchinskii, L.S. Nikolaeva, A.A. Rakhnyanskaya, Yu. A. Kiryanov, E.A. Mezhonova, M.K. Beklemishev, A.M. Evseer, V.S. Pshezhetskii, Koord. Khim. **17**, 1314 (1991) C.A. **115**, 264471 s (1991).

259. B. Vaidya, J. Zak, G.J. Bastians, M.D. Porter, J.L. Hallman, N.A.R. Nabulsi, M.D. Utterback, B. Strzelbicki, R.A. Bartsch, Anal. Chem. **67**, 4101 (1995).
260. K. Hiratani, Bunseki 121 (1992) A.A. **55**, 4A 20 (1993).
261. E. Fritsch, M. Beer, B. Gorski, Z. Chem. **24**, 143 (1984) A.A. **47**, 1B195 (1995).
262. V. Guyon, A. Guy, J. Foos, M. Lemaire, M. Draye, Tetrahedron, **51**, 14065 (1995).
263. K. Saito, Y. Masuda, E. Sekido, Anal. Chem. Acta. **151**, 447 (1983).
264. A. Ohki, M. Takagi, K. Ueno, Anal. Chem. Acta, **159**, 245 (1984).
265. E. Sekido, H. Kawahara, K. Tsuji, Bull. Chem. Soc. Japan **61**, 1587 (1988).
266. B.A. Moyer, C.L. Westerfield, W.J. McDowell, G.N. Case, Sepn. Sci. and Tehnol, **23**, 1325 (1988).
267. Y. Qin, J.Q.C. Yang, Anal. Chem. Acta, **286**, 265 (1994).
268. K. Saito, S. Murakami, A. Muromatsu, E. Sekido, Anal . Chem. Acta. **294**, 329 (1994).
269. K. Saito, I. Taninaka, S. Murakami, A. Muromatsu, Anal. Chim. Acta. **299**, 137 (1994).
270. B.S. Mohite, S.M. Khopkar, Anal. Chim. **59**, 1200 (1987).
271. E. Sekido, K. Saito, Y. Naganuma, H. Kumuzoki, Anal. Sci. **1**, 363 (1985).
272. E. Lachowicz, A. Krajewski, M. Golinski, Anal. Chem. Acta. **188**, 239 (1986).
273. M. Muroi, A. Hamaguchi, E. Sekido, Anal. Sci. **2**, 351 (1986).
274. M. Oue, K. Kimura, T. Shono, Anal. Chem. Actn, **194**, 293 (1987).
275. E. Sekido, K. Chayama, Anal. Sci. **3**, 535 (1987).
276. A. Yu. Nazarenko, V.V. Sukhan, V.M. Timoshenko, V.N. Kalinin, Zh. Neorg. Knim. **35**, 2971 (1990) C.A. **114**, 109634h (1991).
277. K. Chayama, N. Awano, Y. Tamari, H. Tsuji, E. Sekido, Bunseki Kagaku **42**, 687 (1993) A.A. **56**, 5D53 (1994).
278. T. Kumagai, S. Akabari, Chem. Letter 1667 (1989) C.A. **112**, 150553 C (1990).
279. A. Zinke, E. Ziegler, Berch **74**, 1729 (1941).
280. C.D. Gutsche, M. Iqbal, D. Stewart, J. Org. Chem. **51**, 742 (1988).
281. R.M. Izatt, J.D. Lamb, R.T. Hawkins, P.R. Brown, S.R. Izatt, J. Am. Chem. Soc. **105**, 1782 (1983).
282. S.K. Chang, S.K. Kwan, I. Cho, Chem. Letters 947 (1987).
283. R. Ungaro, A. Pochini, G.D. Andreetti, P. Domiano, J. Incl. Pheno **3**, 35 (1985).
284. K.H. Wang, K. Yagi, J. Smid, J. Membrane Bio, **18**, 379 (1974).
285. S. Shinkai, Pure and Applied Chem. **58**, 1523 (1986).
286. J.L. Atwood, A.W. Coleman, H. Zhang unpublished work (cited in ref No. 7).
287. S. Shinkai, H. Koreishi, K. Ueda, T. Arimura, O. Manobe, J. Am. Chem. Soc. **109**, 6371 (1987).
288. R.M. Izatt, J.D. Lamb, R.T. Hawkins, P.R. Brown, S.R. Izatt, J.J. christensen, J. Am. Chem. Soc. **105**, 1782 (1983).
289. S.R. Izatt, R.T. Hawkins, J.J. Christensen, R.M. Izatt, J. Am. Chem. Soc. **107**, 63 (1985).
290. M.A. Mckervay, E.M. Seward, G. Fergusson, B-Ruhl, S.J. Harris, J. Chem. Soc. Chem. Comm. 388 (1985).
291. G. Ferguson, D. Kaitner, M.A. Mckervey, E.M. Seward, J. Chem. Soc. Chem. Comm, 584 (1987).
292. S.K. Chang, I. Cho, Chem. Letter 477 (1984).
293. S.K. Chang, I. Cho, J. Chem. Soc. Perkin Trans. **1**, 211 (1986).
294. R.M. Noyes, J. Am. Chem. Soc. **84**, 513 (1962).
295. S. Shinkai, H. Koreishi, K. Ueda, O. Manabe, J. Chem. Soc. Chem. Comm. 233 (1986).

296. S. Shinkai, H. Koreishi, K. Ueda, T. Arimura, O. Manabe, J. Am. Chem. Soc. **109**, 6371 (1987).
297. M. Takeshita, S. Shinkai, Bull. Chem. Soc. Japan **68**, 1088 (1995).
298. R. Hilgenfeld. N. Saenger, Host Guest complex chemistry II Ed. F. Vogtle Springer Verlag p. 6 (1982).
299. H. Ihm, H. Kim, K. paek, J. Chem. Soc. Perkin Transl, 1997 (1997).
300. S. Shinkai, T. Otsuka, K. Araki, T. Matsuda, Bull. Chem. Soc. Jpn **62**, 4055 (1989).
301. C. Salhi, Z. Asfari, M. Burgard, J. Vicens, Analysis **20**, 351 (1992) C.A. **117**, 179291 (1992).
302. F.A. Neu, S. Fanni, L. Guerra, W.M.C. Gregar, K. Ziat, M.J. S. Weill, G. Barrett, M.A. Mckervey, D. Marrs, E.M. Sewara, J. Chem. Soc. Perkin Trans. II, 113 (1995).
303. F.A. Neu, G. Barrett, S. Fanni, D. Marrs, W.M. Gregar, M.A. Mckervey, M.J.S. Weill, V. Vetragen, S. Wechsler, J. Chem. Soc. Perkin Trans. II, 453 (1995).
304. A.F. Danil de Namor, F.J. Sueros Velarde, A.R. Casal, A. Pugliese, M.T. Goitia, M. Mantero, F.F. Lopez, J. Chem. Soc. Faraday Trans. **93**, 3955 (1997).
305. G. Montavon, G. Duplatre, N. Barakat, M. Burgard, Z. Afsari, J. Vicens, J. Incl. Phenom and Mol. Recog. **27**, 155 (1997).
306. V. Lamare, C. Bressot, J.F. Dozol, J. Vicens, Z. Asfari, R. Ungaro, A. Casnati, Sepn. Sci. and Technol, **32**, 175 (1997).
307. Z. Asfari, C. Bressot, J. Vicens, C. Hill, J.F. Dozol, H. Rouquette, S. Eymard, V. Lamare, B. Tournois, Anal. Chem. **67**, 3133 (1995).
308. F. Arnaud-Neu, V. Bohmer, J.F. Dozol, L. Gruttner, R.A., Jakohi, D. Kraft, O. Mauprivez, H. Rouquette, H. Simon, W. Vogt, M.J.S. Weill, J. Chem. Soc. Perkin Trans. II, 1175 (1996).
309. J.M. Harrowfield, M. Mocerino, B.J. Pecchey, B.W. Skelton, A.H. White, J. Chem. Soc. Dalton Trans. 1687 (1996).
310. P.D. Beer, M.G.B. Drew, M. Kan, P.B. Leeson, M.I. Ogden, G. Williams, Inorg, Chem. **35**, 2202 (1996).
311. R. Ludnig, K. Inoue, T. Yamato, Solvent. Extr, And Ion. Exch. **11**, 311 (1993).
312. P.D. Beer, M.G.B. Drew, A. Grieve, M. Kan, P.B. Leeson, G. Nicholson, M.L. Ogden, G. Williams, J. Chem. Soc. Chem. Commun. 117 (1996).
313. J.F. Malone, D.J. Marrs, M.A. Mckervey, P.O. Hgan, N. Thompson, A. Walker, F.A. Neu, O. Mauprivez, M.J.S. Weill, J.F. Dozol, H. Rouquette, N. Simon, J. Chem. Soc. Chem. Commun. 2151 (1995).
314. H. Boerrigter, W. Verboom, D.N. Reinhoudt, J. Org. Chem. **62**, 7148 (1997).
315. X. Chen, M. Ji, D.R. Fisher, C.M. Wai, J. Chem. Soc. Chem. Commun. 377 (1997).
316. L. Dasaradhi, P.C. Stark, V.J. Huber, P.H. Smith, G.D. Jarvinen, A.S. Gopalan. J. Chem. Soc. Perkin Trans. II, 1187 (1997).
317. S. Shinkai, Y. Shiramama, H. Satoh O. Manabe, T. Arimura, K. Fujimato, T. Matsuda, J. Chem. Soc. Perkin Trans. II, 1167, (1989).
318. T. Nagasaki, S. Shinkai, T. Matsuda, J. Chem. Soc. Perkin Trans. I, 2617 (1990).
319. T. Nagasaki, S. Shinkai, J. Chem. Soc. Perkin Trans. II, 1063 (1992).
320. G. Montavan, G. Duplatre, Z. Asfari, J. Vicens, J. Radioanal. Nucl. Chem. **210**, 87 (1996).
321. K. Yoshifumi, T. Hiroyuki, S. Hirotaka, S. Hideta, and Y. Kimiho, Bull. Chem. Soc. Japan. **69**, 785 (1996).
322. N. Baglan, C. Dinse, C. Cossonnet, R. Abidi, Z. Asfari, M. Leroy, J. Vicens, J. Radioanal and Nucl. Chem. **226**, 261 (1997).

323. O.M. Falana, H.F. Koch, D.M. Roundhill, G.L. Lumetta, B.P. Hay, J. Chem. Soc. Chem. Commun. 503 (1997).
324. T. Nagasaki, S. Shinkai, Bull. Chem. Soc. Japan. **65**, 471 (1992).
325. M. Yilmaz, H. Deligoz, Sepn. Sci. and Technol. **31**, 2395 (1996).
326. Y. Masuda, T. Swano, E. Sekido, Anal. Sci. **7**, 31 (1991).
327. R. Ludwig, S. Tachimori, Solv. Extn. Res. Dev. Jap. **3**, 244 (1996).
328. A.T. Yordanov, O.M. Falana, H.F. Koch, D.M. Roundhill, Inorg. Chem. **36**, 6468 (1997).
329. A.T. Yordanov, D.M. Roundhill, Inorg. Chem. Acta. **264**, 309 (1997).
330. K. Ohto, E. Murakami, T. Shinohara, K. Shiratsuchi, K. Inoue, M. Iwasaki, Anal. Chim. Acta. **341**, 275 (1997).
331. H. Deligoz, M. Yilmaz, Solvent. Extr. Ion. Exch. **13**, 19 (1995).
332. K. Ohto, K. Maruishi, T. Shinohara, K. Inoue, Solv. Extn. Res. Dev. Jap. **3**, 225 (1996).
333. G.G. Talanova, H.S. Hwang, V.S. Talanov, R.A. Bartsch, J. Chem. Soc. Chem. Comm. 419 (1998).
334. A. Gupta, S.M. Khopkar, Talanta **42**, 1493 (1995).
335. V.J. Mathew, S.M. Khopkar, Talanta **44**, 1699 (1997).
336. R.A. Khandwe, S.M. Khopkar, Talanta **46**, 521 (1998).
337. D. Malkhede, P.M. Dhadke, S.M. Khopkar, J. Radio And Nucl. Chem. **241**, 179 (1999).
338. D. Malkhede, P.M. Dhadke, S.M. Khopkar, Candian J. analytical sciences **43**, 143 (1998).
339. D. Malkhede, P.M. Dhadke, S.M. Khopkar, Analyt. Science **15**, 1 (1999).
340. D. Malkhede, P.M. Dhadke, S.M. Khopkar Ind. J, Chem. Technology **7**, 7 (2000).
341. D. Malkhede, P.M. Dhadke, S.M. Khopkar-Ind, J. Chem. **38A**, 1079 (1999).
342. S.M. Khopkar, Madhybharat Journal **35A** 39 B (1995).
343. D. Malkhede, P.M. Dhadke, S.M. Khopkar-Cand J. Anal. Sci. and Spectroscopy (onpress).
344. D. Malkhede, P.M. Dhadke, S.M. Khopkar-Chemia analyticza (warsaw) (on press).

6
Chromatographic Separations

6.1 Introduction

Apart from solvent extraction separations, macrocyclic, macrobicyclic and supramolecular compounds were extensively used in the various kinds of chromatographic separations. They were used in column chromatography, when the researchers had specially synthesised polymeric crown ether resins. They also modified silica support with crown ethers. The monomeric crown ethers were directly used as the stationary phase in the columnar work. The mobile phase at a times consisted of crown compounds. Many interesting separations were effected with polymeric crown ethers as well as cryptands. Only in limited cases mobile phase consisted of crown compounds. On the whole very promising results were obtained.

Further refinement of column chromatography led to use in high performance liquid chromatographic technique. Special kinds of silica bound stationary supports were designed. Vinyl modified silica columns were also used. The reversed phase HPLC work was initiated on extensive scale. Many interesting separations were accomplished by these sophisticated techniques. These investigation was followed by applications of crown compounds in the area of gas chromatography. Instead of use of high boiling solvents as the stationary phase, crown compounds were used in their place. Large size crowns ethers like 30C10 were successfully used for the separation of cations with large ionic size. Even organic compounds were successfully isolated. This work led to use of DB18C16, DB24C8 and DC24C8, which facilitated separation of hydrocarbons. Several experiments were conducted to evaluate McReynold's constant for interpretation of the basis of the separation. This was invariably determined in experimental work.

Extraction chromatography was successfully utilised for the separation of chemically similar metals like gallium, indium, thallium with trioctyl phosphine oxide (TOPO) as the stationary phase. Apart from organophosphorus compounds high molecular weight amines were successfully employed for the purpose of separations. This was followed by exploration of use of crown ethers and cryptands for the purpose of separations. Several alkali metals could be easily separated. The large number of theoretical plates in the column facilitated the efficient separations in extraction chromatography.

Ion exchange chromatography in non-aqueous and organic solvents was used earlier but crown replaced many extractants specially in field of combined ion exchange solvent extraction methods. Addition of crown ethers in mobile phase increased retention behaviour. The anchored cyclic polyethers were comparable to ion exchange resins for separations however due to slow speed of elution methods were not popular with chemist.

The real boost to this technique was given with the discovery of ion chromatography. This was possible due to use of functionalised pellicular resins which could be operated at higher temperatures. The fitment of cavity and ionic size continued to dominate the process of separation. Amino acids were thus separated. Conductivity detectors provided the fast methods of analysis. The choice of heteroatom like oxygen, nitrogen or sulphur in crown compounds determined the selectivity of the separations. Unfortunately cryptands were not so commonly used. Temperature influenced process of separation.

Finally technique of capillary electrophoresis had come to the forefront only recently. The study of factors like mobilities led to the development of newer methods of separations. Lead, thallium, ammonium ion were separated by capillary electrophoresis methods. Calixarenes fortunately found wide scale applications in this technique. They served as selective modifiers. Sulphonic calix(6) arenes gave promising results for separation of various benzenediols and toluendiols.

The crown compounds were extensively used for the chromatographic separations of ionic or polar species. Such techniques in analytical separations included column chromatography, high performance liquid chromatography, gas chromatography, extraction chromatography, ion exchange, ion chromatography and capillary electrophoresis. Each technique is considered in details.

Column Chromatography

Crown compounds were used either as the stationary phase or as the mobile phase in liquid chromatography. Normal monomeric crown compounds like 18C6 were used in both the phases. The water soluble crown compounds were desorbed from column but few ruined the quality of stationary phase. Therefore they were usually immobilised by covalent bonding on stationary phase. Crown containing compounds were bound to the stationary phase, which involved chemical bonding or absorption of crown ethers on suitable supports. Monomeric crown ethers were best as stationary phase, provided they were not very soluble in the mobile phase. Three mode of applications of crown compounds in chromatography consisted of use of polymeric crown resins for the purpose of separation or the use of crown modified as either mobile phase or as the stationary phase or use of the crown compounds as stationary support.

6.1.1 Polymeric Crown Ether Resins

The polymerisation of crown ethers with a suitable functional group was beneficial to immobilize crown compounds on stationary phase [1–10]. The condensation polymer resin also contained cryptands. They were usually polymerised by condensation of crown ether or cryptand with formaldehyde in formic acid [I]. The similar condensation with formaldehyde using phenol, resorcinol, or xylenol as crosslinking agent were also possible. The copolymerisation of crown ether vinyl monomer with DVB also gave cross linked crown polymers.

Such compounds are shown in structures (I) and (II) as follows:

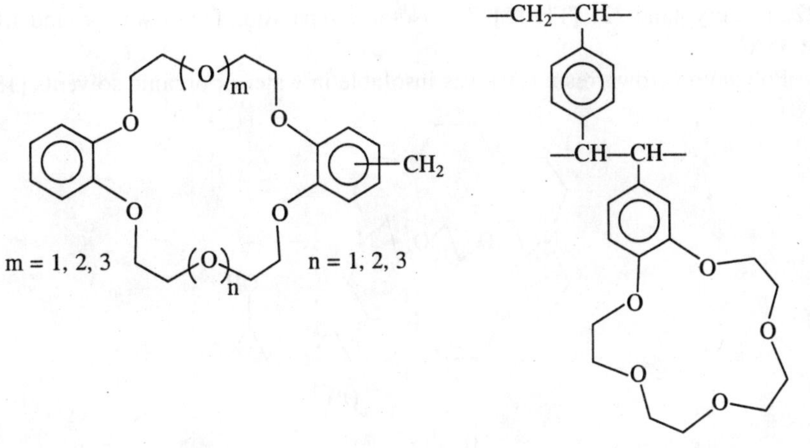

(I)   (II)

m = 1, 2, 3   n = 1, 2, 3

When the cation binding ability of the crown polymer resin to s-block metals was examined it showed that primarily it depended upon kind of crown moiety, metal ion, concentration and solvent employed. The characteristics of these crown polymer resins as stationary phase was studied. Good separation of metals in thiocyante form was effected with DB21C7 on resin as shown above (I) [4]. The retention time of the alkyl ammonium ion descended in the order primary > secondary > tertiary > quaternary amines as expected from the binding of crown ethers. It was possible to carry out anion separations. Such separation was based on the anions of cation binding ability of crown ether, e.g. sodium, chloride, bromide, iodide or thiocyante on B15C5 as shown in (III) structure.

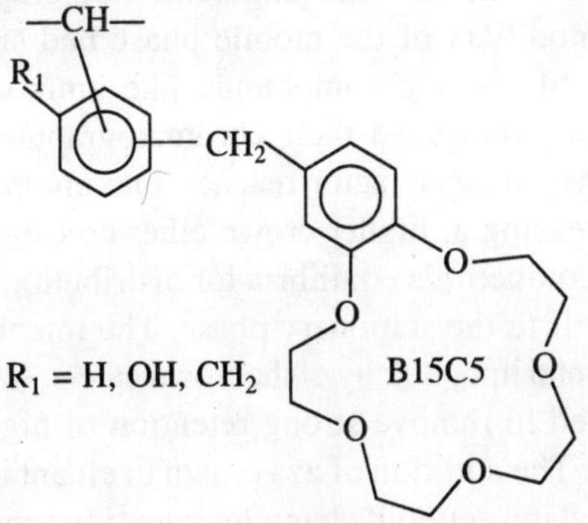

R_1 = H, OH, CH_2

(III)

When ammonium thiocyanate and rubidium chloride were passed on such resin with polymeric crownend (III), ammonium chloride and rubidium thiocyanate was obtained by interconversion on the column. The chromatographic interconversion was useful for many salts [4]. The chromatographic separation of the compounds was dependent on hydrophobic interaction between the crown containing stationary phase and the organic ligands in extraction chromatography. It was possible to separate isotopes of calcium using resins containing cryptand

222 and cryptand 221 [11, 12]. The isotopic separation factor was around 1.02 at 35°C.

Polyamide crown resin (IV) was insoluble in water or organic solvents [13].

(IV)

Therefore it was used in the chromatography separation of alkali and alkaline earths [14, 15]. Such resin exhibited strong attraction for heavy alkali metals showing trend of cation binding ability of crown ether. Polymeric pseudocrown was another compound useful as the stationary phase specially in ion chromatography. Such compounds complexed with metals as real crown ethers. Immobilisation of optically active crown compounds with cross linked polystyrene gave a stationary phase for the separation of aminoacids [16] enantiomers.

6.1.2 Crown Ethers as the Mobile Phase

In liquid chromatography, crown compounds were used directly as the mobile phase. They acted as modifiers of the mobile phase for normal and reversed phase extraction and in the ion exchange chromatography. The use of crown compounds as the modifiers of the mobile phase had significant effect on the retention behaviour of organic compounds like aminoacids [17] in reversed phase technique. They improved their chromatographic behaviour as well as their separations. The capacity factor reached maximum at certain crown ether concentrations, decreasing at higher crown ether concentrations. The excess of crown ether caused competitive equilibria for distribution of organic compounds and crown either itself to the stationary phase. The retention increase depended mainly on cavity size and lipophilicity of the crown ethers. Aza crown or macrocyclic polyamines were used to remove strong retention of highly charged anions on ion chromatography. The addition of aza crown to eluant increased the elution of benzene polycarboxylates without changing retention time of benzoate. This aza crown acted as the ideal modifiers of eluants in anion exchange chromatography.

6.1.3 Crown Ethers as the Stationary Phase

As told earlier these were synthesised by condensation of crown compounds with formaldehyde in formic acid with cross linking by methylene group. With monobenzo crown compounds additional cross linking agent like aromatic hydrocarbon or resorcinol were used. The binding was effected at polystyrene surface through methylene bridges by condensation reaction with formaldehyde.

The thermal stability of such compounds showed a decrease with increase in ring size. The condensation resins were thermally more resistant than polymerised resin. The condensation resins were also resistant to polar solvents. The polymerised resins were resistant to nonpolar solvents. If exchange contained N-in the anchoring group, a protonation of nitrogen atom took place with no complexation. The cyclic polyether containing exchangers were widely used in inorganic chemistry. They were used for the separation of cations, anions, organic compounds and in column, thin layer and high performance liquid chromatography and ion chromatography. During selection of the exchanger the rule to be followed was the diameter of cation/diameter of ring = 0.80. Sodium was separated with B15C5, potassium with DB18C6, rubidium with DB21C7 so also caesium if crown ethers were in polymeric form, silica gel modified by B15C5 was best for ion chromatography. Water determination with DB18C6 with methylene bridge was also possible (18)

Polymers with cryptands as anchoring groups were sold under trade name 'Kryptofix 221B polymer or Kryptofix 222B polymer'. First cryptand was good for complexing sodium while latter was good for entrapping potassium (3, 19) dibenzo21crown7 was best suited for separation of alkali metals specially caesium. The anchoring group was selective for caesium. These exchangers could also be applied as absorbents in thin layer chromatography and electrophoresis poly (ethyleneterephthalate) sheets (8) were coated with suspension of the powdered exchangers in poly (vinylalcohol) solutions, e.g. separation of benzoic, salicyclic and sorbic acid was done by TLC technique [20].

6.1.4 Applications in Column Chromatography

In column chromatographic separation of metal ions with crown compounds either the water or methanol (not exceeding 40%) proved to be better eluant for desorption of weakly bound complex from the stationary support. DB18C6 found extensive applications for the separation of alkali metals [27]. In many instances solvent mixtures were used as the mobile phase during separations [29]. Apart from column, the thin layer and paper chromatographic methods played leading role in separation of fission products including separation of strontium and yttrium [30]. The individual dithiacrown ethers were separated from each another.

In comparison to crown ethers, cryptands were less frequently used. They were used on silica gel plates with ammonical methanol as the solvent for the purpose of separation of selected elements.

The various methods involving columnar technique for the separation of metals as well as organic compounds are summarised in Table 6.1.1.

6.2 High Performance Liquid Chromatography

In HPLC a good separation efficiency and speed was possible using crown compounds as the stationary phase. However all crown polymer were not suitable for HPLC work due to pressure and one needed pressure resistant stationary phase for such job. The use of inorganic materials like silica gel as packing support was useful when crown compounds were immobilised on silica gel by chemical bonding to the silica surface. Copolymerisation of vinyl modified silicas

Table 6.1.1 Application of crown compounds in column chromatographic separations

Element/Compound	Separation Conditions
Alkali metals	Crown ether on silica gel column was used for metal separation [21]. The complexation of crown ether with ammonia on silica gel column was studied in methanol, the mechanism of sorption was then explained [22]. Many crown ether carboxylic acids were used with silica gel column to study the influence of cavity size, position of proton ionisable side arm and capping of residual silanol surface active group with respect to column selectivity [23]. Cyanocobalt salt of potassium chelate was coated on silica or alumina column for the separation of inert gas elements [24].
Amino acids	Chiral crown ether were used as the stationary phase to separate racemic mixture of amino acids, when factors like column height, temperature and extent of crown ether coating influenced the separation [25]. Tetrakiscyclopentano-1, 4, 7, 10 tetraoxacyclodecane on silica gel column was separated with hexane-chloroform ethanol mixture (15:3:2) by gel permeation technique [26].
Alkaline earths	DB18C6 in chloroform on silica gel coated thin layer chromatography plates was impregnated with 2% potassium chloride or bromide with dioxone-pyridine mixture as mobile phase, different from representing various s-block metals complexes were obtained [27]. Also crown ether on silica plates coated with alkali metals was used for their separation [28]. Crown ether polygram SILG/UV impregnateed plate with potassium bromide developer and chloroform-acetone mixture was used for the separation and detection by fluoroscene quenching or with iodine vapour [29].
Y (III)	Symm DB16C5 oxyacetic acid and symm DB16C5 oxysteric acid was used at pH 4.0 and 6.0 respectively in paper chromatographic separation of yttrium and strontium [30]. Thin layer chromatography technique was used for the individual separations of crown ethers.
Crown ethers: Thiacrown ethers	Crown ethers and lysine derivatives were separated on silica gel coated plate with solvent mixture containing benzene, methanol and amines. Color development for visulation was carried out with Dragendorff reagent [31]. Polygram SilG/UV coated with parafin hexane combination was used for separation of crown ethers dissolved in DMF with non-aqueous solvents [32]. The dithiocrown ethers were separated on plates coated with silica gel $F^{60}{}_{254}$ or similar material coated on alumina [33]. Dibenzocrown ether and its ditertiary butyl derivatives in methylene chloride were separated on silica or alumina coated with $^{60}F_{254}$ on TLC plates with solvent mixture containing ethyl acetate and methanol, methylene chloride and chloroform [34]. Dibenzocrown ether in methylene chloride was separated on TLC plates with aqueous methanol acetonitrile or aqueous propanol hexane mixture as the eluant and visualised by ioldine vapour in UV radiation [35] cryptand-222 was separated on silica gel coated plates with methanol ammonia mixture (9:1) and visualised by UV radiation with iodine vapour [36].

was useful in HPLC. The compounds were (V) and (VI) useful for metals separations [37-40]. They have high molecular weight > 3000 [41]

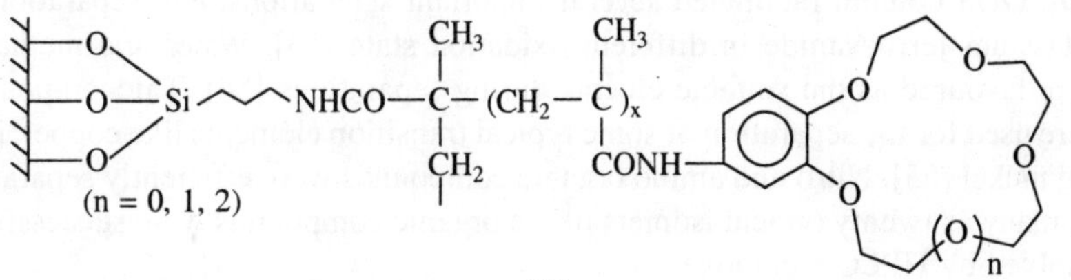

(V)

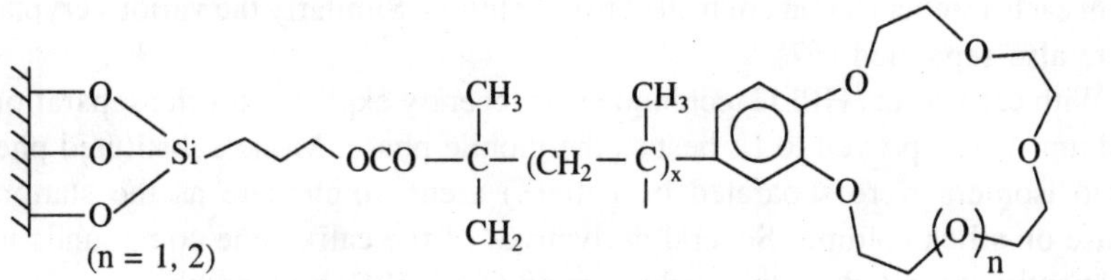

(VI)

The alkali metal halides were separated on poly (18C6), poly (15C5) and poly (12C4) modified silicas. The elution order of alkali metal depended on kind of crown ether used. In ion chromatography such order was not observed, where affinity increased with rise of atomic number. The ion chromatography pattern of poly crown modified silica was influenced by the nature of mobile phase. The separation of anions with common cations was done on crown modified silica (42). The coating silica gel with insoluble crown polymer was another mode to manufacture pressure resistant crown stationary phase for HPLC of ions. Silica gel was coated with polymeric crown ether. Such coated silica was also useful for anion chromatography [43]. HPLC was used for study of coordination compounds [44, 47].

The reversed phase HPLC was employed by Mangia [45] to investigate the chromatographic behaviour of mercuric halide complexes with macrocyclic polyether DB18C6. It acted as hexadentale ligand. Octadecylsilanc (ODS) on silica, i.e. Micropak CH columns were used with mobile phase with phosphate borate buffer of pH 7.0-10.0 and the photometric detector at 254 nm. By this procedure many mercury halides and mixed halides were rapidly separated [45].

The first HPLC technique was applied for study of metal organic system, but now large number of papers have appeared on use of HPLC confirming it as the effective means for the separation and determination of many crown ether complexes, though multielement determinations were some what limited. With availability of sensitive detectors with ICP-AES or AAS or electrochemical technique it had become more reliable tool of separation. Few methods permitted separation of isomers with better resolution and low level of detection of compounds. Only limitations of this method was that many of the methods could not be extended on preparative scale [46].

6.2.1 Applications of High Performance Liquid Chromatography
This technique was not only used for the separation of metals but several organic

compounds. Further not only crown ether, cryptands or calixarenes but also chiral compounds were also utilised for the cleancut separation of aminoacids [50]. ODS column facilitated several important separations, e.g. separation of potassium ferrocyanide in different oxidation state [53]. Water and methanol were favoured as the suitable eluents during separations [54]. Thiocompounds were used for the separation of some typical transition elements like copper, iron and nickel [55]. Nitro and amino organic compounds were efficiently separated. As many as twenty optical isomers of the organic compounds were successfully resolved by HPLC technique.

On bondpacked phenyl column various crown ethers were mutually separated from each another [65] at controlled pH of elution. Similarly the various cryptands were also separated [67].

With calixarene, HPLC technique was recently explored for the separation of sodium. Water proved to be best as the mobile phase. Mono substituted phenol regio isomers were separated by calix(4) arene sulphonate as the stationary phase on silica column. Several derivatives of the calixarene compounds were efficiently separated by this technique on C-60 silica bonded phase.

The various methods for the separation of inorganic and organic compounds using crown ether, cryptands and the calixarenes by HPLC technique are summarised in Table 6.2.1.

6.3 Gas Chromatographic Separations

Most of the gas chromatographic separations were carried out for thermally stable volatile compounds. Interestingly good separations were carried out by this simple technique of gas liquid chromatography exploiting the difference in partition coefficient of individual compounds on the column. The variety of variation in the flow rate as well as programming the temperature of the column has assisted chemist to carry out several novel separations. In most of the cases inert gas like helium or nitrogen was used as the carrier gas. The compound being analysed were volatile and formed vapour phase along with carrier gas which in turn was carried down the column. Usually variety of materials were used as the stationary support. They consisted of chromsorb, silica, etc. The particle size of these supporting materials greatly influenced the partition behaviour of solute on stationary phase.

As a rule stationary phase was composed of high boiling liquid. During chromatographic separation, the volatile components, partitioned between gas phase and high boiling liquid phase hence the term gas liquid chromatography was used. Few attempts were made by chemist to change this stationary phase. Instead of high boiling liquid alternatively better compounds were used. For instance with trifluoroacetylacetone metal complexes were resolved by gas chromatography.

The applications of crown ethers as well as cryptands blended on stationary support as stationary phase are discussed in this section. On sorption of metals depending upon the cavity of crown and ionic size of metal they were complexed and retained on column which in turn were eluted with suitable eluant. Generally capillary columns were preferred in these investigations.

Table 6.2.1 Separations on HPLC with crown compounds

Elements/ Compounds	Separation Conditions
Orgnometallic complexes	High performance liquid chromatography technique was extensively used for the separation of metal crown ether complexes [47]. The organic compounds and the individual complexes of crown ethers with metals were also separated [48]. The chiral crown ethers were used for the separation of enantiomeric aminoacids and amines on HPLC [49]. The inorganic as well as organic compounds were isolated with crown compounds as the stationary phase [50].
Alkali metals	Dodecyl 18C6 on octadecylsilanised C-18 silica column was used for their resolution with aqueous methanol (40%). Potassium was strongly sorbed but lithium was weakly sorbed [51]. With crown ethers they were separated as complexes with controlled equilibration or time by HPLC method [52] with DB18C6 on column ODS-silica two complexes like potassium ferrocyanide and ferricyanide were separated with 5% aqueous methanol as mobile phase and analysed by photometric detector at 767 nm and 249 nm respectively [53].
Alkaline earths	On silica gel column packed in stainless steel tubes with poly B15C5 and bis B15C5 was used to separate alkaline earths with aqueous alcohol mixture. The alkali metals were eluated with water while the alkaline earths were removed with aqueous methanol [54].
Co (III)	Benzocrown ether separated it on alkali nitrile bonded Zorbex CN column with triethylamine in hexane and ethyl acetate with photometric detector. Copper, iron nickel were also separated with diethyldithiocarbamate [55].
Hg (II)	On Micropak CH column (i.e. octadecyl silane on silicagel) with DB18C6 in methylene chloride, the halide salts were separated with methanol borate buffered at pH 8.8 and detected by photometry. Potassium iodide was added to acetate solution to give better resolution [56].
Organo-nitro compounds	With polyvinyl B18C6 on bonded silica nitro compounds were separated with acetate buffer at pH 5.0. Addition of potassium chloride improved the retention behaviour of 4-nitro compounds on the column [57] with sil-18C6 such separation was also fissible for amino acid for structure identification [58]. With crown ethers in silica column twenty optical isomers were separated with THF methanol mixture [59]. With polyvinyl B18C6 on silica gel column separated with methanol buffer mixture for amino compounds at pH 5 or pH 8.0 with UV detection due to hydrophobic interaction [60]. Polycyclic organic compounds were separated on chloropropyl silane bonded phase in stainless steel column. B15C5 derivative of crown ether were separated with methanol aqueous solution from aromatic derivatives [61].
Crown ethers thiaderivatives	With-cyclodextrin on stainless steel column with methanol buffered at pH 4-in the presence of triethylamine was used to separate enantiomers of amine acids crown ethers and fused ring compounds [62]. On column of Wakas 15C8 they were separated by gradient elution with aqueous acetonitrile within short period on various column [63]. N-substituted dibenzocrown ether on dynamically modified silica gel with aqueous methanol were separated [64].

Elements/ Compounds	Separation Conditions
Azacrown compound	1-aza 15C5, 1aza 18C6 di and thia aza 15C5 and several analogues compounds were separated on bond packed phenyl column with methanol containing α-naphthalene 1-sulphonic acid at pH 2-2 and analysed by photometric detector, diaza compounds were eluated first then mono aza compounds were removed [65].
Thiacrown	Tetra thia ether were separated on μ-bond pack CN column with aqueous methanol and analysed by photometric UV detector [66]. They were also separated after synthesis by HPLC [67]. Cryptand-222 was eliminated from 2-decyl-2 (F^{18}) fluoro-D-glucose on Al tech carbohydrate column with 83% acetonitrile as the mobile phase. The separation of cryptand from other products were possible by HPLC method [68].
Sodium (I)	With calix(4) arene ester as stationary phase on silica gel column sodium was separated with 30% aqueous methanol as mobile phase [69] with calix(4) arene tetradiethyl amide on silica gel separated it with water as the mobile phase. The conductivity detector was used. The method was selective for separation of sodium from calcium and magnesium [70].
Phenols/regio isomers	Calix (6) arene p-sulphonate on silica column with methanol acetate buffer was used for separation of monosubstituted phenol regioisomer and some positional isomers [71] calixarene on silica gel HPLC column were also used with chloroform heptane mixture as the mobile phase [72] to separate 4-alkyl phenols.
Calix(n) arene:	Calixarene isomers of tetramethyl calix(4) arene octal and octacetate and octamethyl ether were separated with 50% acetonitrile [73]. The diastero isomers of calix(4) arene and cavitand were separated on Si-60 or RP-18 column with aqueous acetonitrile or methylene chloride acetone mixture. Such separation was possible on TLC [74] p-tert butyl calix(n)arene and p-t butyl-dihomox calix(4) arene on C18 column were separated with solvent mixture containing acetonitrile, methylene chloride and acetone on silica gel column [75]. With toluene methanol (2:3) mixture p-t butyl calix (4, 6, 8) arene were separated on C-60 silica bonded column [76] calix(n) arenes and their phosphoryl derivatives were separated on Li chromsorb RP-18 column with 85% aq acetonitrile solution [77] p-t butyl calixarene derivatives were separated on sphersab ODS column with acetonitrile ethyl acetate and acetic acid mixture (8:2:1) and compared with supercritical fluid extraction chromatography method [78] calixarene and sulphonated derivatives were separated on C-60 bonded silica phase and results were similar to one done on ODS silica column [79] p-t butyl calix(n) arene (n = 4-10) were separated on sphersorb ODS column with acetonitrile methanol ethylacetate mixture in short interval gradient elution [80].

Since crown ethers formed stable complexes with metal cations they found wide scale applications in analytical chemistry. They were used in several separation techniques such as solvent extraction, ion selective electrodes and liquid chromatography. Blasius [81] first reported the use of crown ethers in chromatography. It became popular specially in liquid chromatography. Only in recent years chromatographers have paid more attention to use of these crown compounds in gas chromatography [81]. The packed column coated with DB24C8, and DB18C6 were used for the separation of hydrocarbons, alcohols, amines. Graphitized thermal carbon blacks modified with crown ethers was used for same purpose [82]. The crown ethers were coated on glass capillary columns and characterised [83]. 18C6 substituted poly siloxane was used as the stationary phase [85].

A typical application was that of 4, 4 dipentadecyl or 4-3 dipentadecyl dibenzo 30C10 (VII) and 3-penta decyl benzo 15C5 (VIII) coated on glass capillary column or fused silica capillary columns. Their chromatographic behaviour was investigated.

DSU30C10 $C_{15}H_{31}$

SU15C5 $C_{15}H_{31}$

(VII) (VIII)

Fused silica column had better efficiency [84]. The selectivity of crown ether depended mainly on the relative size of the solute and crown ether cavity, kind of heteroatoms and conformational flexibility of crown ether ring. These compounds were versatile as stationary phases. Very good separations of organic compounds with hydroxyl group was fissible, due to deactivation of the residual silanol groups on the surface of column wall by crown rings.

A systematic investigation was carried out on use of several crown ethers as the stationary phase in gas chromatography [86]. For example DB18C6, DB24C8, DC24C8 were studied with reference to polarity, selectivity and stability. The methods were extended for the separation of nitrophenol and nitroaniline isomers without any derivatisation [87]. Blasius [1–10] and Sousa [88] used the crown ethers in chromatography. They were subsequently used in HPLC [89] and ion pair chromatography [90]. However by and large use of crown ether in gas chromatography was rare. The chromatographic characteristics of 18C6 was first studied [91] DB18C6 coated on the fire brick C_{22} was used for separations of dichlorophenol [92]. Several hydrocarbons were also separated by GLC [93, 94]. Oligo (ethyleneoide) substituted polysiloxanes with polyethylene glycol as stationary phase had good selectivity due to matching of size and shape of crown ether cavity.

In all GLC work usually McReynold's constant was evaluated [95] to interpret theoretically several separations.

6.3.1 Applications of Gas Chromatography with Macrocyclic and the Supramolecular Compounds

Most of these technique involved use of above mentioned compounds as the stationary phase, involving separation of array of organic compounds like aldehyde, ketones and polyhydric alcohols. The separations were invariably carried out on glass capillary columns. B15C5 and DB24C8 were explored for separation of polar organic mixtures [101]. Programme temperature gas chromatography was utilised for isolation of 18C6, 21C7, 24C8 and similar crown compounds [104]. Argon was preferred as the carrier gas while dibutyl pthalate was used as the internal standard. Nitrophenols and nitroanilines were thus separated. Various separations of organic compounds using gas chromatography are best summarised in Table 6.3.1. They were mostly used for the separation of various organic compounds.

Table 6.3.1   Gas chromatography separations of organic compounds with crown ethers

Compounds	Separation Conditions
Aliphatic compounds	Crown ethers and acyclic polyethers were used as the stationary phases [96, 97]. It was used for the separations of hydrocarbons, alcohols, aldehydes, ketones amines and several aromatic compounds and for the mixture of compounds with high boiling point [98]. Crown ether coated glass capillary columns were used for the separation of solution [99]. The McReynold's constant was studied in gas chromatography methods for eight crown ethers [100]. Crown ether coated on fused silica capillary column were used with alkyl B15C5 and DB24C8 for the separation of polar organic mixtures [101].
Nitrophenols	Crown ethers polysiloxane on fused silica capillary column cured with aza tert butane as initiater was used for the separation of phenolic and ethyl substituted diphenyl ketones [102]. Similar stationary phase was used with 18C6 for separation of solutes [103] DC18C6 in tetrachloroethane was used in GC-MS separations of radioactive products [104]. Various crown ethers like 18C6, 21C7, 24C8 were isolated on supports like OVS-225 by programmed temperature gas chromatography. Dibutyl pthalate as an internal standard and argon as carrier gas [105] on fused silica column coated with BP-5 compound B15C5, B18C6, DC18C6 were separated at pH 6.8 with acetonitrile [106]. DB18C6, DB24C8, DC24C8 were used as the stationary phase for the separation of nitrophenols, nitroaniline with good thermal stability. DB18C6 provided the best stationary phase [107] with B15C5 coated on support span 80 in stainless steel column were used for the separation of various organic compounds like hydrocarbons, alcohols, ketones, esters, ethers [108].
Glycols & enantiomers	Chiral crown ethers were used at pH 10 on crown pak CR column for separation of the enantiomorphous compounds and drugs with water as eluant [109] 5% crown ethers in dichloromethane with the stationary phase on fused silica capillary coulmn was used for the separation of glycols [110] chiral crown ethers on crown pak CR at 25°C was used to ascertain the purity of optically active compounds [111].
Ketones	On fused silica capillary column coated with crown ether polysiloxan with cross linkage of dicymyl peroxide as an indicator was used with efficiency [112] for separation of ketones. An enactive isomers was separated on silica column [113] at pH 3.0 when binding ratio to active sites was greater [113].
Amino acids	Various compounds like chiral phenyl alkanols phenyl, phenylamines and amino acids were separated on chirasilmon 18C6 and chiralaza 15C5 in dichloromethane by temperature programme chromatography [114].

(Contd)

Compounds	Separation Conditions
	A polyhydromethyl siloxane derived chiral 18C6-25 in dichloro methane on fused silica column purged with nitrogen as carrier gas was used at 260°C. The enantioselective stationary phase facilitated the separation of racemic mixture containing amino hydroxyl group compounds [115].
Phenols	The phenols were separated on propyloxymethyl 15C5 polysiloxane coated fused silica capillary column also used for the separation of phenols chlorophenols, nitrophenols and substituted derivatives [116] calixarene crown siloxane telomers were used as the stationary phase [117].
Oligomers	p-t butyl calix(4) arene, calix(8) arene were used to separate oligomers on silica column like xylene, ethyltoluene, diethylbenzene, and butylbenzene. The efficiency of separation was low due to poor solubility in the stationary phase [118]. Calix crown compounds were used on polysiloxane in dichloromethane as the stationary phase. Calixcrown compounds were used for separation of isomers of aromatic compounds [119].
Hydrocarbons	Calix(4) arene and its derivatives were used as the stationary phase for the separation of aromatic hydrocarbons and analogues compounds due to steric hindrance formation of inclusion compounds and presence of hydrogen bonding and dispersion of similar forces [120]. Calixresorcinarene with cyclodextrin as the stationary phase was used to study of reaction mechanism with guest molecule indicating presence of synergic effect [121].

6.4 Reversed Phase Extraction Chromatography

This was one of the technique which had come to forefront in recent years. The main advantage was need of very small volume of extractant as well as eluant during separation. As many as seven components were separated by proces of elution in few minutes with small volume of an eluant. The usual mode involved the utilisation of neutral stationary support like silica gel or Kel F. The stationary support was made hydrophobic by coating it with dimethyldichlorosilane in the rotary vaccum evaporator for few hours. This hydrophobic stationary support was subsequently coated with stationary phase consisting of suitable extractant. Good amount of work was carried out for the metal separations involving TBP [122] TOPO [123] and high molecular weight amines [124] such as amberlite LA-1, LA-2, and aliquat 336. With last compound separation of the several anionic complexes of metals with organic carboxylic acids such as malonic, citric or tartaric acid was also carried out.

The monomeric crown ethers were best reagents as the stationary phase in extraction chromatography, but use of monomeric crown ethers had restricted applicability on account of their solubility in the aqueous phase. However this problem was circumvented by coating them on hydrophobic stationary support. It was not necessary even to use polymeric crown compounds in liquid chromatographic work. The solid crown compounds were directly useable. A column packed with solid B18C6 was used for separation of lanthanide series of elements [125] but direct packing of crown ether in column had several problems as the stationary phase was likely to decompose due to its solubility in an eluant.

Table 6.4.1 Extraction chromatograpic separations with crown ethers

Element	Separation Conditions
Li (I)	With B15C5, 15C5 and DC18C6 as the stationary phase with perchlorate as counter ion it was separated. The different isotopes were separated in aqueous solution after complexation [132].
K (I)	Diaza 18C6 or parent compound was used to separate various isotopes such as K^{39} and K^{41} [133].
Cs (I)	With 12C4, 15C5, 18C6 and nitro substituted DB18C6 in chloroacetic acid with crown ether coated on chromsorb column caesium was separated, it was stripped with dilute mineral acids and trichloroacetate [134], DB18C6 and mesitylene coated on silicone treated kieselguhr column separated them with sodium perchlorate buffered at pH 7.0 alternatively ethanol perchlorate solution at pH 4.0 was used, s-block metals were not extracted by the stationary phase and were thus separated [135].
Mg (II)	Various crown ethers such as DB18C6, DC18C6 coated on silica gel which was rendered hydrophobic with picric acid as counter anion were used to separate various isotopes and analysed them. The addition of 1.5M sulphuric acid increased the extractibility of Mg^{25} and Mg^{26} [136].
Sr (II)	At pH 2-5 crown 24C8 or DC18C6 was used to separate it. The metal was stripped with 1M hydrochloric acid from the column [137].
Ba (II)	Crown ether DB18C6 coated on silica gel was used to extract it and subsequently strip it with 1M acetic acid. The coextracted uranium (VI) molybdenum (VI) were subsequently stripped with 1M perchloric acid and 0.5M ammonia respectively. Barium was determined colorimetrically [138, 139].
Alkaline earths and alkali metals	B15C5, DB18C6, DC18C6 and 1, 4, 7, 10, 13 pentaoxotri decane – 1, 3 diyldiquionoline in chloroform was coated in hydrofluoron. It gave good separation with 0.1M sodium thiocyanate at pH 7.0 – 12.0 for alkali metals and 0.1M sodium perchlorate at pH 7.0 for alkaline earths. Mesitylene could also be used with some advantage [140].
Co (III)	DB18C6 in dichloromethane was coated on silica gel column and cobalt was extracted quantitatively from 1M thiocyanate solution when latter acted as the counter anion. The extracted metal was stripped with 0.5-2M of hydrochloric or sulphuric acid and determined spectrophotometrically as nitro R salt. The separation of cobalt from chromium, molybdenum, nickel and copper was possible [141].

Therefore solvent extraction using crown compounds for cation separation was most popular. The crown compounds were dissolved in water immiscible solvent, to extract cations selectively. The multi-step solvent extraction was accomplished in the technique of extraction chromatography. Apart from silica, diatomccous earth was made hydrophobic by treatment with dimethyldichlorosilane as stationary support and it was coated by mixing silicanised support with crown ethers dissolved in chloroform or suitable solvents followed by evaporation of excess of solvent. Usually DB18C6, DC18C6 or B15C5 were used for such purpose of coating. The alkali and alkaline earths were separated with thiocyanate of sodium or ammonium as the eluting agents. The kind of crown ether used, the nature and composition of the solvent and pH of extraction influenced the chromatographic behaviour on the column [126-129]. Crown compounds were used for counter current chromatography, a typical example was that of extraction chromatography. Centrifugal counter current chromatography with 18C6 in chloroform as stationary phase permitted cleancut separation of sodium and potassium [130]. Several solvent extraction systems involving crown compounds were extended to extraction chromatography without significant modification in conditions. The hydrophobic interaction was beneficial for immobilizing crown compounds on stationary support. For instance alkyl silicanised silica had hydrophobic surface and crown compound which was highly lipophillic was immobilised through hydrophobic interaction. Octadecylsilanised silica (ODS) coated with crown ether was best stationary phase in HPLC. High separation efficiency was encountered [131] due to the existence of the large number of theoretical plates. The high elution rate of mobile phase also facilitated the separate ion of the metal ions.

6.4.1 Applications of Extraction Chromatography for Separations
Several metals like alkali metals, few alkaline earths, and some transition metals were separated involving the use of crown ethers as the stationary phase in reversed phase extraction chromatography. Crown compounds like B15C5, 15C5, DC18C6, 12C4, 18C6, 24C8, DB18C6 invariably provided the excellent separations. The counter anion used for the purpose of ion pair complexation consisted of perchlorate, picrate, thiocyanate etc. The method facilitated the separation of isotopes of lithium [132] or potassium [133]. Usually stripping of metals from the column was carried out with dilute mineral acids. Silica, kieselghur, chromosorb were tried as the stationary support for organic coating. The isotopes of magnesium could be separated with clarity [136]. Many complex mixtures of transition metals were isolated with ease [138, 139]. A novel process was developed for the separation of cobalt from other interfering ions with DB18C6 coated on silica gel with thiocyanate as the counter anion and dilute hydrochloric acid as the stripping agent [141].

The interesting separations of various metals involving crown ethers are summarised in Table 6.4.1 using extraction chromatography technique.

6.5 Ion Exchange Chromatography
Ion exchange is one of the most poweful tool of separation. The technique came in to fore front due to impressive separations of the lanthanide series of elements. This was followed by deionisation or demineralisation of water. Subsequently all chemically similar elements like zirconium hafnium, niobium tantalum or gallium, indium thallium, belonging to the same group of periodic classification were separated from each another by cation exchange chromatography.

One could witness spectacular progress in metal separations with the discovery of anion exchange chromatography utilising anionic or negatively charged complexes of metals, with common mineral acids. Good amount of work was carried out on the separation of anionic chloride or sulphato complexes of metals. For example separation of uranium (VI) or zirconium (IV). This field became more popular with the applications of organic ligands like dicarboxylic organic acids which formed anionic complexes specially with transition metals at the constant pH. The complexes of organic complexing agents had an edge over mineral acids because one could control pH of complexation, reagent concentration and solubility of complex in organic or aqueous phase to facilitate separations.

The solubility of complexes of metals led to the discovery of use of nonaqueous and organic extractant in metal separation. For instance organic polar solvents like acetone, dioxane, tetrahydrofurane were used in combination with mineral acids as the eluants. Subsequently more powerful extractants like TOPO, TBP and TBPO were extensively used in combined ion exchange solvent extraction technique (CIESE).

The use of macrocyclic and macrobicyclic polyether was of recent origin. Very limited number of supramolecular compounds have been so far used for the purpose. In this section we explore the possibility of using crown ether cryptands and calixarene in ion exchange separation.

The conventional ion exchange chromatographic separation of metals or other ionic compounds was carried out by ion exchange methods with common eluting agents. The effect of crown on ion exchange chromatographic behaviour of alkali metal ions was extensively investigated. In typical separation addition of crown ethers to mobile phase increased the retention time of amino acids during separation by encouraging ionic interaction between the resulting ammonium cation and anionic site of ion exchanger due to the crown complex formation [142]. Same time addition of crown ethers and crown complexing metal ions to the mobile phase nullified the increase in retention due to the existance of competitive equilibria between ammonia and metal ion. Crown compounds themselves were individually separated using metal ion loaded cation exchange and nonaqueous solvent [143]. Such separation was based on ion dipole interaction between crown compounds and the cationic site of the stationary phase. Those of compounds which were complexed were eluted slowly in separations on the chromatographic column.

It was interesting to compare naturally available exchanger with one anchored with cyclic polyethers. The latter were high resistant to chemicals and temperature. They could take up both cations as well as anions. The stripping of the solvent shell of ions was possible. The binding of ions was assisted in solvents less polar than water. The stability was dependent on solvent, nature of crown compound and position and number of heteroatoms like oxygen, sulphur or nitrogen. The elution was possible by pure solvents. The thermal stability had decreased with number of rings. The condensation resin were thermally more resistant than copolymerised resins. They were resistant to polar solvents. The complete protonation of N-atoms in anchoring group took place at pH < 2.0 in absence of complex formation. The cyclic polyether bound to silica gel or polystyrene were used in ion chromatography. They were used for the separation of cations, anions, neutral salt and organic compounds, trace analysis and for analysis of water.

Another typical example was the efficient chromatographic evaluation of crown ether complexation with alkali metal ions [144]. This method primarily dwelt on determination of complexation constant. This method was comparable to one for evaluation of the complexation constant by conductometry [145] or potentiometry [146] and nuclear magnetic resonance technique [147]. Several crown ethers like B15C5, B18C6, DB18C6, DB21C7, DB24C8 and DB3010C were explored as the suitable complexing agents of potassium ion which formed 1:1 complex, other alkali metals were also studied. The match of size of metal ion with the cavity size of the crown ether determined the complexation selectivity [148].

6.5.1 Applications of Ion Exchange Chromatography for Separations

Very limited work was carried out on ion exchange column with crown compounds. The metals studied consisted of alkali metals, alkaline earths, trans uranic elements, cadmium and copper. The mixed solvents like THF and methanol were used as the eluants [148]. Salts of the alkali metals were also used as the eluants [152]. With carboxamide resins, methanol was used as the eluant. The methods were used for the separation of isotopes of alkaline earths. The polycrown ethers could separate uranium (VI) from titanium. Cadmium was separated with B15C5 on polymer based exchanger in the presence of potassium iodide with dilute solution of hydrochloric acid [159]. This facilitated its separation from chemically similar metal like zinc. p-tert butyl calix (8) arene [160] was successfully used for the separation of copper, zinc and nickel with dilute hydrochloric acid as the eluant.

Various ion exchange separations using crown ether and calixarenes are summarised in Table 6.5.1.

6.6 Ion Chromatography Separation

Although greater strides were made in inorganic chemistry with chromatography for the separation of cations as well as anions, these methods found less applications in industry primarily because of speed of analysis. In any typical separation, sorption of ion its elution and subsequent analysis took quite some time. There were no rapid methods available to ascertain the concentration of the species directly on the column or in the solution immediately after separations.

In comparison with ion exchange, partition chromatography principle involving gas chromatography or high performance liquid chromatography became popular as tool of the separation due to availability of versatile detectors. For instance in gas chromatography electron capture (ECD) flame ionisation (FID) or conductivity (CD) detectors proved to be most valuable for the detection and determination of volatile component on the chromatographic column. Similarly in HPLC methods, refractive index (RI) photometric (UV) or amperiometric, atomic absorption (AAS) detectors proved to be most versatile for the characterisation and analysis of nonvolatile organic compounds on the column.

In recent time, ion exchange chromatography was oriented as instrumental technique with discovery of detectors involving measurement of conductivity or absorbance or diffusion generated current. Since kind of reaction involved in such of separations was not only confined to ion exchange but also to the phenomena of ion exclusion and ion pair formation, the technique as a whole was termed as 'Ion Chromatography'. With the discovery of pellicular resins containing labile cation or anion it was easy to carry out separation under high pressure. But most impressive part of instrumentation was the existance of suppressor column which

Table 6.5.1   Ion exchange separations with crown compounds

Element	Separation Conditions
Alkali metals	1, 4, 7, trioxaheptolyne or 1, 4, 7, 10 tetraoxadecylene or 1, 4, 7, 10, 13 pentaoxatridecylene compound on IEX- 5105P (sulphonate form) resin in ammonical form was used to separate individual elements with mixed solvents like methanol or tetrahydrofurane [148]. Various isotopes were separated with crown ethers [149]. The crown ethers themselves were also separated on cation exchange resin [150]. Radio tracer alkali metals like sodium or potassium were used to study distribution pattern of crown ether on tungstophosphate resin [151]. Cation exchange resin in the presence of crown ether were eluted with 0.1M potassium solution on TKS gel IC-column. The elution profiles were varied with change of resin with varying capacity [152]. Ten crown ether carboxylate resins were used with methanol eluent. They were separated with varying conditions like change in cavity of crown ether, length of carbamide chain, concentration of methanol as well as metal ion concentration to achieve good clean out separation [153].
Ca (II)	Isotopes of calcium like Ca^{46}, Ca^{44} were separated using 18C6 [154]. Use of DMSO increased capacity of the resin. The stability of calcium crown complex however was increased by addition of DMSO in acetic acid and chloroform mixture [155]. Crown ether on diphonix resin bonded to DVB or BioRad AG, Dowex-X8 was used to separate calcium, strontium and radium with propanol water mixture when the presence of hydrophobic crown ether showed decrease in the cation sorption [156]. The selectivity was enhanced by synergic action of cation exchanger and the presence of functional ionic group of the resin like sulphonic or diphosphoric acid [156].
Sc (III)	On polymeric ionisable resin with DB16C5 oxyacetic acid separated it from strontium [157].
U (VI)	Poly (methylene 3 3' dibenzo) 18C6 with 0.02M picric acid had separated it in ethanol from titanium with 0.1M hydrochloric acid while titanium was released with picric acid. Ethanol picric acid mixture offered better separation [158].
Cu (II)	With p-t-butyl calix (8) arene on XAD-4 resin at pH 3-9 the metals like copper, iron, zinc, lead, nickel and cobalt were sorbed and subsequently eluted with 0.1M hydrochloric acid. Uranium (VI) and thorium (IV) could be complexed at pH 4.5 [160].
Cd (II)	With crown B15C5 on polymer based exchanger with 0.1M potassium iodide, it was separated with 0.5M hydrochloric acid and separated from zinc and analysed by atomic absorption spectroscopy [159].

predominantly assisted to convert weak eluate to form compound which would have better conductivity for measurement. Dionex Co. made significant contribution to render this technique attractive world wide amongst analytical chemist. Several macrocyclic compounds had cyclic structures with heteroatom such as oxygen, nitrogen or sulphur which provided electron rich surrounding for cation which fitted in cavity of the molecule forming stable complexes. In comparison, cryptand provided a three dimensional cavity for metal cation entrapment. These compounds were used extensively in ion chromatographic separations [161-194]. The selectivity for cation was decided by the ability of the cation to fit in to cavity of macrocycle. Since these macrocycles were neutral molecules, anion was coupled to form electrically neutral ion pair complex in low dielectric media. Such macrocycles were hydrophobic hence were useful to extract anion from hydrophillic aqueous solutions. It was thus possible to isolate cations with common anions or anions associated with common cations.

6.6.1 Advantages of Ion chromatography

Cram [165-166] first used these macrocyclic compounds for the separation of amino acids. Usually macrocycles were incorporated into the stationary phase. It had several advantages. Since these macrocycles had no charge cations were eluted from the column without the use of ionic eluant providing low back ground conductivity and high sensitivity. One could have different selectivity with these macrocycles as the stationary phase was changed from one macrocycle to all together other macrocycle to influence the selectivity. Large number of crown ether were used as the stationary phase to improve the selectivity. Since cation attraction was in narrow range, group I and II metals were eluted isocratically with conductivity detector. Another advantage of macrocycle was it would separate cations and anions, because cations were complexed and remained on the stationary phase. On the contrary the anions were retained along with crown compound. The small cations were easily plugged in to hole of crown ether while large cations formed sandwich type complex between two rings of crown compounds. Usually better retention was demonstrated by the cation which had ionic radius similar to cavity of crown ether (preferred ratio was 0.80-1.0 for cation: cavity). In general in macrocycle based chromatography, cations were eluted in order of their increasing binding constants. In such phenomena the mechanism of retention was based on ligand exchange instead of ion exchange [166]. The binding constants were evaluated [167-168]. Some "reverse" observation had been noted, viz. 12C4 showed great affinity for sodium than potassium but in usual ion exchange potassium had greater affinity [169]. Further when no one cation showed an ideal fit ligand cavity then several other factors predominated to determine the selectivity. If ion was very large which could not fit in cavity it formed 2:1 complex, e.g. caesium which was more stable [170] formed sandwich complex.

6.6.2 Separations with Cryptands and of Anions

Very limited experiments were carried out in ion chromatographic separations using cryptand as the stationary phase. As a rule cryptands encapsulated cation better than crown ethers. The binding constant were large than those for crown ethers. Cryptand 222 was better than 18C6 for separation of alkali metals with highest binding constant for potassium [171].

It was not necessary to add salt to mobile phase during elution if macrocycle was used as the stationary phase. Pure water served as the best eluant. The retention of cation diminished on addition of diluent to the mobile phase in few systems [172]. The complexation reaction was exothermic, hence complexes had better stability at lower temperature. At high temperature retention showed decrease due to weaking of cation macrocycle interaction [173]. In nonpolar solvents the positively charged crown complex remained with anions to maintain electrical neutrality. This had facilitated separation of anion in the presence of common cations. In such case retention was not dependent on type of crown ether used.

By ion exchange process anions were separated with macrocyclic compounds. With hydroxide of cation in mobile phase separation was possible. This was because mobile phase cation formed complex with macrocycle in stationary phase, resulting in anion exchange site with eluant ratio in the mobile phase [174, 175]. Thus macrocylic compounds were introduced in ion chromatography as stationary phase or even as mobile phase. Crown ethers were impregnated in stationary phase by physical trapping on solid support, or polymerisation or covalent bonding to silica or polymer support. Much of the work had been described in earlier sections. Crown ether coating on silica based C_{18} or styrene-DVB gave best results [176]. The preparation of column was simple. A water methanol solution containing adequate concentration of crown or cryptand compound was passed through column for several times when capacity depended upon ligand concentration. Only with time ligand was lost from column due to washing with eluant when eluant other than water was used [177]. The polymeric macrocycle as stationary phase was used by several workers [178-179]. Such phases were better as they were stable towards different solvents and such phases were obtained by coating macrocycle polymer on the solid support poly (DB18C6) was best polymeric resin for coating (IX). It was used for the separation of hydrochloric acid and sodium chloride or hydroxides of sodium and potassium. Such resins were then coated on solid support [180].

(IX)

Finally monomeric crown ethers were covalently bound to silica or polymeric packings [181-182]. Such phase showed high efficiency and easy to synthesise then above kinds of phases. Thus most of resins were based upon benzo derivative of crown ethers. Benzo crowns had bound cation weakly in comparison to that the binding by analogue. The cation complexation was best in aqueous solution. The oxygen atom complexes s-block metal, nitrogen atom in macrocycle was best to complex silver mercury while sulphur containing crown compounds complexed gold and palladium [183-187]. The diaza crown ethers were also used as stationary phase [188] for separation of the transition metals. Covalently bound macrocycles attached to silica yielded better results than polymeric bound resins [189-191].

6.6.3 Applications of Ion Chromatography for Metal Separations

So far this technique was extended for the separation of mainly alkali and alkaline earths, but most significant use of ion chromatography was for separation of anions [192]. With water as the mobile phase B15C5 was first tried but polycrown ether like poly (DB18C6) found extensive applications for anion separations. Crown compounds coated with piezoelectric crystal on DVB resin were also used to separate various anions.

The alkali metals were separated on polyamide resin coated with crown ethers with water as the eluant [196]. Carboxylated crown ethers also provided useful separations. 12C4 on polymer based silica gel with methanol as mobile phase was used for the separation of sodium [198]. Cryptand 222 on poly (styrene divinyl benzene) column at pH 2.0 was used for the separation of halide anions [201]. Various ion chromatographic methods for the separation of anions and s-block metals are summarised in Table 6.6.1.

6.7 Capillary Electrophoresis with Crown Compounds

It was also called capillary zone electrophoresis. This technique could separate small amounts of substance in very short time with good resolution. This method permitted analysis to nanogram (10^{-9}) concentration with detector sensitivity of 10^{-18} attovel. In this process charged species were separated based on the difference in migration rate under the influence of the electrical potential. The method resembled HPLC, and these methods were not a real chromatographic method as it did not exploit the difference in the migration rate of the constituent on stationary support.

The separation medium was fused silica capillary tube (10 μm internal diameter and 1m-length) containing an appropriate electrolyte, with one of capillary diped in better solution. Direct current was employed with average current of 250 μa with 1000-30, 000 volts potential as was done in the conventional electrostatic precipitator. The peak volume used was 1μl. With laser beam d-electron could be excited upto 10^{-21} zeptomoles. The phenomena of electroosmosis operated, silica capillary had silicates as ionisable group (silanol). At pH 2.0 $SiOH^{3+}$ generated negative charge zeta potential, leading to accumulation of cations at cathode leaving behind neutral molecules, while anions were detected last. In 16 mts, 18 amino acids were separated. This technique was invariably used for the separation of amino acids proteins peptides; nucleic acids and biopolymers. Inorganic cations like alkali metals were also separated in 10 minutes as many as at 10^{-21} gram concentration.

6.7.1 Use of Crown Ethers in Capillary Electrophoresis

In isotachophoretic determinations crown ethers were used [202]. For such purpose it was necessary to adjust effective mobilities of analyt ions which had similar ionic mobilities, e.g. s-block metals were separated with 18C6 in leading electrolyte the complexing agent for these ions [203-209]. Thallium, lead, silver formed inclusion complexes with crown ethers in chromatographic column with favourable formation constant [210] and that their effective mobilities decreased [211]. It was not possible to have separation of ions as ammonium and thallium due to

Table 6.6.1 Ion chromatographic separation with crown compounds

Elements/anion	Separation Conditions
Anions	Crown ethers and cryptands were coated with silica based resins with water as mobile phase to get separation of several anions from river water [191]. B15C8 was used for separation of salt [192]. Extensive applications of macrocyclic ligand in ion chromatography technique was feasible. Poly (DB18C6) on silica gel coated on ion exchange resin was used [172, 193] to separate variety of anions like chloride, nitrate, bicarbonate, sulphate, thiocyanate etc. with water as eluant. From rain water [194] crown ethers or cryptands coated with piezoelectric crystal on DVB resin was used to separate anions the water was used to separate anions while methanol was used to separate cations [195].
Alkali metals	The conductivity detector was used. Polyamide resin packed with crown ethers were used to separate with water as an eluant then it was noted that the presence of anionic species influenced their separations [196]. 0.01M carboxylate crown ethers in chloroform was used for complexation of alkali metals with 0.01M hydrochloric acid as an eluant [197]. Silica gel modified by crown ethers like poly and bis (B18C6) or polyform was used with 90% methanol as an eluant at 40°C. An increase in concentration of the methanol influenced the selectivity of separation. Alkaline earths could be similarly separated [198].
Na (I)	With 15C4 on polymer based silica gel with methanol aqueous solution as the mobile phase was separated. Spherical resin with propylene coated hydroxysityl groups gave better separation [199]. Elution order was increasing with increasing atomic number [199].
NH_4^+	Tetradecyl 18C6 on Dionex MPIC column separated it with 0.25M methane sulphonic acid on the mobile phase by technique of gradient elution [200]. Inorganic anions were also separated with cryptand-221 on polystyrene divinyl benzene resin at pH 2.0. It had protonated and acted like anion exchanger. Water was used as mobile phase for halide ions beyond pH > 10.0 after deprotonation resin could be used as cation exchanger for the separation of alkali metals with methanol as the mobile phase [201].

similar ionic mobilities. The 18C6, 15C5 complexes created difference in ionic mobilities on complexation with these metals. One generally evaluated potential unit (PU) [212] for identification of the analytical ions. The stabilities of inclusion complexes of analyt ion with crown ether depended upon the kind and location of heteroatom in crown structure, number of such heteroatoms, the interrelationship between diameter of cation and the cavity of crown, charge of cation, flexibility of crown ring and the intensity of interaction between the heteroatom and the solvent [213] employed during separation.

6.7.2 Calixarenes in Capillary Electrophoresis

The selectivity in capillary electrophoresis was easily modified by the use of calixarene [214]. The use of the calixarene as additive was interesting in capillary electrophoretic separations because its addition changed the effective mobilities of the solutes reflecting changes in selectivity of the separation. Crown ethers were earlier used for such purpose [215]; due to availability of cavity for host guest interactions with the solutes. Since calixarene were macrocyclic molecules made up of units metal linked by methylene bridge and were stable with substituent p-methyl, p-tert butyl or p-phenyl on upper rim rendering them insoluble in water. The substitution of upper rim with sulphonic group in calix (6) arene made them water soluble and selective complexing ligand [216] (X). This facilitated complexation of several ions [217-220]. The separation of the organic molecules was also facilitated by its use, e.g. benzdiols and toluidines. This led to full scale separation of three of the isomers. Usually elongated solutes could easily penetrate the calixarene cavity [211]. Thus is short calixarene could be used as additive in capillary electrophoresis to adjust selectivities. The separation of chlorophenols, benzonediols and toluidines was fissible by addition of 3.4 mM of p-sulphonic calix (6) arene with buffers of pH 7.0.

6.7.3 Misceller Electro Kinetic Capillary Chromatography (MECC)

The neutral molecules migrated electro-osmotically. The addition of detergent led to formation of miscelles in mobile phase. This tended to separate neutral molecule. The detergent used was sodium dodecylsulphate (SDS). The micelles absorbed nonpolar compounds in to the hydrocarbon interior of the particle, i.e. solubilising nonpolar species. The separation of phenols was done by this technique [221-222] with SDS misceller solution in borate phosphate buffer. The separation in MECC was due to liquid-liquid partition with miscelles as the pseudostationary phase. This was environmentally important [223, 224] contribution.

$$\left[\begin{array}{c} SO_3 \\ \bigcirc \\ OH \end{array} CH_3 \right]_6$$

(X)

6.7.4 Applications of Capillary Electrophoresis with Crown Compounds

Excellent separation of ions with similar mobilities such as potassium and thallium, ammonia and silver lead and thallium was achieved by capillary electrophoresis [202]. Alkali and alkaline earths were separated on silica capillary with 18C6 at pH 4.8 [225]. Although in comparison to crown ether and calixarene macrobicyclic compound like cryptand-222 was not that extensively utilised in such separation there was reference for its use for the separation of several anions by varying pH and electro-osomotic flow of the solution [226]. Cryptand-222 was utilised as the bifunctional modifier at neutral pH for the separation of ammonium, sodium and potassium. Several anions were separated by protonated cryptand. Such anions included carboxylate anions and the oxyanions [229]. The phenols were separated with p-sulphonic calix(4) arene on fused silica capillary with phosphate buffer at pH 5.0 [230].

The various methods for the separation of inorganic anions cations as well as organic molecules like nitrocompounds phenols, hydrocarbons with capillary electrophoresis are listed in Table 6.7.1.

Table 6.7.1 Separation by capillary electrophoresis with crown compounds cryptands

Element/Compounds	Separation Conditions
TI (I)	With 15C5 and 18C6 on PTFE column electrode containing Triton X-100 and crown ether with 0.1M histadine at pH 4.0 was separated by isotachophoretic determination. It facilitated separation of the potassium and thallium; silver and ammonia or lead and thallium [202].
Transition metals	With 18C6 on fused silica capillary with 0.5M 18C6 and 5M lactic acid buffered at pH 4.8 several metals like alkali alkaline earths and transition elements were separated [225].
Anions	With cryptand-222 and 18C6 on fused silica, capillary, several anions were separated such separation depended on pH and electro-osmotic flow [226]. Separation of benzoderivatives in the presence of tetraethylammonium chloride and triethylamine was also effected. The separation of dinitrobenzoylpolyoxyethylenedodecylether was also separated [227]. With cryptand-222 used as the bifunctional modifier at pH 7.0 were separated by capillary electrophoresis. Also several anions were separated [228] with protonated cryptand-222 with acetic acid as the eluant for carboxylate anions and oxyanions.
Organic compounds	p-sulphonic calix (6) arene was used at certain pH [214, 229]. p-sulphonic calilx(6) arene on fused silica capillary with phosphate buffer at pH~5.0 separated and determined in UV at 275 nm (23). p-carboxyethyl calix (7) arene buffered solution was separated where cavity size and geometry of molecule influenced complexation [231]. p-sulphonated calix(4) arene (n = 4, 6, 8) and 4-hydroxybenzene sulphonate were separated on fused capillary at room temperature, with borate buffer at pH 8.3 and determined in UV region. The addition of magnesium chloride reduced migration time [232].

Conclusion

We can summarise that amongst chromatographic techniques crown compounds specially polymeric crown ether have been extensively utilised in liquid column chromatography. Same methodology was extended to the technique of high performance liquid chromatography. With silica support monomeric crown compounds were used as the stationary phases. Crown compounds were also used as mobile phase in columnar work.

In gas chromatography involving several separations, crown compounds were impregnated on the stationary support like chromsorb. Large size crown like 30C10 or 24C8 were used on capillary columns. Several organic compounds were separated by the gas chromatographic methods.

The extraction chromatography had many practical applications. Monomeric crown ether were used for the purpose. Many difficult chemically similar elements were separated by this technique.

Ion exchange was used recently. The technique mainly exploited differences in solubilities of complexes. However most spectacular applications were noted in field of ion chromatography. Many separations were carried out including those of amino acids. Cryptands also furnished good results. Best contribution of this method was for the separation of anions which was not possible by any other methods.

Finally in capillary electrophoresis crown ethers and calixarene were used mainly as the modifiers to change ionic mobilities of the cations to promote separation. Misceller elctrokinetic capillary chromatography provides a scope for metal separations with use of sodium dodecylsulphate (SDS) for micelle formation. It is expected that this technique would demonstrate astonishing progress in this area of analytical chemistry in future years.

References

1. E. Blasius, W. Adrian, K.P. Janzen, G. Klautke, J. Chromatography **96**, 89 (1974).
2. E. Blasius, K.P. Janzen, Chemi. Ing. Tech. **47**, 594 (1975).
3. E. Blasius, P.G. Maurer, J. Chromatography **125**, 511 (1976).
4. E. Blasius, K.P. Janzen, W. Neumann, Mikrochim Acta II **279**, (1977)
5. E. Blasius, K.P. Janzen, W. Adrian, G. Klautke, R. Lorscheider, P.G. Maurer, V.B. Nguyen, T. Nguyen-Tien, G. Scholten J. Stockemer, Z. Anal. Chem. **284**, 337 (1977).
6. E. Blasius, K.P. Janzen, H. Luxenburger, V.B. Nguyen, H. Klotz, J. Stockemer, J. Chromatography **167**, 307 (1980).
7. E. Blasius, K.P. Janzen, M. Keller, H. Lander, T. Nguyen-Tien G. Scholten, Talanta **27**, 107 (1980).
8. E. Blasius, K.P Janzen, W. Klein, H. Klotz, V.B. Nguyen, T. Nguyen-Tien, R.P. Peifter, G. Scholten, H. Simon, H. Stockemer, A. Toussaint, J. Chormatography **201**, 147 (1980).
9. E. Blasius, H. Lander, Z. Anal. Chem. **303**, 177 212 (1980)
10. E. Blasius, M. Keller, Z. Anal. Chem. **318**, 390 (1984).
11. K.G. Heumann, H.P. Schiefer, Angew Chem. **318**, 390 (1984).
12. S. Fujine, K. Saito, K. Shiba, J. Nucl. Sci. Tech. **20**, 439 (1983).
13. W.M. Feigenbaum, R.H. Michel, J. Poly. Sci. A-1 **9**, 817 (1971).
14. M. Igewa, M. Tanaka, Y. Abe, M. Yamaguchi, T. Yamobe, Bull. Soc. Seawater Sci. Japan **33**, 331 (1980).
15. M. Igawa, I. Ito, M. Tanka, Bunseki Kagaku **29**, 580 (1980).
16. G.D. Y. Sogan, D.J. Cram, J. Am. Chem. Soc. **98**, 3038 (1976).
17. T. Nakagawa, H. Mizunuma, A. Shibukawa, T. Ueno, J. Chromatograpy **211**, (1981).
18. E. Blasius, K.P. Janzen, Topics in Current Chemistry **98**, 163 (1981).
19. E. Blasius, P.G. Maurer, Makromol. Chem. **178**, 649 (1977).
20. J.W. Copius-Peereboom, H.W. Beekes, J. Chromatography **14**, 417 (1964).
21. K. Kimura, T. Shono, J. Liq. Chromatogr. **5**, 223 (1982).
22. O. Tetsuo, U. Toshinori, J. Chem. Soc. Faraday Trans. **92**, 4977 (1996).
23. M.G. Hankins, T. Hayashita, S.P. Kasprzyk, R.A. Bartsch, Anal. Chem. **68**, 2811 (1996).
24. S.J. Carter, L.S. Stuhl, J. Chromatography **291**, 348 (1984).

25. T. Shinbo, T. Yamaguchi, K. Nishimura, M. Sugiura, J. of Chromatography **405**, 145 (1987).
26. I.A. Vakhitina, N.P. Samoilova, T.I. Kostenko, O.G. Tarakonov, V.S. Lebedev, Zh. Prikl. Khim **58**, 713 (1985) AA **47**, 12C63 (1985).
27. J.F. Biernat, T.Z. Wilczewski, Pol, J. Chem. **53**, 513 (1979).
28. J. Kostrowicki, E. Luboch, B. Makuch, A. Crygan, J.F. Biernat, A. Horbaczewski, J. Chromatography **454**, 340 (1985).
29. E. Blasius, W. Klein, Z. Anal Chem **322**, 348 (1985).
30. C.M. Wai, H.S. Du, Anal. Chem. **62**, 2412 (1990).
31. L. Trezl, P. Bako, L. Fenichel, I. Ruszhak, J. Chromatography **269**, 40 (1983).
32. T. Czerhati, M. Szogyi, L. Gyorfi, Chromatographia **20**, 253 (1985).
33. P. Kus, J. Chromatography **436**, 338 (1988).
34. P. Kus, J. Plannar chromatography **8**, 472 (1995).
35. P. Kus, J. Plannar Chrom. **9**, 293 (1996).
36. T. Chaty, J.R. Dahl, Nucl. Med. Biol. **16**, 385 (1989) AA **53**, 79132 (1991).
37. K. Kimura, M. Nakajima. T. Shono, Anal. Letters **A13**, 741 (1980).
38. M. Nakajima, K. Kimura, T. Shono, Anal. Chem. **55**, 463 (1983).
39. M. Nakajima, K. Kimura, T. Shono, Bull Chem. Soc. Japan **56**, 3052 (1983).
40. M. Nakajima, K. Kimura, E. Hayata, T. Shono, J. Liquid Chromatography **7**, 2115 (1984).
41. K. Kimura, M. Nakajima, T. Shono, J. Poly Sci. **23**, 2327 (1987).
42. K. Kimura, H. Harino, M. Nakajima, T. Shono, Chem. Letter 747 (1985).
43. E. Blasius, K.P. Janzen, J. Zender, Z. Anal. Chem. **325**, 126 (1986).
44. B.R. Willeford, H. Veening, J. Chromatography **251**, 1 (1982).
45. A. Mangia, G. Parolari, E. Gaetani, C.F. Laureri, Anal. Chim. Acta. **92**, 111 (1977).
46. S.J. Carter, L.S. Stuhl, J. Chromatography **291**, 348 (1994).
47. B.R. Willeford, H. Veening, Chromatography Rev. **26**, 61 (1982), AA **43**, 3C5 (1982).
48. Y. Jin: R. Fu, Hauxue Shiji **11**, 339 (1989), AA **53**, 4B2 (1991).
49. T. Tamaguchi, Kagaku to Kogyo **42**, 255 (1989), AA **52**, 7C39 (1990).
50. L.A. Kartsova, B.V. Stolyarov, Zh. Anal. Khim. **48**, 582 (1993), AA **56**, 1B41 (1994).
51. K. Kimura, E. Hayata, T. Shono, J. Chem. Soc. Chem. Commun. 271 (1984).
52. A. Kaufmann, Chromatographia **29**, 76 (1990).
53. I.N. Papadoyannis, G.K. Sarns, Anal. Letters **21**, 21 (1988).
54. M. Nakajima, K. Kimura, T. Shono, Anal. Chem. **55**, 463 (1983).
55. P. Zhu, D. Wang, J. Jin, S. Cao, Chromatographia **25**, 419 (1988).
56. A. Mangia, G. Parolari, E. Gaetani, C.F. Laureri, Anal. Chim. Acta **92**, 111 (1977).
57. K. Kimura, H. Harino, M. Nakajima, T. Shono, Chem. Lett. 747 (1985) AA. **47**, 12J31 (1985).
58. T. Nakagawa, A. Shibukawa, A. Kaihara, H. Tanaka, Nippon Kagaku Kaishi 1001 (1986) CA. **107**, 7541p (1987).
59. T. Shinba T. Yamaguchi, K. Tachibana, Eur. Patent Ref. C.A. **113**, 78433p (1990).
60. K. Kimura, M. Tanaka, T. Shono, J. Chromatography. **522**, 107 (1990).
61. N.X. Ye,: W. Xu,: S.L. Da,: Z.H. Wang, Fenxi Huaxue **25**, 857 (1997) AA **60**, 2D155 (1988).
62. H.L. Jin, A.M. Stalcup, D.W. Armstrong, J. Liquid Chromatography **13** 473 (1990).
63. T. Okada, Anal chem **66** 2163 (1994).

64. D. Shila, W. Zhao, L. Liu, P. Sheng, C. Wu, Wuhan Daxue Xuebao, Ziran Kixueban 81 (1986) CA **108**, 31022 m (1988).
65. Y. Chung, D. W. Kim, K. Lee, C.S. Kim, Y.L. Cha, K.Y. Hong, Mikro Chem. J. **53**, 454 (1996).
66. Y. Li, B. Chen, J. Yang, O. Peng, Chem. Huaxue Xuebao **42**, 701 (1984) AA **47**, 4C43 (1985).
67. J. Jurczak, R. Ostaszewski, J. Coordi. Chem. **27**, 201 (1992).
68. S.M. Moerlein, J.W. Brodack, B.A. Siegel, M.J. Welch, Appl. Radiat. Isot. **40**, 741 (1989) AA **52** 8E35 (1990).
69. J.D. Glennon, K.O. Connor, S. Srijaranani, K. Manley, S.J. Harns, M.A. Mckervey, Anal. Letters **26**, 153 (1993).
70. J.D. Glennon, E. Horne: K. Hall, D. Cocker, A Kuhn S.J. Harris, M.A. Mackervey, J. Chromatography A. **731**, 47 (1996).
71. Y.K. Lee, Y.K. Ryu, J.W. Ryu, B.E. Kim, J.H. Park, Chromatographia **46**, 507 (1997).
72. F.J. Ludwig, A.G. Bailie. Jr. Anal. Chem. **56**, 20810 (1984).
73. G. Man, T. Lippmann, K. Muller, H.V. Tittlebach, Chromatographia **34**, 453 (1992).
74. T. Lippmann, G. Mann, GIT, Fachz Lab **39**, 203 (1995) AA **57**, 10D213 (1995).
75. F. Vocanson, R. Lamartine, Chromatographia **41**, 204 (1995).
76. Y. Saito, H. Ohta, H. Tarasaki, Y. Katoh, H. Nagashima, K. Jinno, K. Itoh, R.D. Trengove, J. Harrowfield, S.F. Y. Li, J. High Resol. Chromatography. **19**, 475 (1996) AA **85**, 12D220 (1996).
77. O.I. Kalchendo, J. Lipkowski, R. Nowakowski, V.I. Kalchenko, M.A. Visotsky, L.N. Markovsky, J. Chromatography Sci. **35**, 49 (1997).
78. B.F. Graham, J.M. Harrowfield, R.D. Trengove, I. Rodrigues, S.F.Y. Li, J. Chromatography Sci. **35**, 232 (1997).
79. K. Jinno, K. Tanabe, Y. Saito, H. Nagashima, R.D. Trengove, Anal. Comm. **34**, 175 (1997).
80. L. Rodriguez, S.F. Y. Li, B.F. Graham, R.D. Trengove, J. Liquid. Chrom. **20**, 1129 (1997).
81. R. Li, Wuhan Daxue Xuehao Ziran Kexueban **4**, 121 (1985).
82. E.V. Zagorevskaya, N.Y. Kovaleva, J. Chromatography 365 7 (1986).
83. D.D. Fine, H.L. Gearhart, H.A. Mottola, Talanta **32**, 751 (1985).
84. Y. Jin, R. Fu, Z. Huang, J. Chromatography **469**, 153 (1989).
85. C.A. Rouse, A.G. Finlinson, B.J. Tarbet, J.C. Pixon N.M. Djordjevic K.G. Markides, M.L. Lee, Anal. Chem. **60**, 901 (1988).
86. N.R. Ayyangar, A.S. Tambe, S.S. Biswas, J. Chromatography **543**, 179 (1991).
87. M.N. Gandhi, S.M. Khopkar, Crown ethers and cryptands in solvent extraction (unpublished work)
88. L.R. Souso, D.H. Hoffman, I. Kaplan, D.J. Cram, J. Am. Chem. Soc. **96**, 7100 (1974).
89. T. Nakagawa, H. Murata, A. Shibukawa, K. Murakami H. Tanaka, J. Chromatography **330**, 43 (1985).
90. A. Shibukawa T. Nakagawa, A. Kaihara, K. Yogi, H. Tanaka, Anal. Chem, **59** 2496 (1982).
91. R.V. Vigalok, L.F. Bubachinkova, Usp. Gaz. Khrumatogr (Kazon) **6**, 191 (1981).
92. A. Ono, Analyst **108**, 1265 (1983).
93. R. Li, Sepu- **4**, 304c (1986).
94. C.J. Pederson-J. Am. Chem. Soc. **89** 2405 (1967).
95. W.O. McReynolds, J. Chrom. Sci. **80**, 685 (1970).

96. X.C. Zhou, C.Y. Wu, X.R. Lu, Y.Y. Chen, J. Chromatography. A. **662**, 203 (1994).
97. P. Jing, R.N. Fu, Fenxi Huaxue **23**, 104 (1995) AA **57**, 7B7 (1995).
98. R. Li, Sepu. **4**, 304 (1986) CA **106**, 112869d (1987).
99. S.M. Khopkar, Basic Concepts of Analytical Chemistry 2nd Edition New Age International Publisher Ltd. (1999).
100. J. Shi, G. Lu, W. Guo, D. Wang, H. Hu, Gaodeng Xuexiao Huaxue Xuebao **10**, 647 (1989) CA 112 191083t (1990).
101. A. Zhang, N. Ge, Z. Guan, J. Deng, H. Liu, J. Zhu, R. Fu, Z. Huang, B. Zhang, J. Chromatography **521**, 128 (1990).
102. C. Wu, C. Wang, J. Cheng, X. Lu, Sepu **8**, 355 (1990) CA **114**, 135327c (1991).
103. C. Wu, H. Li, Y. Chen, X. Lu, J. Chromatography. **504**, 279 (1990).
104. T.G. Mayasoedova, G.V. Dmitrieva, Khim. Vys Energy 25 244 (1991) CA **115**, 123587x (1991).
105. Y. Inoue, G.W. Gokel, Cation Binding by macrocycles. Marcel Dekker Publisher (1990).
106. T.S. Fedorova, A.A. Karnishin, T.I. Kostenko, Zh. Anal. Khim. **39**, 2222 (1984) AA **47**, 12C65 (1985).
107. Y. Yamini, M. Shamsipur, Talata, **43**, 2117 (1996).
108. Y. Liu, D. Qiao, M. Li, Sepu. **10**, 33 (1992) AA **54**, 10B28 (1992).
109. O. Kamato, K. Takahashi, T. Doi, J. Chromatography A. **675**, 244 (1994).
110. L.S. Cai, C.Y. Wu, H.M. Han, J. Microcolumn Sep. **7**, 137 (1995) AA **57**, 7B17 (1995).
111. S.C. Lin, W.J. Maddox, J. Liquid Chromatography **18**, 1947 (1995).
112. C. Wu, C. Wang, J. Cheng, X. Lu, Sepu. **8**, 355 (1990) AA **54**, 1B17 (1992).
113. R.A. Thompson, Z. Ge, N. Grinberg, D. Ellison, P. Tway, Anal Chem. **67**, 1580 (1995).
114. H. Yan, X.C. Zhou, Y.Y. Chen, C.Y. Wu, X.R. Lu, J. Chromatography A **753**, 269 (1996).
115. X.C. Zhou, C.Y. Wu, H. Yan, Y.Y. Chen, J. High. Resol. Chromatography **19**, 643 (1996) AA **59**, 4B16 (1997).
116. C.M. Wang, C.Y. Wu, Huaxue **25**, 25 (1997) AA **57**, 847 (1997).
117. Z.L. Zhang, C.P. Tang, C.Y. Wu, Y.Y. Chen, J. Chem. Soc. Chem. Commun. 1737 (1995).
118. P. Mnuk, L. Felt, V. Schurig, J. Chromatography **732**, 63 (1996).
119. W.Y. Zhang, C.Y. Wu, J.L. Wang, S.W. Zhong, Sepu. **15**, 204 (1997) AA **60**, 1D121 (1998).
120. S. Zhang, W. Zhang, C. Wu, Z. Zhang, Y. Chen, Fenxi Kexue Xuebao **14**, 14 (1998) CA **128**, 200300w (1998).
121. Y.X. Wen, D.Q. Xio, Y. Ling, R.N. Fu, J.L. Gu, R.J. Dou, A.Q. Luo, Chromatographia **46**, 177 (1997).
122. P. Narayan, Extraction Chromatographic separation of some elements with high molecular amines from malonate solution Ph. D. Thesis I.I.T. Bombay (1985).
123. R.B. Heddur, Extraction chromatographic separation of main group elements with Trioctyl phosphine oxide as an extractant Ph. D. Thesis I.I.T. Bombay (1984).
124. Chhaya Dixit, Reversed phase extraction chromatographic separation of elments with tributyl phosphate from mixtures and environmental samples Ph. D. Thesis I.I.T. Bombay (1984).
125. R.B. King P.R. Heckley, J. Am. Chem. Soc. **96**, 3118 (1974).
126. W. Smulek, W.A. Lada, Radiochem Radioanaly Letter **30**, 199 (1977).
127. W.A. Lada, W. Smulek, Radiochem. Radioanaly Letters **34**, 41 (1978).

128. W. Smulek. W.A. Lada, J. Radio andl. Chem. **50**, 169 (1979).
129. T. Kimura T. Tshimori T. Hamada, Anal. Chem. **54**, 1129 (1982).
130. T. Araki, Y. Kubo, T. Toda, M. Takata, T. Yamashita, W. Mureyama, Y. Nunogaki, Analyst **110**, 913 (1985).
131. K. Kimura, H. Harino, E. Hayata, T. Shono, Anal. Chem. **58**, 2233 (1986).
132. A. Yu. Tsiradze, S.V. Demin, A. V. Levkin, V.I. Zhilov, S.F. Nikol'skii D.A. Knyazev, Zh. Neorg. Khim. **35**, 2158 (1990) AA **54**, 1D29 (1992).
133. A.V. Levkin, V.V. Basmanov, S.V. Demin, V.N. Mikhailov, Zh. Fiz. Khim. **65**, 1987 (1991) CA **115**, 197003a (1991).
134. E. Peimli, J. Radioanal. Nucl. Chem. **144**, 9 (1990).
135. W.A. Lada, W. Smulek, Radiochem. Radioanal. Lett **34**, 41 (1978).
136. A.V. Lerkin, V.V. Basmanov, S.V. Demin, A. Yu. Tsivadze, Zh. Fiz. Khim. **64**, 1376 (1990) AA **53**, 4D38 (1991).
137. R. Suess, H. Bruchertseifer, S. Fischer, K. Walter, N. Trautmann, ZFl-Mitt **63**, (1991) AA **55**, 6D53 (1993).
138. B.S. Mohite, C.D. Jadage, S.R. Pratap, Analyst. **115**, 1367 (1990).
139. B.S. Mohite, D.N. Zambare, B.E. Mahadik, Analyst. **119**, 2033 (1994).
140. W. Smulek, W.A. Lada, J. Radioanal. & nucl Chem. **50**, 169 (1979).
141. Yi. Yu. Vin, S.M. Khopkar, Indian J. Chemistry, **27**A, 458 (1988).
142. S. Iwagami, T. Nakagawa, J. Chromatography **369**, 49 (1986).
143. S. Aoki, M. Shiga, M. Tazaki, H. Nakamura, M. Takagi, K. Ueno, Chem. Letter 1583 (1981).
144. B.S. Mohite, Ph. D. thesis Solvent extraction of alkali & alkaline earth elements with crown ether, I.I.T. Bombay (1986).
145. E.V. Zagorevskaya, N.V. Korateva, J. Chromatography **365**, 7 (1986).
146. S.M. Khopkar, M.N. Gandhi, J. Sci. & Ind Re. **53**, 630 (1994).
147. C.Y. Wu, C.M. Wang, Z.R. Zeng, X.R. Lu, Anal. Chem. **62**, 968 (1989).
148. B.S. Mohite, Anal. Chem. **66**, 4097 (1994).
149. S.A. Katalinkov, I.A. Mushletsov, Trud Inst. Misk Khim Tekhol Inst. im D.E. Mendleeva 3 (1989) A.A **53**, 8D4 (1991).
150. T. Okada, T. Usui, Anal. Chem. **66**, 1654 (1994).
151. K.K. Luk, M.L. Miles, L.H. Bowen, J. Radioanal Nucl. Chem. **82**, 55 (1984).
152. T. Okada, T. Usui, Anal. Chem. **66**, 1654 (1994).
153. T. Hayashita, K. Yamosaki, J.C. White, S.P. Kasprzyk, R.A. Bartsch, Sepn. Sci. and Technol. **31**, 2195 (1996).
154. B.E. Jepson, M.R. Clager, J.L. Green, Pure and Applied Chem. **65**, 489 (1993).
155. B.E. Jepson, N.A. Novotny, W.F. Evans, Sepn. Sci. and Technol. **28**, 507 (1993).
156. M.L. Dietz, R. Chiarizia, E.P. Harwitz, V. Talanov, R.A. Bartsch, Anal. Chem. **69**, 3028 (1997).
157. D.J. Wood, S. Elshani, C.M. Wai, R.A. Bartsch, M. Huntley, S. Hartenstan, Anal. Chim. Acta. **284**, 37 (1993).
158. D. Shen, He Huaxue Yu Fangshe Huaxue **5**, 53 (1983) A.A **46**, 5B117 (1984).
159. D. Shen, S. Liu, T. Le, Fenxi Huaxe **12**, 934 (1984) A.A **47**, 6B47 (1985).
160. R. Pathak, G.N. Rao, Anal. Chim. Acta **335**, 283 (1996).
161. E. Blasius, K.P. Janzen, Top. Current Chem. **98**, 163 (1981).
162. M.N. Gandhi, S.M. Khopkar, Ind. J. Chem. **30**, A 706 (1991).
163. T. Shono-Bunseki Kaga Ku, i.e. Jpn. Soc. Anal. Chem **33**, E 449 (1984).
164. M. Takagi, H. Nakamura, J. Coordin Chem **15**, 53 (1986).
165. G. Dotsevi, Y. Sogah, D.J. Cram, J. Am. Chem. Soc. **92**, 1259 (1975).
166. G. Dotsevi, Y. Sogah, D.J. Cram, J. Am. Chem. Soc **98**, 3038 (1976).
167. R.M. Izatt, J.S. Bradshaw, S.A. Nielsen, J.D. Lamb J.J. Christensen, Chem. Rev. **85**, 271 (1985).

168. A. Bajaj, N. Poonia, Cordi. Chem. Rev. **87**, 55 (1988).
169. M. Nakajima, K. Kimura, E. Hayata, T. Shono, J. of Chromatography **7**, 2115 (1984).
170. M. Nakajima, K. Kimura, T. Shono, Bull. Chem. Soc. Japan **56**, 3052 (1993).
171. K. Kimura, H. Hayata, T. Shono, J. Chem. Soc. Chem. Comm. 271 (1984).
172. J.D. Lamb, R.B. Smith, J. Chromatography **546**, 73 (1991).
173. T. Iwachido, H. Naito, F. Samukawa, K. Ishimaru, Bull. Chem. Soc. Japan **59**, 1475 (1986).
174. J.D. Lamb, P.A. Drake, J. Chromatography **482**, 367 (1989).
175. J.D. Lamb, P.A. Drake, K.E. Wolley, P. Jadik and R.M. Cassidy (Ed), Advance in Ion Chromatography Vol. 2 215 (1990).
176. K. Kimura, H. Harino, E. Hayata, T. Shono, Anal. Chem. **58**, 2233 (1986).
177. R.G. Smith, P.A. Drake, J.D. Lamb, J. Chromatography **546**, 139 (1991).
178. J.P. Joly, B. Gross, Tetrahedron Lett. **26**, 4231 (1989).
179. E. Blasius, K.P. Janzen, Isr. J. Chem. **26**, 25 (1985).
180. M. Igawa, K. Saito, T. Yamabe, M. Tanaka, Anal. Chem. **53**, 1942 (1981).
181. T.G. Waddell, D.E. Leyden, J. Org. Chem. **46**, 2406 (1981).
182. M. Nakajima, K. Kimura, T. Shono, Anal. Chem. **55**, 2406 (1981).
183. J.S Bradshaw, R.L. Breuning, K.E. Krakowiak, J.B. Tarbet, M.L. Breuning, R.M. Izatt, J.J. Christensen, J. Chem. Soc. Chem. Comm. 812 (1988).
184. R.M. Izatt, R.L. Bruening, M.L. Bruening, B.J. Tarbet, K.E. Krakowiak, J.S. Bradshaw, J.J. Christensen, Anal. Chem. **60**, 1825 (1988).
185. J.S. Bradshaw, K.E. Krakowiak, B.J. Tarbet, R.L. Bruening, J.F. Biernat, R.M. Izatt, J.J. Christensen, Pure and applied chem. **61**, 1619 (1989).
186. R.M. Izatt, R.L. Bruening, B.J. Tarbet, D. Griffin, R.L. Bruening, K.E. Krakowiak, J.S. Bradshaw, Pure and applied chem. **62**, 1115 (1990).
187. J.S. Bradshaw, K.E. Krakowiak, R.M. Izatt, M.L. Bruening, B.J. Tarbet, J. Heterocyd. Chem. **28**, B47 (1990).
188. V. Dudler, L.F. Lindoy, D. Sallin, C.W. Schlopfer, Aust. J. Chem. **40**, 1551 (1984).
189. N.Y. Kremliakova, A.P. Novikov, B.F. Mayasoedov, J. Radio and Nucl. Chem. 145 (1990).
190. E. Peimli, J. Radio and nucl. Chem. **144**, 9 (1990).
191. M. Igawa K. Saito, M. Tanaka, T. Yamabe, Bunseki Kageku **32** E 137 (1983), AA **46**, IJ 38 (1984).
192. S. Quici, P.L. Anelli, Chim. Oggi. **7**, 49 (1989) AA **52**, 12A8 (1990).
193. R.G. Vibhute, S.M. Khopkar-J. Radioanal and Nucl. Chem, **152**, 487 (1991).
194. E. Blasius, K.P. Janzen, J. Zendev, Z. Anal. Chem. **325**, 126 (1986).
195. Y.S. Jane J.S. Shin, Analyst **120**, 517 (1995).
196. M. Igawa, I. Ito, M. Tanaka, Bunseki Kagaku **29**, 580 (1980) AA **40**, 5B 23 (1981).
197. R.A. Bartsch, W. Walkowiak, T.W. Robinson, Sepn. Sci. and Tech. **27**, 989 (1992).
198. M. Nakajima, K. Kimura, T. Shono, Bull. Chem Soc. Japan **65**, 3052 (1983).
199. M. Nakajima, K. Kimura, E. Hayata, T. Shono, J. Liquid chromatography **7**, 2115 (1984).
200. B.R. Edward, A.P. Giauque, J.D. Lamb, J. Chromatography **706**, 69 (1995).
201. S.J. Chen, J.S. Shin, J. Chin. Chem. Soc. **38**, 211 (1990) AA **54**, 8B 37 (1992).
202. K. Fukushii, K. Hiro, J. Chromatography **523**, 281 (1990).
203. M. Tazaki, M. Takagi, K. Ueno, Chem. Letter 639 (1982).
204. F.S. Stover, J. Chromatography **298**, 203 (1984).
205. F.S. Stover, J. Chromatography **368**, 476 (1986).

206. K. Fukushi, K. Hiro, Talanta **35**, 55 (1988).
207. K. Fukushi, K. Hiro, Z. Anal. Chem. **332**, 115 (1988).
208. K. Fukushi, K. Hiro, Talanta **35**, 55 (1988)
209. A.A. G. Lemmnes, F.M. Everaerts, J.W. Venema, H.P. Jonkar, J. Chromatography **439**, 3459 (1986).
210. R. Oda, T. Shono, I. Tabushi (Ed.), Crown ether chemistry crown ether no. Kageku Kagaku-Dojin Kyoto (1978).
211. M. Tazaki, T. Hayashita, Y. Fujino, M. Takagi, Bull. Chem. Soc. Japan **59**, 3459 (1986).
212. H. Miyazaki, K. Kato, Tosoku-Denki-Eido Ho Kodanshi Science p. 30 (1980).
213. K. Takemoto, M. Miyata, K. Kimura, Hosetsu. Kagobutsu Tokyo Kagaku Dojin Tokyo p. 18 (1989).
214. I. Shohat, E. Grushk, Anal. Chem. **66**, 747 (1994).
215. R. Stoklin, Erni F. Chromatographia **33**, 32 (1992).
216. S. Shinkai, S. Mori, H. Koreishi, T. Tsubaki, O. Manabe, J. Am. Chem. Soc. **108**, 2409 (1986).
217. S. Shinkai, S. Mori, T. Arimura, O. Manabe, J. Chem. Soc. Chem. Comm. 238 (1987).
218. S. Shinkai, S. Mori, K. Araku, O. Manabe, Bull. Chem. Soc. Japan **60**, 3679 (1987).
219. S. Shinkai, H. Koreishi, K. Uedo, T. Arimura, O. Manabe, J. Chem. Soc. Chem. Comm. 233 (1986).
220. S. Shinkai, H. Koreishi, K. Uedo, T. Arimura, O. Manabe, J. Am. Chem. Soc. **109**, 6371 (1987).
221. K. Otsuka, S. Terabe, T. Ando, J. Chromatography **9**, 348 (1985).
222. S. Terabe, K. Otsuka, K. Ichikawa, A. Tsuchiya, T. Ando, Anal. Chem. **56**, 113 (1984).
223. M.N. Gandhi, S.M. Khopkar, Anal. Chim. Acta. **270**, 87 (1992).
224. S.M. Khopkar, Environmental Pollution Analysis Wiley Eastern Publishers Ltd. (1994).
225. C. Francois, P. Morin, Drenx J. Chromatography **717**, 393 (1995).
226. J.D. Lamb, B.R. Edwards, R.G. Smith, R. Garrick, Talanta **42**, 109 (1995).
227. T. Okada, J. Chromatography A. **695**, 309 (1995).
228. C.S. Chiou, J.S. Shih, Analyst. **121**, 1107 (1996).
229. D. Malkhede, P.M. Dhadke, S.M. Khopkar, J. Radioanal. and Nucl. Chem. **241**, 175 (1999).
230. X.B. Hu, X.R. Lu, T. Zho, J.K. Cheng, Guodency Xuexiao Huaxne Xuebao **18**, 1616 (1997).
231. S. Sun, M.J. Sepaniak, J.S. Wang, C.D. Gutsche, Anal. Chem. **69**, 344 (1997).
232. Y. Zhang, I.M. Warner, J. Chromatography **688**, 293 (1994).

7
Extractive Spectrophotometry

7.1 Introduction

The use of organic reagent, in the inorganic and analytical chemistry was known since ages. With discovery of colour forming reagents also called chromogenic ligands this field took great strides towards identification and determination of elements at low concentrations. As a matter of fact entire qualitative analysis schemes were based on the use of chromogenic ligands for the detection of cations. The large number of elements were identified with plethora of organic ligands. Such ligands consisted of β-diketones, oxines, oximes, naphthols azonaphthols, dithiozone, diethyldithiocarbamates, dithiol which contained a donor atom such oxygen or nitrogen or sulphur. These methods not only provided a tool for mere identification of these metals but also furnished reliable methods for quantitative analysis. Such quantitative analysis was predominated by colorimetry or the photometric methods of analysis with chromogenic ligands.

With the discovery of selective, sensitive and specific reagents, these methods were not only used for quantitative analysis involving spectral methods but also were used for the separation of elements. Such separation techniques largely became popular with discovery of solvent extraction methods. The methods proved boon for many reasons. When metal formed complexes with these chromogenic ligands, they were primarily soluble in the aqueous phase with limitation of their overall stability, but when such species were transferred to hydrophobic phase such as organic solvent, their stability was remarkably enhanced providing better method not only for separation but also for quantitative determination. The major share of success of such methods goes to extractive photometric determination. Several transition and main group elements were analysed by these simple but reliable methods of analysis.

However, such analytical chemistry was restricted to d and p-block elements. While s-block alkali and alkaline earths were totally neglected. So also anions were neglected. Fortunately with the discovery of crown ethers the analytical chemistry of s-block metal received great boost for further activities. Not only novel methods for separation of these metals were developed but even attempts were made to discover newer methods for their simultaneous extraction and direct spectrophotometric determination.

This concept led to discovery of chromoionophores. They were composed of chromophores, i.e. one which gave colour reaction which in turn was measured by spectral methods and another part was ionophore. Now this ionophore could be cationic or anionic however generally one was concerned in cation analysis with latter. Both of these together really constituted chromoionophores. As told these chromoionophores were either neutral, cationic or anionic. From the view point of coordination complexation anionic or what were called monoprotonic ionophores were most useful. This was because the proton was ionisable it gave away and offered place for cationic metal to form complex. Such complexes were soluble in hydrophobic media like organic solvents permitting liquid-liquid partition separation. Chromoionophores were composed of two indispensable component but depending heavily upon each another for complexation. Ionophores recognised ions while chromophore converted chemical reaction to measurable optical signal as in spectrophotometry. The anionic chromophore really did the trick of analysing cationic species.

Now the solvent extraction process proceeded with the formation of uncharged species. The ionophores like crown ethers generally reacted with anionic chromophores to form neutral or uncharged complex species promoting extraction. Due to the presence of chromophoric group extracted species was coloured which in turn was used for the direct photometric analysis.

Such photometric analysis was carried out in one of the following ways:

(a) One could either select a crown ether which itself possessed a powerful chromophoric group. Such compounds included thia crown ethers or aza crown ethers. They not only formed complex and extracted metals but also developed colour in hydrophobic phase which was directly measured. Such informarion is mainly included in this chapter.

(b) The second category of extractions involved the use of common crown ethers however the counter anion for process of ion pair formation was selected in such fashion that it not only extracted metal as neutral species but also provided colour reaction. The cationic crown metal moiety reacted with anionic counter ions like eosine, erythrosine metanill yellow, tropelion-000 to form coloured ion pair complexes which were directly measured spectrophotometrically. Few process involved also synergic extractions.

(c) The third kind of compounds consisted of use of normal crown ethers with use of any counter anion irrespective of wheter they had chromo-ionophoric, group to form ion paired complex followed by its stripping to the aqueous phase to undertake colorimetric estimation by external chromogenic ligand in the aqueous phase. Several of such extractions are covered in previous chapter.

In the course of this discussion most of the methods which involved solvent extraction followed by any nonspectral method of determination have been included in earlier chapter. While those of the extractive procedure which necessarily involved subsequent spectrophotometric analysis either in the organic or the aqueous phase are included in this chapter. Finally those techniques involving extraction followed by measurement of metallic species by technique of molecular luminescene method or atomic emission methods are discussed at length in the next chapter.

7.2 Classification of Chromoionophores

The discovery of crown ethers boosted the use of organic reagents in metal analysis. This was true specially for use of metal selective reagents for s-block elements. Chemist obtained significant information from colour reaction; a change in colour indicated progress of reaction. Chromoionophores were coloured ionophores. A chromophore was introduced at any point in ionophores. Such substitution did not always lead to change in the spectra. However the chromophore should be ideally located in ionophore in such position so as to facilitate the complexation. A chromophore was composed of a ionophore which could detect ions, had a chromophore component which converted interaction to give optical signal. Thus in molecular construction of two parts, viz. ionophore and chromophores could not remain independent of each another.

Thus chromoionophores promoted complexation and interaction leading to spectral variation. This permitted identification of ionic species. It helped liquid-liquid partition facilitating direct spectrophotometry. Anionic ionophores were very useful in solvent extraction. Large number of reviews [1–3] have appeared. These chromophores have been classified in two classes as per their molecular charge. They are:

(1) Neutral uncharged chromophores
(2) Anionic proton dissociable chromophores.

The interaction of metal with anionic chromophore was more favoured as compared to reaction with neutral chromophores. The latter were usually studied in nonaqueous solution favouring solubilization of metal salts. Anionic proton dissociable chromophores were extensively investigated in aqueous organic and liquid-liquid extraction. Since they carried dissociable protons in their uncharged form and they in turn complexed with metal ion in exchange with protons. The undissociated chromophores were lipophillic and soluble in organic solvents. Few anionic chromophore metal complexes dissolved in water also. The electronic spectra of anionic chromophore, e.g. phenolate derivative were not sensitive to the difference between metal species in coordination with shift of about 10 nm-30 nm. This thus did not facilitate spectrophotometric analysis in water inspite of having a spectral shift.

7.3 Neutral Chromoionophores

The ionophore function was effected by crown ether ring and chromophoric function was provided by intramolecular or intermolecular charge transfer absorption [4]. The electronic donor end of chromoionophore was directed to the centre of crown ether cavity so that the donor end was permitted to react directly with incoming metal ion. The light absorption caused electronic charge transfer from the donor end to the acceptor end in the chromophore. The increased energy needed for the electronic excitation, i.e. light absorption led to hypsochromic or blue shift. The donor and acceptor components were situated at two sites and were covalently connected. By binding metal ions ionophores brought chromophore donor acceptor together to develop new absorption band by intermolecular type charge transfer. The absorption band shifting was governed by following factors:

(a) Size fitness between metal ion and crown ether.
(b) Surface charge density of metal cations and
(c) Nature of the solvents in influencing solvatochromism

7.4 Monoprotonic Crown Ether Dyes

A monoprotonic chromophore was introduced into crown ether structure, a nearness to ethereal function led to dissociation of the chromophoric proton by complexation of positively charged metal ions with crown ether macrocycle [5]. The role of metal binding and role of its detection were carried out individually by crown ether and protonic chromophore respectively. Anionic chromophore helped metal binding ability of crown ether and as such these two roles could not be separated apart. The proton dissociable chromophores were useful in extraction photometry. Such group was introduced on the edge of crown ether via a linker arm on which chromophore was attached. The proton dissociable group of chromophore reacted with metal leading to change in absorbance value. If anion generated on deprotonation was singly charged ionophore formed monovalent complex extractable into organic solvent. However if two anionic side arms were added on crown ether there new molecule extracted divalent metal [6, 7]. The crown ether, linkage arm and anionic group, i.e. donor atom were important from point of solvent extraction process. Such chromoionophores are depicted in structure (I) and (II) below:

(I)   (II)

The structure of crown ether, the steric orientation and length of linker arm, nature of side arm chelate ring and anion group were important in extractions. The nature of interaction was included coordination to ion pairing depending on basicity of anionic group and acidity of metal ions.

7.4.1 Metal Extraction Ability and Selectivity

A monoprotonic ligand on dissociation was distributed between aqueous and organic phase when ligand inorganic phase was maximum. A 1:1 complex formed due to special structural form of anionic group. A phenolic group was common, it was hydrophobic and the anionic oxygen was monodentate. Interaction was restricted in molecule. If side arm contained carboxylate which was hydrophillic and bidendate interaction was not restricted to single molecule. There was competition between proton and metal ion for extraction into the organic phase. Hence extraction constant was measure of extractability. If it was too large extraction occurred at lower pH. The extraction constant and ratio of extraction constant of two metals decided separability of two metallic species. The picrylamino-substituted B15C5 extracted metal in 1:2 form but 18C6 type dye extracted metals in 1:1 form [8, 9] (see structure III). The pH control was also of great importance to get reproducible results. Crown ethers obtained from the benzo crown ethers were useful in extractive photometry, but the extraction efficiency was low [10]. This was related to intramolecular ion pair structure in organic phase, the anionic charge was not stabilised in nonpolar solvents. A good extraction was fissible due to the effective delocalization of anionic charge in chromophore. A steric congestion altered hydration around anionic amino nitrogen. This was not true when anion was phenolate [10] as shown in (III) and (IV).

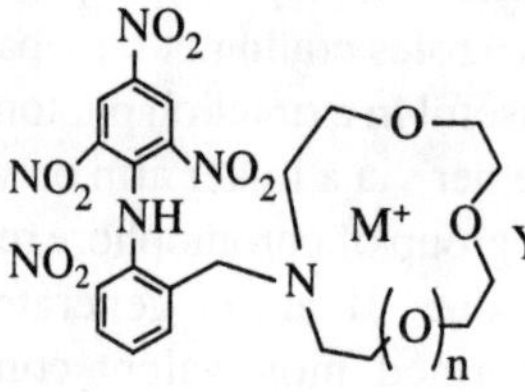

(III) 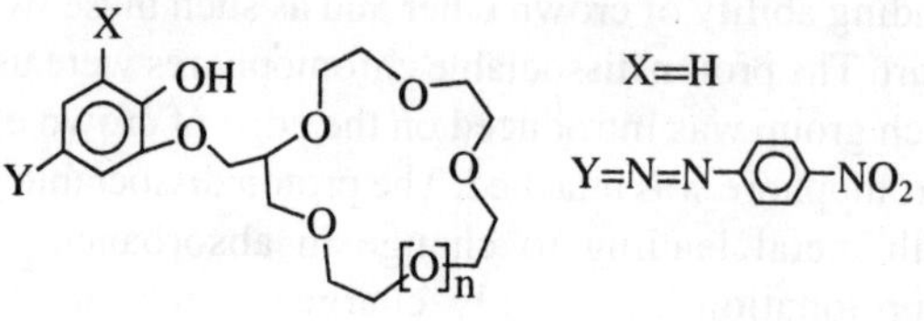 (IV) X=H Y=N=N–⟨O⟩–NO$_2$

7.4.2 Side Armed Monoaza Crown Ethers

To enhance metal extraction efficiency or increase K_{ex} or D one could introduce a chromophoric side arm into crown ether structure so that deprotonated anionic group could interact directly with metal ion which was embedded in cavity of crown ether [11, 12]. These side armed crown ether dyes (III) (IV) were prepared via synthetic reactions starting from glucolglycerol ether. These compounds (IV) were soluble. The high selectivity for extraction of s-block metals was determined by size selectivity, the coordination interaction and ion pair interaction of anionic species with crown ether bound metal ion. A fixed negative charge increased coordination reaction improving extractibility of heavy atoms. The extraction photometry was modified in homogeneous solution by using surfactant miscelles as a pseudophase for extraction was useful for such job. The lengthening of linker arm by two methylene units led to new compound [13] which gave eight membered chelate ring a loose chelate in real sense.

7.4.3 Nature of Anionic Group on Extraction

It was necessary to calculate proton dissociation and metal extraction constants of typical crown ether chromoionophore with different side arms of 15C5 [14-17]. In the following structures (V) and (VI), the compound (V) contained

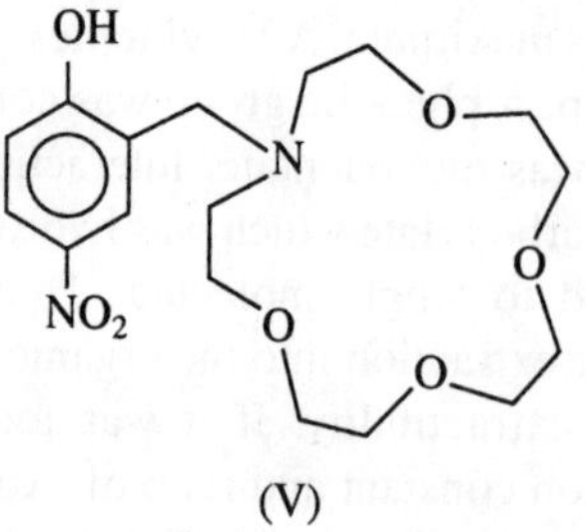

(V) 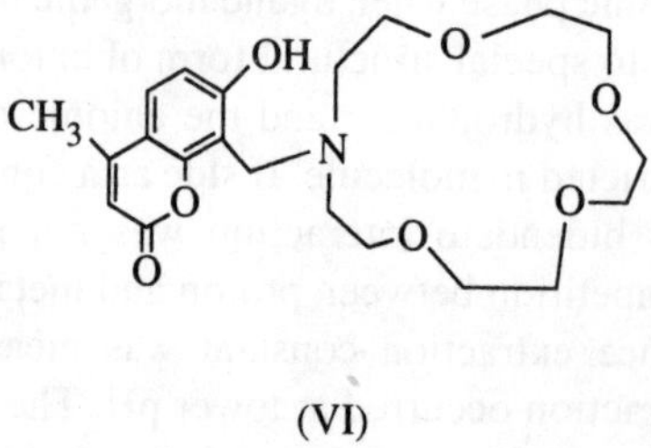 (VI)

monoaza 15C5 rings and the anionic side arm formed the same O-and N as coordinating six membered metal chelate. The combined proton affinity of phenolate (O⁻) and tertiary amine (N) was given by (pKa(OH) + pKa(NH⁺)) the values were 15.5. The greater was proton dissociation consetant (Ka) consequently the greater was metal extraction constant K_{ex}. See Table 7.1 and see structures (V) to (VIII)

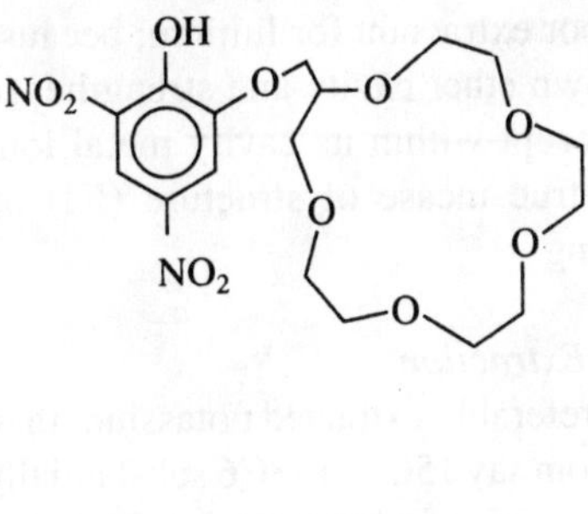

(VII)

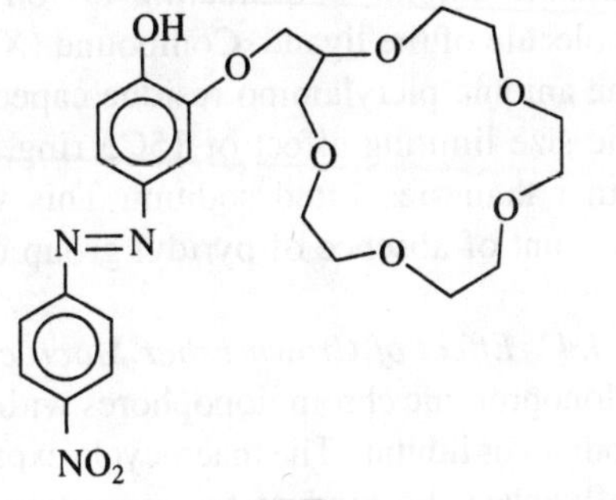

(VIII)

Table 7.1 Proton dissociation and metal extraction constant of 15C5 derivatives

Structure	pK_a		pK_{ex}		Metal selectivity
	OH	NH⁺	Na⁺	K⁺	Na⁺/K⁺
(V)	5.8	9.7	9.8	10.4	3.3
(VI)	6.3	10.5	10.3	11.5	15
(VII)	3.2	–	3.6	4.2	3.5
(VIII)	7.5	–	8.4	9.2	6.3

In two proton dissociate constant [pK_a(OH) and pK_a(NH⁺)] of the four phenolic aza crown ether, the dissociation constant of [pK_2OH] correlated with metal extraction constant showing significance of phenolate anion metal ion interaction for extracting the complex in the organic phase. The phenolic proton was ultimate proton to be knocked out to facilitate complexation and extraction of metal. The proton affinity (or basicity) of the anion in ligand was demonstrated in metal selectivity. For ligands of proton affinity (VI) the selectivity was high.

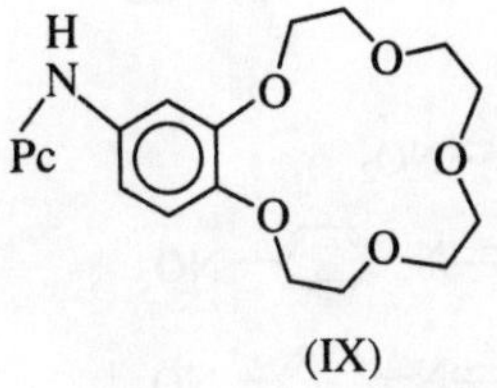

(IX) 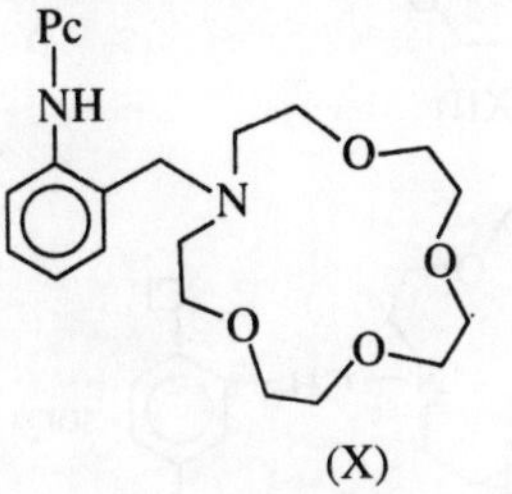   (X) 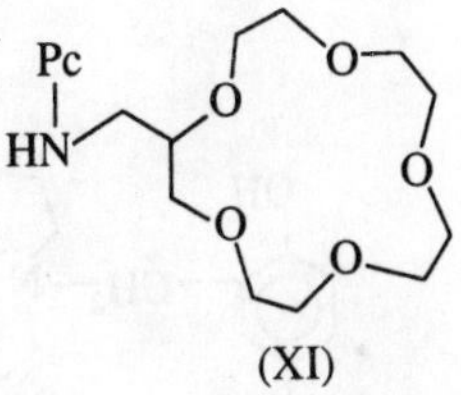   (XI)

The compound (IX) and (X) both contained picrylamino group as proton dissociating chromophore. An anionic nitrogen formed upon deprotonation of picrylaminoproton was sterically affected as the adjacent two ortho nitro groups and the hydration could effectively stabilise the charge on the amino nitrogen. It extracted alkali metals forming [MR. HR] kind of complexes. Other anionically armed crown ethers did not behave like this. This was due to rigid linker arm in (IX). For purpose of the extraction hydrated water molecule in coordination sphere of the metal ion should have been replaced by etherial oxygen or a suitable anionic coordinating or ion pairing group with the help of another molecule of the ligand. Compound (X) was poor extractant for lithium; because the anionic picrylamino residue caped the crown ether cavity and strengthened the size limiting effect of 15C5 ring not to accept within its cavity metal ions other than size fitted sodium. This was not true incase of structure (XI) on account of absence of pyridyl group on the ring.

7.4.4 Effect of Crown Ether Macrocycle on Extraction

Monoprotonic chromoionophores with 18C6 preferably extracted potassium than sodium or lithium. The macrocycle expansion from say 15C5 to 18C6 substantially influenced the kind of the extraction. The basicity of phenolate ion had also influenced the extraction. Similarly for changing from 12C4 to 15C5 the extractibility of sodium showed decrease. Thus reduction of crown ether ring size to 9C3 led to disappearance of extractibility for alkali metals. A simple monoprotonic phenol (including 9C3) was not sufficient to extract small ions like lithium but atleast four neutral donor atoms were essential apart from anionic phenolate donor to effectively extract lithium.

7.5 Diprotonic Crown Ether Chromoionophores

The introduction of two proton dissociable chromophores into the crown ether ring led to those dyes which were selective for divalent metal ions, e.g. alkaline earths [10]. For the purpose such of the crown ethers could be considered where in two anionic chromophoric groups were capable of reacting straight with metal ion from the axial directions. These kind of ligands were obtained from DA18C6 and similar diazacrown ethers structures (XII) and (XIII) represented as follows.

(XIV)

Compound (XII) was bad extractant for alkaline earths due to the low coordinating ability of the amide nitrogen in crown ether ring but compound (XIII) was good extractant [18, 19]. The metal extraction selectivity was Ca > Sr > Ba > Mg. The compound (XIII) existed as zwitter ionic species but in methylene chloride they were present as neutral aminophenol structure. The fluoroscent form of the reagent (XII, XIII) could be obtained via a Mannich reaction of diaza crown ether with 4-methylumbelliferone to compound (XIV).

The ligand needed more alkaline condition for better extraction. The diprotonic [12] crown ethers could also be obtained from DA 24C8.

Thus in summary we note that main feature of extraction capacity and selectivity in the divalent metal diprotonic chromoionophore was more or less same as that of monovalent nonprotonic chromoionophore. The cation-anion charge interaction dominated over cation dipole action in nonpolor solvents. Small variation in size of crown ether did matter but change in kind and position of anionic side arms determined the course of extraction with a need of definite coordination stereochemistry around central metal ion (better six). The location of N-atoms in diazacrown ether ring was also of significance from the view point of extraction of metals, e.g. in DA 15C5 the placement of two nitrogens in adjacent position decreased the extraction of calcium but similar placement of N-atom in 18C6 increased the extraction of calcium. This showed that possibility of ligand to adjust to typical coordination stereospecificity needed by a metal was of significance in extraction.

7.6 Other Protonic Chromoionophores

All these chromoionophores were useful largely in the extraction photometric determination of the specially alkali metals and bivalent metal ions, e.g. compounds (V) (VII) (IX) and their analogues were used for photometry. These [14, 20-22] reagents were commercially available in the market. There was an introduction of variation of picrylaminochromophore where three nitro groups in picryl moiety was replaced cyano or trifluoromethyl group [30, 31, 32, 33]. This improved spectral properties separating [HR] and [MR] species in organic solvent with better selectivity for potassium extraction. The crowned phenols (XV) were studied, containing built in phenolic hydroxyl group with macrocyclic structure [23]. They contained 2-4 dinitro phenylazo groups. The change in colour of these crowned phenols was due to deprotonation of the phenolic hydroxyl group. Usually the diprotonic crown ether chromoionophores were commonly used for the extraction photometry or fluorimetric analysis of bivalent metal ions. The thiacrown ether based chromoionophores were also used for extraction photometry of soft metals like silver and copper [24-26].

A typical bicyclic crown phenol is shown in structure (XV) [27, 28]. The colouration in organic phase was related to deprotonation of the phenolic proton under the influence of proton removing amine and metal cation stabilizing resultant phenolate anions.

(XV)

Finally benzocrown ethers with azolinked side arms were also used [29]. These compounds changed their complexation and extraction selectivity and responded to light. In such system usually ion pair type association complex was usually formed.

The crown ether dyes as chromogenic reagents was still in stage of the development. A proton dissociable and neutral crown ether dyes were therefore synthesised. Molecular absorption spectroscopy was well suited for flow analysis [30] due to speed simplicity and stability.

Ionophores were one which promoted transfer of ions from aqueous phase to hydrophobic phase. Such transfer could take place by ion extraction or ion channel formation. Chromoionophores were best for ion extraction. The chemistry encompassed metallochromic extractants. The development of chromoionophore was confinded to reaction with cations but not concept of cell biology. A chromoionophore provided with optically sensing function and ion binding capacity.

In addition to molecular absorption photometry involving above mentioned chromoionophores, an alternative technique was widely used. This method was based on ion pair extraction of crown ether and cryptands which were used while counter anions like bromocresol green, resazurin, picrate, eosin, xylenol orange, erythrosine and thiocyanate were used for typical of pairing anions. These crown ethers or cryptands were not coloured the experimental procedure was more or less same as those used for chromogenic crown ethers [33].

7.7 Analytical Applications of Crown Ethers in Extraction Photometry

Apart from utilisation of crown ethers for the separation of elements by solvent extraction, they were also used for extractive photometric analysis of metals. Such process involved two stages, viz.

1. The stage involving chromoionophoric crown ethers like monoprotonic chromoionophores, chromogenic proton ionisable crown ether, thia and aza crown ethers.
2. The second category included usual crown ethers like 15C5 (XVI), 18C6 (XVII), 24C8 (XVIII) for the process of extraction accompanied by

utilisation such of the chromogenic counter anions like eosine, erythrosine, thiocyanate, metanil yellow which could give coloured reaction which could be directly measured spectrophotometrically.

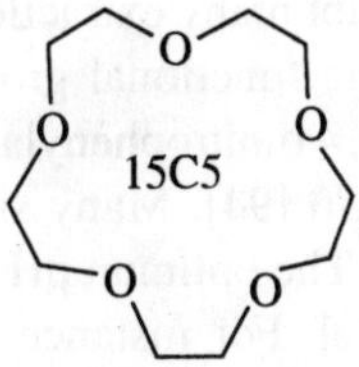

15C5 (XVI)   18C6 (XVII)   24C8 (XVIII)

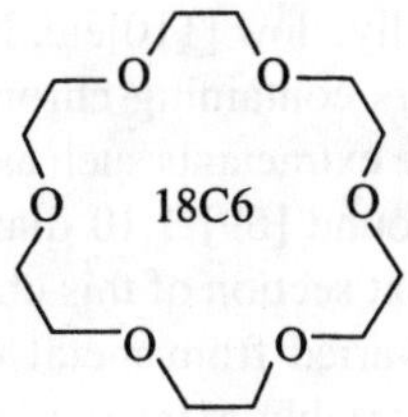

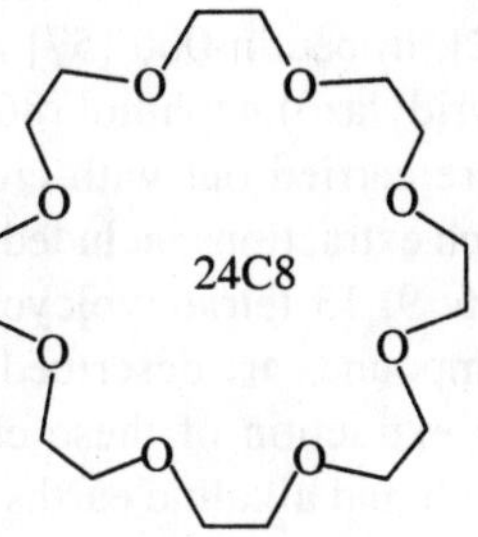

Of these first category of crown ethers was already discussed earlier. Now we would consider second category of ligands which use chromogenic counter anion for formation of ion pair complexes. The principal substituted crown ethers used are shown in structures from (XIX) to (XXI) as shown below:

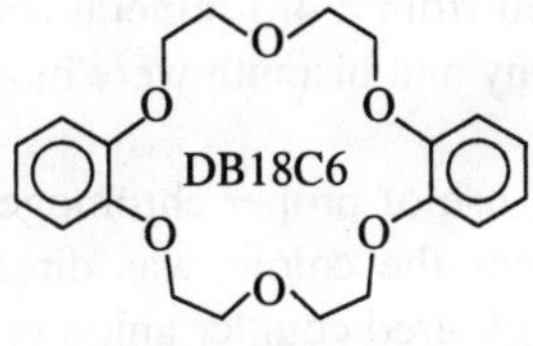

DB18C6 (XIX)   12C4 (XX)   DC18C6 (XXI)

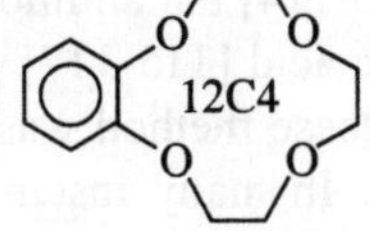

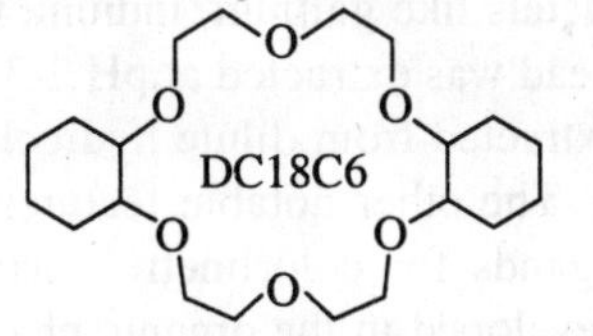

Most of the metals in periodic classification of elements were successfully extracted by various crown ethers. They largely included alkali metals and alkaline earths lanthanides, some actinides, zirconium, niobium, molybdenum, some platinum group metals the coinage metals, gallium subgroup elements, lead and antimony and bismuth. Variety of crown ethers were employed for such purpose. They largely included 15C5, B15C5, 18C6, DC24C8, DB24C8, however most intensively studied crown ether was 18C6 and its dibenzo and dicyclohexy derivatives. As regards diluents used for dissolution of these crown ethers the selection was restricted to dichloroethane, dichloromethane, chloroform nitrobenzene, benzene, butanol, but most commonly used diluent were methylene or ethylene chloride. As discussed in earlier chapter on extraction equilibria solvents with high dielectric constant favoured ion pair formation. As regards concentration of the extracting agent used generally was low as compared to conventional extraction procedure. It generally varied from 0.01M [57], 0.02 [69], 0.001M [73], 0.5M [93] etc. Usually 0.01-0.05M of extractant was quite adequate for the quantitative extraction of metals. Only in exceptional cases 0.5M reagent was used. In order to form electrically neutral complex during ion pair formation, the concentration of the counter anion used was kept more or less equivalent to that of the extractants.

Although picric acid was most popular counter ion for the extraction of alkali and alkaline earth elements subsequently an attendation was focussed on such of the counter anions which could give coloured reactions. Such counter ions mainly consisted of the use of bromocresol green [51] dipicrylamine [49] bromothymoblue [52], tropaelin-000 [57] zircon [65] arsenazo [74-79], xylenol orange [84] 1-2 (pyridylazo) naphthol [86], metanillyellow [110]etc. No doubt many extractions were carried out with crown ethers containing chromogenic functional group. Such extractions included use of the extractants such as 16 (2, 4 dinitrophenylazo) 3, 6, 9, 13 tetraoxybicyclic compound [69] 1.10 diazo 18C6 [94]. Many such compounds are described in the next section of this chapter. The optimun pH for the extraction of these elements varied from metal to metal. For instance the alkali and alkaline earths were invariably extracted in the pH region of pH 3-7. Most of the extraction centering around pH 3.0 while inner transition metals like lathanides were extracted in alkaline region with pH extending from pH 9.0 [77, 78].

Uranium or thorium were extracted from mild acidic solution while middle group transition metals like molybdenum, zirconium, vanadium, were extracted from 6-8M of the common mineral acids. Platinum group metals were extracted in region of pH 3.0-8.0, copper was extractable at pH 13.0 [96]. The p-block metals like gallium, indium, thallium were extracted from 2-4M mineral acids. Lead was extracted at pH 2-3 [110-114] but antimony and bismuth were mostly extracted from dilute hydrochloric acid [115-117].

The other notable features of these method was use of proper chromogenic ligands for colorimetric analysis. In many instances the colour was directly developed in the organic phase with use of either coloured counter anion or by adding external chromogenic extractable reagent. Alternatively metals were stripped to aqueous phase by mineral acid and susequently analysed spectrophotometrically by addition of external chromogenic ligand [74-84]. The colouration with counter ion was useful, e.g. cobalt with thiocyanate as counter anion [93] or silver with dipicrylamine [99] or mercury with dithiozone as counter anion [100]. In few instances the complex with counter anion and K-crown complex favoured encapsulation of the metal ion with colour forming reaction (see XXII). For instance extraction of indium with potassium bromide [103 or potassium iodide [104] or extraction of thallium [105] (see structures XXIII).

(XXII)

Some of the significant observations on the extraction of these metals followed by spectrophotometry either in the organic phase or even in the aqueous phase were as follows:

1. The novel methods for the separation of chemically similar elements like zirconium and hafnium [83] or the isolation of niobium and tantalum [86] or sequential separation of gallium, indium and thallium [102-105].
2. Many of the counter anions acted as the synergistic extractant facilitating thereby quantitative extraction, e.g. use of dithiozone in the extraction of mercury [100], or lead [114] or extraction of uranium with 4-(2-pyridylazo) resorcinol (PAR) it formed mixed complexes [82] with DC18C6.
3. The counter anion permitted direct colorimetric determination after extraction, e.g. vanadium [84] extracted with DB18C6 with potassium thiocyanate as counter ion.
4. In large number of cases stripping was followed by colorimetric analysis with external chromogenic reagent added to the aqueous phase.

The various parameters influencing the extraction of metals with different crown ethers followed by photometric analysis by appropriate ligand are summarised in Table 7.7.1.

(XXIII)

7.8 Application of Cryptands in Extractive Photometry

In comparison to crown ether cryptands, had distinct advantage as the complexing ligand for metals primarily for two reasons, viz. they formed three dimensional complexes which were relatively stable. Further since the heteroatom was nitrogen and oxygen, the bond formation led to stable complex formation specially when they were transferred to hydrophobic phase. Finally very low reagent concentration was essential for quantitative extraction of the metals.

(XXIV)   (XXV)   (XXVI)

The kind of metals which were extracted by cryptands consisted of alkali metals alkaline earths and transition metals [118-135]. Most of the extraction which were carried out were quantitative in the pH region of pH 6.0-9.0. The diluents used for the dissolution of the cryptands consisted of chloroform,

Table 7.7.1 Applications of extractive spectrophotometry with crown ether

Element	Extraction Conditions
Alkali metals	At pH > 6.7, 0.01M DB18C6 in benzene extracted it from picrate media [34, 35]. Usually 1:1 complex was formed. The method was promising for the separation of potassium [36]. B15C6, B18C6 in chloroform extracted them as 2:1 complexes [36] cis and trans bis (crown ethers) containing B15C5, B18C6 in picrate media formed 2:1 complex in chloroform or acetonitrile, cis isomer had better extractivity than trans form [37]. Crown ether carboxylate derivative extracted various alkali metals in chloroform from chloride or sulphate solution at pH 6-7 for potassium and pH 8-12 for sodium it was used in ion chromatographic studies [38]. A dialkyl derivative of DB18C6 and its other derivatives extracted alkali metals from picrate solutions in chloroform [39]. Lariat ether such as 2, 4, 6 trinitroanilio methyl substituted 15C5 and analogue had extracted alkali metals from chloride solutions [40]. Chromogenic crown ether containing pendant nitrophenolic coumarin and picrylamino moieties in dichloroethane were used for the extraction of alkali metals [41]. Derivatives of monobenzo 15C5 and DB 18C6 were also studied and effect of ionic size and cavity in the crown ether was investigated [42]. B15C5, 18C6, B18C6 and acyclic B18C6 in chloroform were utilised [43] when shift in absorption maxima was noted.
Li (I)	0.02M of crowned dinitrophenol aza compounds were used to extract it and determine it at 565 nm [44] however several alkali metals showed strong interference with the exception of sodium. Chromogenic crown ether derivatives of 14C4 in dichloromethane extracted it and was determined at 464 nm [45]. 14C4 containing phenylazo group was also used [46]. Spirobenzopyrane derivative in dichloroethane extracted it from the picrate solution [47] and analysed at 543 nm. With 18C6 or DB18C6 in dichloroethane conductance was measured and were compared with spectrophotometric data [48].
Na (I)	DC18C6 in water methylene chloride mixture was extracted with dipicrylamine or picric acid as anion [49]. From 12C4 in dichloromethane and picrate solution extraction was carried out and then determined photometrically at 375 nm [50]. At pH 3.0-4.7, B15C5 in chloroform extracted it with bromocresol green as the counter anion and was determined colorimetrically at 410 nm [51]. B18C6 in the presence of potassium salt of tetrabromophenol pthaleinethyl ester extracted it and used in flow injection analysis [52]. 0.5% B15C5 in chloroform with bromothymol blue as the counter anion in ethanol was extracted at pH 5.1 and analysed at 415 nm [53]. At pH 3.4, 15C5 extracted it from bromophenol blue in chloroform and analysed at 407 nm [54]. Alkyl derivative of DB18C6 in dimethyl formamide extracted it from thiocyanate solutions [55].
K (I)	A 0.015M of 18C6 in benzene selectively extracted it from picric acid solution and analysed at 425 nm [56] however rubidium and thallium showed strong interference. At pH 3.0, 0.01M DB18C6 in chloroform extracted it as mixed ligand complex with tropaelin-000 as the counter anion and determined at 405 nm [57]. Potassium from water was thus analysed. At pH 7.0, 0.01M DB18C6 in chloroform extracted it with

(Contd)

Element	Extraction Conditions
	bromothymol blue and determined at 412 nm but sodium and caesium showed interference [58] 0.5% of 18C6 in benzene with eosin (04%) extracted and determined at 543 nm [59] 18C6 in water methanol mixture extracted it from picrate media and subsequently studied by high performance liquid chromatography [60].
	Dibromo DB18C6 in chloroform with tropaelin-000 or bromocresol green- and bromothymol blue as counter anion was extracted and analysed at 405-420 nm [61] B15C5 in presence of dipicrylamine extracted it and analysed at 420 nm [62]. 18C6 in chloroform extracted potassium from solution containing methyl orange in chloroform and analysed at 415 nm, but strontium, barium and lead showed strong interferences [63]. B18C6, 18C6 and similar crown ethers like DB18C6, B15C5 with dipicrylamine as counter anion extracted it, when it was noted that dyes containing sulphonic group were more effective [64] B18C6 in benzene or chlorobenzene extracted it with novel counter anions by flow injection colorimetry [65] 10% DB18C6 at pH 5.1 extracted it in chloroform with 0.04% bromothymol blue in ethanol as counter ion and analysed at 445 nm [66]. 18C6 with tropaelin-000 extracted it but halides showed interference [68]. Also extracted at pH 4.5 wim 18C6 [67].
Rb (I)	0.025M of 16 (2-4 dinitrophenylazo) 3, 6, 9, 13 tetraoxybicyclic compound extracted it in chloroform and determined at 575 nm [69] in triethanolamine. If caesium was coextracted when intramolecular ion pair was formed [70].
Ca (II)	0.02M DC24C8 in chlorobenzene extracted it with 0.05M propylene orange and analysed at 420 nm [71].
Sr (II)	At pH 3.0 0.01M 18C6 in toluene extracted it from picric acid and stripped it with 1M hydrochloric acid and analysed at 670 nm as its complex with chlorophosphonazo [72].
Ba (II)	At pH 3-9, 0,001M DB18C6 in nitrobenzene extracted it from 0.01M picric acid, stripped with 1M nitric acid and measured spectrophotometrically as its sulphonazocomplex at 640 nm. It was possible to separate it from other alkaline earths [73].
Sc (III)	At pH 2.5-4.5 it was extracted with 0.04M 18C6 in dichloroethane with 0.4M picric acid as counter anion and was subsequently stripped with 0.01M nitric acid and determined spectrophotometrically as its arsenazo complex at 650 nm when, zirconium, hafnium and vanadium showed strong interference [74].
Ln (III)	15C5, 18C6 extracted cerium (III) along with prosodemium from picrate solution [75] as 1:2 sandwich complex.
La (III)	At pH 9.0 diazo derivative of crown in the presence of 2-thenoyl trifluoroacetone extracted it [76]. At pH 9.0, 0.02M DB24C8, or DC24C8 in dichloroethane extracted it from 0.0002M picric acid and determined photometrically as its arsenazo (III) complex at 650 nm after stripping it with 0.5M perchloric acid DB24C8 was superior to dicycloanalogue. The complex was 1:1:3 for lanthanum: crown ether: picrate anion. Azacrown in chloroform extracted it at pH 9.0 and determined spectrophoto-metrically at 372 nm [77]. Inductively coupled plasma atomic emission spectroscopy (ICP-AES) was also used for such analysis [78].

(Contd)

Element	Extraction Conditions
Ce (IV)	At pH 6-8 it was extracted with 0.05M, 15C5 in dichloromethane from 0.01M picric acid. It was subsequently stripped with 1M perchloric acid and analysed as its arsenazo-III complex at 655 nm. It was separated from large number of metals in multicomponent mixture [79].
Th (IV)	At pH 2.0-3.5, 0.065M 18C6 extracted it from 0.04M nitric acid and was back washed with 0.5M nitric and determined photometrically as its complex with arsenazo-III at 655 nm [80].
U (IV)	From 6-8M hydrochloric acid it was extracted with 0.02M DC18C6 in chloroform and was subsequently stripped with 0.05M hydrochloric acid and was determined as its arsenazo-III complex at 665 nm [81] 0.01M of DB24C8 in nitrobenzene extracted it from 6-10M hydrochloric acid stripped with 2M nitric acid and analysed as its 4 (2 pyridylazo) resorcinol (PAR) complex at 530 nm [82].
Zr (IV)	From 8.5M hydrochloric acid it was extracted with 0.025M of DC18C6 in methylene chloride, it was stripped with 0.5M hydrochloric acid and determined as its arsenazo-III complex at 860 nm [83].
Hf (IV)	While hafnium was extracted from 9M hydrochloric acid with 0.0075M of DC18C6 in dichloromethane, it was stripped with 0.1M. Perchloric acid was determined as its complex with xylenol orange at 540 nm [83]. This excellent method provided sequential separation of zirconium and hafnium [83].
V (V)	From solution containing 0.003M hydrochloric acid and equimolar concentration of potassium thiocyanate it was extracted with 0.001M. DB18C6 in benzene and was directly measured spectrophotometrically however lead cadmium, titanium, cerium showed interference [84].
Nb (V)	From 6M hydrochloric acid containing 3M potassium thiocyanate it was extracted with 0.5M DB18C6 in benzene [85]. Also with 02M DC18C6 in methylene chloride it was extracted from 7-10M hydrochloric acid, stripped with 0.5M sulphuric acid and measured as its complex with 4 (2pyridylazo) resorcinol at 540 nm. It was thus possible to separate niobium from tantalum [86] from each other.
Mo (VI)	From 8M hydrochloric acid, it was extracted with 0.01M DC18C6 in nitrobenzene, stripped with 2M nitric acid and determined photometrically as its complex with tiron at 390 nm [87]. From 3-8 hydrochloric acid also it was extracted with 0.01M DB18C6 in dichloromethane stripped with 0.05M hydrochloric acid and measured spectrophotometrically as its tiron complex at 390 nm [88].
Mn (II)	At pH 7-11, 0.15M macrocyclic Schiffs base extracted it in nitrobenzene and determined directly at 435 nm [89]. On the extraction at pH 3-6 a coloured complex was easily measured. Such complex at this pH was quite stable [89].
Re (VII)	With 0.01M DC18C6 in ethylene chloride it was extracted and determined as basic green complex at 640 nm [90].
Ru (III)	With 18C6 it was extracted due to second sphere coordination and simultaneous adduct formation [91].
Os (VIII)	At pH 3-8 with 0.1M DB18C6 in dichloromethane extracted it and measured as its aminophenol complex photometrically at 420 nm [92].
Co (III)	At pH 1.0-7.0 it was extracted with 0.5M 18C6 in methylene chloride in the presence of 0.2M ammonium thiocyanate enabling direct colorimetric

(Contd)

Element	Extraction Conditions
	determination at 623 nm [93] other crown ethers such as B15C5, DC18C6, DB18C6, diazo 18C6 in dimethylsulphoxide formed 1:1 complex quite efficiently [94].
Pd (II)	DB18C6 ditertiarybutylbenzo 18C6, dibenzyl derivative of 18C6 in butanol extracted it quantitatively [95] and determined colorimetrically.
Cu (II)	At pH13.0 it was extracted with DB18C6 in butanol with the zincon as the counter anion to form ion paired complex and estimate it directly at 610 nm (96), but nickel and iron showed interference.
	At pH 5-6 the macrocyclic Schifs base with picrate as anion extracted it in chloroform [97]. Macrocyclic Schifs base in chloroform also extracted it from 6M nitric acid and analysed photometrically at 435 nm [98].
Ag (I)	At pH 6-9, nitrogen and sulphur containing macrocycles in chloroform or methylene chloride extracted it with dipicrylamine as the counter anion and simultaneously analysed at 380 nm. It was also possible to extract it at pH 5-8 and estimate photometrically at 360 nm [99].
Hg (II)	1-3 diaza 2-thiabenzo 15C5 in chloroform extracted it with dithizone as the synergistic agent [100] also macrocycle formazane in chloroform extracted it from picrate solution and the analysis was carried out at 568 nm [101]. The coextracted copper was analysed by atomic absorption spectroscopy (AAS).
Ga (III)	From 6M hydrochloric acid it was extracted with 0.02M 18C6 in methylene chloride, stripped with 1M acetic acid and determined as its 1 (2-pyridylazo) naphthol (PAN) complex at 545 nm [102]. Gallium was thus separated from several metals.
In (III)	From hydrochloric acid containing either sodium bromide, or potassium bromide or iodide it was extracted with B15C5 or DB18C6 in dichloromethane and analysed as its complex with the 4-(2 pyridylazo) resorcinol [103]. From 4M hydrobromic acid it was quantitatively extracted with 0.02M 18C6 in methylene chloride, then stripped with 1M perchloric acid and analysed at 545 nm as its complex with 1-(2 pyridylazo naphthol) (PAN) [104].
Tl (III)	From 4M sulphuric acid containing 2M potassium iodide, it was extracted with 0.05M of 18C6 in dichloromethane and was determined as its iodo complex at 400 nm [105].
Ge (IV)	From 8M hydrochloric acid it was extracted with 0.02M DC18C6 in methylene chloride, stripped with 2M hydrochloric acid and [106] determined photometrically as its complex with phenylfluorone at 530 nm.
Sn (VI)	From 1.5M sulphuric acid containing 1.60M potassium iodide it was extracted with 0.04M DC18C6 in dichloromethane and determined directly as its iodide complex at 410 nm. An excellent separation from associated elements as antimony and bismuth was possible [107]. It was also extracted from tetrabromophenol with 0.2M DC8C6 in chloroform and analysed at 607 nm [108].
Pb (II)	At pH 3-5 it was extracted with 0.1M of 18C6 in chloroform with tropaelin-000 as counter anion and determined at 410 nm [109]. It was also extracted with 18C6 in the presences of metanil yellow [110]. About 0.5M of 18C6 in chloroform extracted and separated it from many metals at milligram concentrations [111]. With tropaelin-000 as counter ion it was also extracted

(Contd)

Element	Extraction Conditions
	with 0.01M 18C6 in chloroform from 0.1M hydrochloric acid and analysed at 490 nm [112]. At pH 2-0 to 3.0 about 0.02M DC18C6 in chloroform extracted it in the presence of 0.06M picric acid. Lead was stripped with 0.5M perchloric acid and analysed as its complex with arsenazo-III at 635 nm [113]. The extraction was also fissible with 0.5M DC18C6 in chloroform in the presence of 0.0025 of dithizone and analysed photometrically at 514 nm [114].
Sb (III)	From 1M sulphuric acid containing 0.1M potassium iodide and 0.25M ascorbic acid as the complexing agent, it was extracted with 0.2M of 18C6 in dichloromethane and analysed as its iodocomplex directly at 495 nm. The method was extended for the analyses as its iodocomplex directly at 495 nm. The method was extended for the analysis of antimony in real samples [116]. The method permitted sequential separation of the three metals [115].
Bi (III)	From 1M sulphuric acid containing 0.75M potassium iodide it was extracted with 0.05M of 18C6 in dichloroethane and determined directly as its iodide complex at 495 nm [116]. 0.1M crown derivative was used for extraction and determination of few aliphatic amines at 520 nm [117].

dimethylsulphoxide, toluene, nitrobenzene, dichloroethane, or mixture of solvents [125]. Variety of counter anions were used for the purpose of extraction. They consisted of 1-2 (pyridylazo) naphthol (PAN), methyl orange, picric acid, erythrosine, eosine, or dithizone. However eosine and erythrosine were most popular as they permitted simultaneous extraction and direct spectrophotometric determination of metals. Another attractive feature of these methods was concentration of cryptand used during extraction was really very small ranging from .001M-0.1M of cryptand concentration. The commonly used cryptands consisted of cryptand-222 (XXIV), cryptand-221 (XXV), cryptand-211 (XXVII), however best results were obtained with cryptand 222. One of the interesting aspect of cryptand chemistry was the possibility of separating metal pollutants from aquatic environment and waster water effluents.

For instance Gandhi and Khopkar [131] developed excellent methods for the separation of nickel, cadmium, manganese, lead, thallium and copper. These metals were extracted with either cryptand-222, 221 or 222B in mixture of solvents like nitrobenzene, toluene or dichloromethane. The extension of these studies to solvent extraction of zirconium [129] and uranium [130] provided rapid methods for extractive separation followed by extractive photometry involving the use of 8-hydroxyquinoline as the chromogenic ligand. The chemistry of the extraction of calcium was still interesting as such extraction was possible in the presence of water soluble ethanol along with toluene with erythrosine as counter anion which also acted as the chromogenic ligand [169]. In few instances instead of spectrophotometry the technique of atomic absorption spectroscopy was used for the rapid analysis of metals from real samples. This was true for few separation of the metal pollutant from aquatic environment by solvent extraction [131].

The various methods used for the solvent extraction and direct photometric determination of metals with cryptands spelling out reagent concentration, kind of solvent used pH of extraction, nature of counter anion, and optimum wave length use for the measurement are summarised in Table 7.8.1.

Table 7.8.1 Application of extractive photometry with cryptands

Element	Extraction Conditions
Na (I)	With cryptand-222 in chloroform it was extracted and measured at 520 nm wtih 4 (2-pyridylazo) resorcinol (PAR) as the counter anion[118].
K (I)	Cryptand-222 in chloroform extracted it in the presences of methyl orange as counter anion and analysed at 415 nm [119] but the alkaline earths showed interference. This extraction was analogues to one with 18C6. With the chromogenic cryptands in ethoxy ethanol in the presences of the surfactant was extracted and analysed at 540 nm [120]. The cryptohemic sphere, i.e. complex in diethylene glycol monoethyl ether was extracted [121].
Alkaline earths	All cryptands, viz. cryptand-222, 221, 211 in dimethyl sulphoxide extracted them in the presence of mureoxide as a chromogenic ligand as 1:1 complex [122]. Cryptand-222 in chloroform or dichloromethane extracted them from picric acid as counter anion. The best solvent was dichloromethane [123].
Mg (II)	At pH 9.5 it was extracted with 0.0138M cryptand-221 in triethanolamine with metaphthalein as a counter anion the determined photometrically at 568 nm [124]. Sodium was used as the scavanger and calcium was simultaneously determined.
Ca (II)	Cryptand-221 in dimethylsulphoxide with mureoxide as the indicator for kinetic studies was investigated [125]. It was also possible to extract it with cryptand-222 with mixture of solvent with erythrosine as anion [169].
Sr (II)	At pH~9.0, 0.02M cryptand-222 in a mixture of toluene and nitrobenzene was extracted with erythrosine-B as the counter anion and was determined at 547 nm [126]. Cryptand 222 in nitrobenzene toluene mixture in the ratio of 4:1 was used for extraction with erythrosine as counter anion and also chromogenic ligand for photometric analysis at 467 nm. The method was effective even in the presences of barium, calcium, iron and phosphate [127].
Ba (II)	With cryptand 222 it was photometrically titrated when lead perchlorate was used as the standard solution [128].
U (VI)	At pH 6.0, 0.01M cryptand-222 in chloroform or nitrobenzene extracted it with 0.005M eosine as counter anion. It was stripped with 0.1M perchloric acid and determined at 430 nm [129] as oxinate complex at pH 4.5. It was separated from fission products.
Zr (IV)	With cryptand-222 in toluene it was quantitatively extracted at pH 5.0 and was stripped with mineral acid and was determined spectrophotometrically as its coloured complex with oxine in chloroform [130].
Ni (II)	At pH~7.5 with 0.01M cryptand-222 in dichloroethane it was extracted in the presence of erythrosine as counter anion, stripped with 1M hydrochloric acid and determined by both spectrophotometry or by atomic absorption spectroscopy. However metals like chromium, iron, manganese, cadmium, lead showed strong interference [126, 131].

(Contd)

Element	Extraction Conditions
Cd (II)	At pH 6.0, 0.08M cryptand-221 nitrobenzene toluene mixture (1:4) was extracted with trichloro or triiodo fluoroscene as the counter anion [132] and was analysed at 545 nm; but lead interfered [132].
Pb (II)	With 0.1M cryptand -222 in chlorobenzene it was extracted in the presence of 0.04M dithiozone and was analysed photometrically at 545 nm [133, 134]. Lead from high purity alloys was also analysed. It was also extracted with cryptand-222B and analysed by atomic absorption [170] spectroscopy.
NH_4^+	About 0.01M cryptand in chloroform extracted it in the presence of 4.0 μm dithiozone and analysed spectrophotometrically at 496 nm but heavy metals showed strong interference [135].

7.9 Application of Thia and Aza Crown Ethers in Extraction Photometry

The field of extraction photometry received real boost on account of the discovery of several thia and aza crown ethers. These were monoprotonic chromoionophores capable of forming distinctly coloured complexes with transition metals. In comparison to normal crown ether thia compounds contained sulphur as the donor atom (see structures XXVII to XXIX) while aza crown ethers had nitrogen as the donor atom in their cyclic structure facilitating formation of relatively stronger bonds with the metals leading to the complexes with better stability (see structures XXX to XXXII). This helped extraction as in the presences of the (–N ≡ N–) aza group it permitted direct colorimetric determination of the metals. The metals which were generally extracted by them consisted of alkaline earth elements, lanthanides, copper, silver, cadmium and mercury. So far they were not used for the extraction of less familiar elements. The most preferred solvent was chloroform but dichloroethane was used only in the exceptional cases [140]. The counter anion used for the formation of ion pair complex consisted of picric acid (XXXIII), erythrosine [141] (XXXVII), and eosine (XXXVI) and bromocresol green (XXXV), dipicrylamine (XXXIV), and xylenol orange (XXXVIII). The average pH for the quantitative extraction was ranging from pH 4-8 and very limited extraction were carried out below pH 3.5 [142] or above pH 8 [140]. To prevent hydrolysis of interfering ions sulphosalicyclic acid (for iron) or EDTA (for nickel) was used as the sequesting agents. The principal aza crown ethers used consisted of A12C4, DA18C6, A18C6, DADB 18C6 etc. While common thia crown compounds were generally synthesised like thia crown ether, crown ether etc. (structures XXV).

The various methods utilised for the extractive photometric determination of coinage metals with thia crown ethers indicating optimum condition of extraction are summarised in Table 7.9.1.

The structures of some typical thia and aza crown ethers which are commonly used for extraction photometry are depicted below:

Some of the chromogenic planner counter anions used in the extraction along with crown ethers and cryptand are shown in structures (XXXIII) to (XXXVIII).

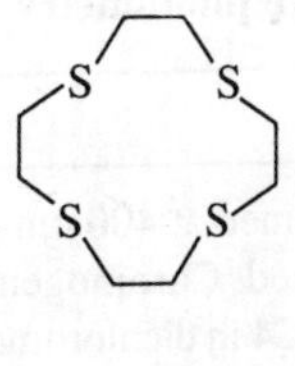

(XXVII) 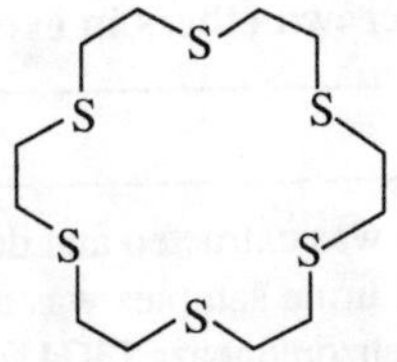 (XXVIII) 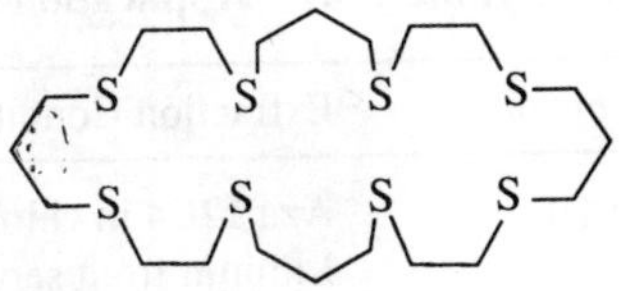 (XXIX)
Thia crown ethers

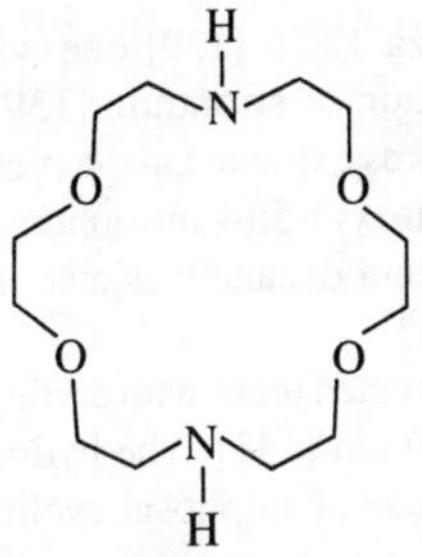

(XXX) 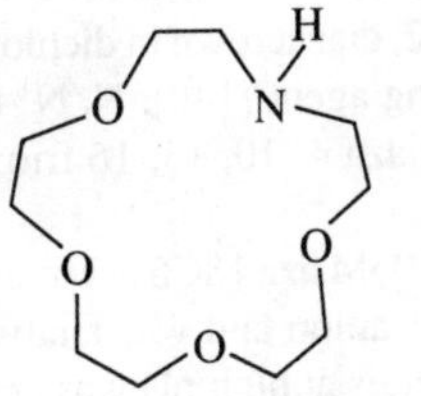 (XXXI) 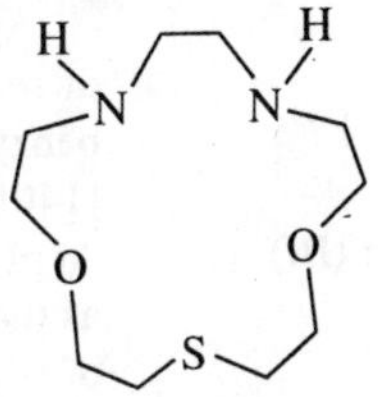 (XXXII)
Azacrown ethers

7.10 Applications of Calixarene in Extractive Photometry

In comparison to crown ether and cryptands calixarene were discovered recently and as such they had limited applications in extraction photometry. These compounds were better complexing ligands because they permitted effective substitution in upper or lower rim of the structure. For instance in structure (XXXIX) one could replace R-with tertiary butyl or tertiary octyl group or

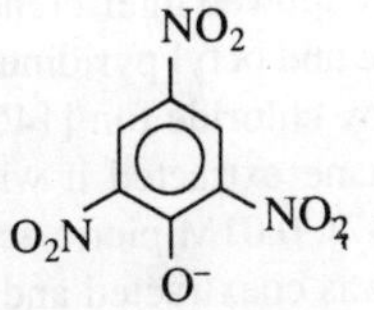

Picric acid (XXXIII)

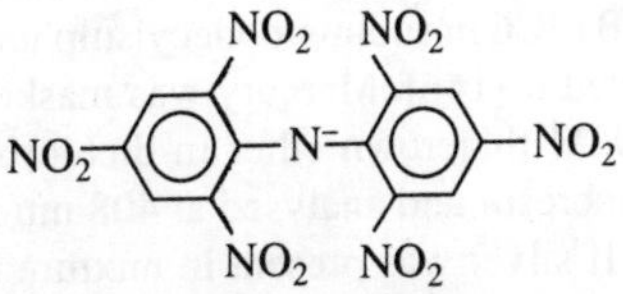

Dipicrylamine (XXXIV)

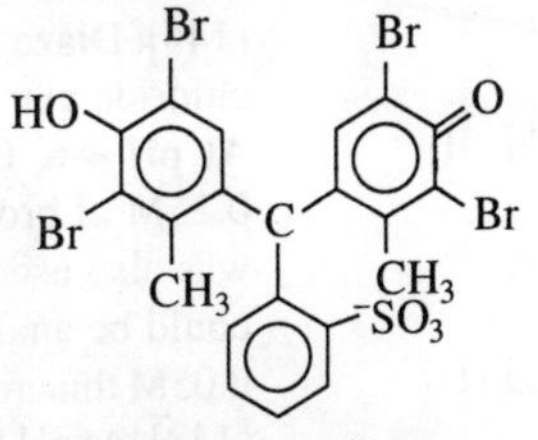

Bromocresol green (XXXV)

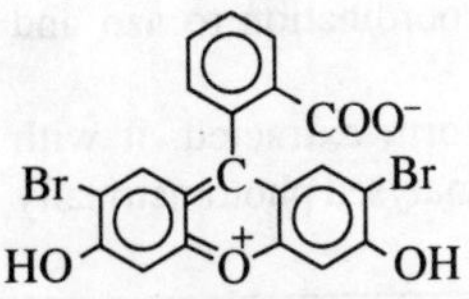

Eosine (XXXVI)

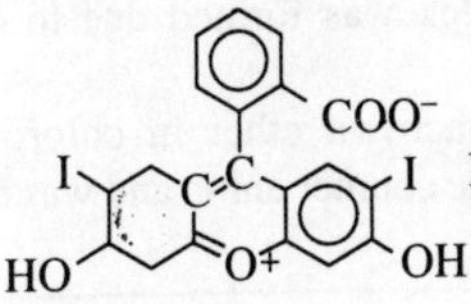

Erythrosine (XXXVII)

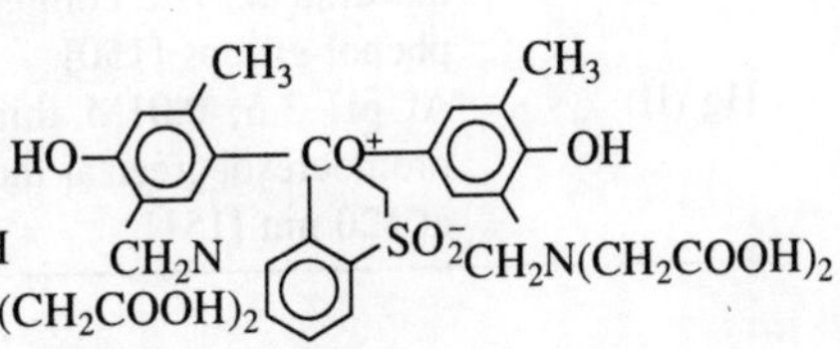

Xylenol orange (XXXVIII)

Table 7.9.1 Application of azacrown ethers in extractive photometry

Element	Extraction Conditions
Li (I)	Aza 12C4 in chloroform was extracted and determined at 400 nm [136]. Lithium from serum and urine samples was analysed. Chromogenic aza 12C4 and chromogenic benzomonoaza 13C4 and 14C4 in dichloromethane extracted it [136, 137] quantitatively. N (2 hydroxy 5 nitrobenzyl) monoaza 12C4 was best for the extraction [136]. The other effective extractant was 13C4 [138].
Ca (II)	The alkaline earths were extracted with diaza 18C6 [139] one with 2-pendent anion side arm was ideal for extraction of strontium [139]. At pH 5-12, diazacrown in dichloroethane was extracted in the presence of sequestering agent [140]. N, N^1 bis (2 hydroxy)-5 (4-nitrophenylazo) benzyl 1, 7 diaza 4, 10, 13, 16 trioxacyclopenta decane was also useful [140].
Ln (III)	At pH 11.0, 0.05M aza 18C6 in chloroform extracted them with erythrosine as the counter anion and was analysed at 550 nm [141]. The hydrolysis of s-block metals at high pH was arrested by use of sulphosalicyclic acid [141].
Cu (II)	At pH 1.7, 0.07M tetraaza macrocyclic ligand extracted it in chloroform with metanil yellow as the counter anion and was analysed photometrically at 406 nm [142], but cobalt and nickel showed strong interference [143].
Ag (I)	A ligand 8, 9, 17, 18 tetrahydro 7H dibenzo (1, 4, 8, 12) dioxadiazo cyclopentadecane in chloroform and dimethylsulphoxide mixture was extracted with 0.04M picric acid as the counter anion and EDTA (disodium salt) as the sequestering agent [144]. The complex was photometrically analysed at 368 nm.
Cd (II)	0.01M of 1, 4, 8, 11 tetraazatetramethylcyclotetradecane in chloroform extracted it with either eosine or erythrosine as counter anion and determined at 550 nm but copper, silver and mercury showed interference [145]. Diaza DB18C6 in sodium dodecylsulphonate and octyl pyridinum chloride extracted it [146]. Mercury was masked by chloride ion [146].
Cu (II)	At pH 4-6, 0.05M thiacrown ether in dichloroethane extracted it with 0.25M of bromocresol and analysed at 408 nm [147]. 0.01M picric acid was also used. If silver was present in mixture it was coextracted and it could be analysed at 378 nm by simultaneous spectrophotometry [147].
Ag (I)	0.05M thiacrown ether in chloroform extracted it from picric acid solution [148]. At pH 8.0, 4 picrylamino benzene 1, 4, 8, 11 tetrathio cyclopentade 1, 3 ene i.e. thiacrown ether in ethylene chloride was extracted and analysed photometrically at 450 nm [149]. In the presence of copper extraction was carried out at pH 7.4 instead of pH 8.0 [150]. Benzothiocrown in dichloroethane also extracted it with amylalcohol as the diluent. The complex was formed due to coordination to azo and phenol groups [150].
Hg (II)	At pH 3.5, 0.01M thiacrown ether in chloroform extracted, it with bromocresol green as the counter anion and was analysed photometrically at 420 nm [151].

common sulphonic acid (HSO_3) group to improve its solubility in water. Similarly in the lower rim if the OH-group could be easily replaced by keto group or acetyl group it facilitated its complexation with several metals (see structure XL). In addition to these two sites of substitution one can change cavity size or the annular space to accommodate large size metal ion. This was done by changing moiety (n) from 4, 6, 8 onwards phenyl rings in the structure.

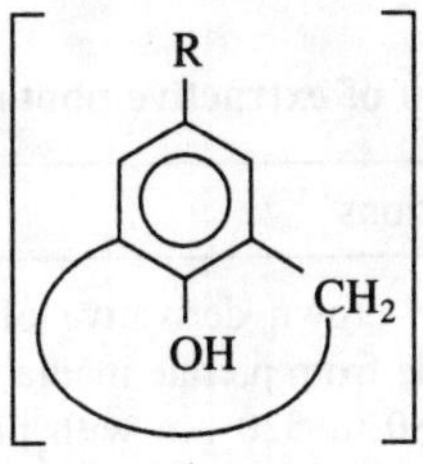

(XXXIX)

So far calix (6) arene had proved to be best extracting agent when its lower rim was substituted with acetyl group. As such in very few instances it was possible to accomplish direct spectrophotometric dermination. Alkali metals were extracted with tertiary butyl derivative of calix (5) arene in methylene chloride with picrate as the complexing anion [152]. A bathochromic shift of 140 nm was noticed with nitrophenyl azophenyl analogues [153]. The calix(4) arene diamide was successfully used for extraction of lanthanum from the picrate solutions [154]. The upper rim substituted calix (6) arene with sulphonic group permitted extraction of cerium [155]. While uranium was extracted with tetra (N-p-chlorophenyl) hydroxamate phenyl tetrahydroxy calix(4) arene with ethyl acetate [157]. Silver was extracted with hexahydroxy (4 phenyl aza) calix (6) arene in chloroform [158]. While fulerene was extracted with calix (8) arene [162].

However most outstanding contribution in the chemistry of calixarene was the application of hexa-aceto calix (6) arene (XL).

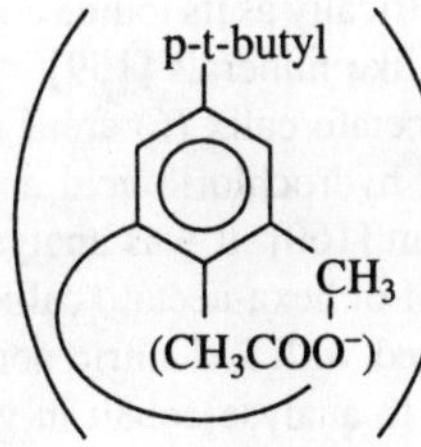

(XL)

This ligand had proved to be most selective for the solvent extraction of palladium [162], cobalt [161], iron [160]. In all these cases metals were extracted around pH 7.0-7.5 with 0.001M of the ligand invariably in toluene as the diluent. These metals were stripped with dilute mineral acids and were determined in the aqueous phase with suitable chromogenic ligands, e.g. cobalt with Nitroso R-salt [161], palladium with stannous chloride, [162], iron (III) with thiocynate [160].

In addition to these metals other transition metals [163-167] like thorium [156], chromium [158], manganese, [159], copper [163] as well as the main group elements like lead [166], bismuth [167] and thallium [165] were quantitatively extracted with hexa-acetato calix (6) arene.

The details of the conditions for the extraction, stripping and the determination of various metals with different calix(n)arene are best summarised in Table 7.10.1

Table 7.10.1 Application of extractive photometry with calixarenes

Element	Extraction Conditions
Alkali metals	Tetrahydroxybutyl crown derivative of calix (5) arene extracted it in methylene chloride from picrate media [152]. A bathochromic shift was observed from 380 to 520 nm with nitrophenyl azophenol derivative [153].
La (III)	With calixarene (4) diamide it was extracted as the 1:1 complex from picrate media [154].
Ce (IV)	With 0.025M calix(4) arene 5, 11, 17, 23 tetrasulphonate extracted and determined photometrically at 417 nm [155].
Th (IV)	At pH 7.5 it was extracted with hexa-acetatocalix (6) arene, it was then stripped with 0.05M nitric acid and determined spectrophotometrically at 545 nm with Thoron [156].
U (VI)	At pH 6.0 about 0.2% of 5, 11, 17, 23 tetra (N-p-chlorophenyl) hydroxamate-i-phenyl 25, 26, 27, 28 tetra hydroxycalix(4) arene extracted it quantitatively in ethylacetate and washed with potassium thiocyanate before colorimetric determination [157]. Uranium from environmental samples was analysed.
Cr (III)	At pH 6.5 it was extracted with 0.001M acetyl derivative of calix (6) arene in toluene, stripped with 4M hydrochloride acid and determined as its xylenol orange complex at 520 nm. Chromium from ores was analysed [158].
Mn (II)	At pH 6.0, 1×10^{-4}M of the hexa-acetato derivative extracted it quantitatively in hexane and was stripped with 2M sulphuric acid and analysed photometrically as its iodide complex at 545 nm. It was analysed from real sample like minerals [159].
Fe (III)	At pH 7.5 hexa-acetato calix (6) arene extracted it quantitatively it was stripped with 1M hydrochloric acid and determined as its thiocyanate complex at 480 nm [160]. It was analysed from drugs.
Co (II)	At pH 7.0, 0.075M of hexa-acetato calix (6) arene in toluene extracted it then it was stripped with 2M nitric acid and analysed at 500 nm. The method was used to analyse cobalt in vitamin B12 preparation [161].
Pd (II)	At pH 7.5 acetyl derivative of calix (6) arene in toluene extracted it quantitatively and was subsequently stripped with 2 M nitric or perchloric acid and analysed as its stannous chloride complex after reduction at 635 nm [162]
Cu (II)	Copper was extracted with 0.0001M hexacetato calixarene with phosphate buffer solution of pH 9.2, it was stripped with 1-3M mineral acid and was determined spectrophotometrically at 685 nm as its complex with rubeanic acid [163].

(Contd)

Element	Extraction Conditions
Ag (I)	37, 38, 39, 44, 42 hexahydroxy-5, 11, 17, 23, 29, 35 hexakis (4 phenyl azo) calix (6) arene in chloroform extracted it from picrate media and analysed photometrically at 450 nm [164].
TI (I)	It was quantitatively extracted at pH 3.0 with 0.001M reagent hexa-acetato calix (6) arene in toluene, it was stripped with 2M mineral acid and analysed spectrophotometrically with PAR at 530 nm. Thallium from sediments and other real samples was analysed [165].
Pb (II)	At pH 6.0 lead was quantitatively extracted with 1×10^{-4}M hexa-acetato calix (6) arene, stripped with 6M nitric acid and analysed spectrometrically as its complex with 4-(2 pyridylazo) resorcinol PAR [166]. Lead from samples of easily fusible alloys was analysed.
Bi (III)	From 0.1M hydrochloric acid it was quantitatively extracted with 1×10^{-4} hexa-acetato calix (6) arene in toluene and was subsequently stripped with 2M nitric acid and was then analysed spectrophotometrically as its complex with thiourea at 470 nm. Bismuth from drugs was then analysed [167].
Fulerene	C-60 buckiminister fulerene was quantitatively extracted with calix (8) arene as its complex derivative and analysed spectrophotometrically at 440 nm in solution containing water [168].

Conclusion

Crown ethers as the chromogenic extractant for the metals are in still early stage of development but fundamental concept of molecular recognisation by them has been established. Molecular absorption spectroscopy was well suited not only for laboratory scale analysis but also for continous flow injection method because of simplicity of operation and stability of the complexes.

Crown ether or similar compounds were ionophores which promoted transfer of ions from aqueous to organic phase involving process of ion extraction or favouring complextion and lipophilisation of metals. The progress was restricted to interaction with cationic species by anionic or monoprotonic chromoionophores. Those reaction with anions if discovered could compete with ion chromatography and ion selective electrodes. The basic idea of the chromoionophore was to furnish optically sensing function in addition to complexing properly [170].

The progress is slowly made in that direction by utilising crown ethers, cryptands, and calixarene in the technique of extraction absorption spectroscopy. Many more thrilling discoveries are expected to be known in next century.

References

1. M. Takagi, K. Ueno, Topics in current chemistry, Springer Verlag **121**, 39 (1984).
2. M. Takagi, Cation binding by macrocycles-Complexation of cationic species by crown ethers Ed. Y. Inoue, G.W. Gokel Marcel Dekker, New York (1990).
3. T. Kaneda, J. Synth. Org. Chem. **46**, 98 (1988).
4. H.G. Loehr, F. Voegtle, Acc. Chem. Res. **18**, 65 (1985).
5. M. Takagi, Analytical Letters **10**, 1115 (1977).
6. Y. Katayama, K. Nita, M. Ueda, H. Nakamura, M. Takagi, K. Ueno, Anal. Chem. Acta, **173**, 193 (1985).

7. M. Takagi, J. Coord. Chem. **15**, 53 (1986).
8. J.D. Lamb, Coordination Chemistry of macrocyclic compounds Ed. G.A. Melson Plenum Press p. 143 (1979).
9. N.K. Dalley, Synthetic multidentate macrocyclic compounds Ed. R.M. Izatt, J.J. Christensen Academic Press p. 207 (1978).
10. T. Yamashita, H. Nakamura, M. Takagi, K. Ueno, Bull. Chem. Soc. Japan **53**, 1550 (1980).
11. H. Nakamura, H. Sakka, M. Takagi, Chem. Letter 1305 (1981).
12. H. Nishida, Y. Katayama, H. Kalsaki, H. Nakamura, M. Takagi, K-Ueno, Chem. Letter 1853 (1982).
13. Y. Katayma, R. Fukuda, M. Takagi, Anal. Chim. Acta. **185**, 295 (1986).
14. M. Takagi, H. Nakamura, K. Ueno, Analyt. Letter **10**, 1118 (1977).
15. Y. Katayama, H. Nakamura, M. Takagi, Anal. Sci. **1**, 393 (1985).
16. H. Nakamura, H. Nishida. M. Takagi, K. Ueno, Bunseki Kagaku **31**, E131 (1982).
17. B.P Bubnis, G.E. Pacey, Chem. Ber. **114**, 638 (1982).
18. H. Nishida, M. Tazaki, M. Takagi, K. Ueno, Mikro Chim. Acta. **1** 281 (1981).
19. M. Shiga, H. Nishida, H. Nakamura, M. Takagi, K. Ueno, Bunseki Kagaku **32**, E293 (1981).
20. H. Nakamura, M. Takagi, K. Ueno, Anal. Chem. **52**, 1668 (1980).
21. H. Nakamura, M. Takagi, K. Ueno, Talanta **26**, 921 (1979).
22. H. Nakamura, H. Nishida, M. Takagi, Anal. Chim. Acta, **139**, 219 (1982).
23. S. Misumi, T. Kaneda, Inst. Sci. ind. Res. Oska. Uni. **44**, 29 (1987).
24. M. Muroi, A. Hamaguchi, E. Sekido, Anal. Sci. **2**, 351 (1986).
25. E. Sekido, K. Chayama, J. Chem Soc. Japan **907**, (1986).
26. E. Sekido, K. Chayama, Talanta **32**, 797 (1985).
27. J.P. Dix, F. Voegtle, Chem. Ber. **113**, 457 (1980).
28. J.P. Dix, F. Voegtle, Chem. Ber. **114**, 638 (1981).
29. S. Shinkai, J. Am. Chem. Soc. **104**, 1967 (1982).
30. K. Kimura, S. Iketani, H. Sakamoto, T. Shono, Anal. Sci. **4**, 221 (1988).
31. M. Takagi, Anal. Chim. Acta, **126**, 185 (1981).
32. H. Sumiyoshi, Talanta **24**, 763 (1977).
33. G.E. Pacey B.P. Bubnis, Y.P. Wu, Analyst. **106**, 636 (1981).
34. Y. Katayama, M. Takagi, Senryo to Yakuhin **33**, 65 (1988) C.A **109**, 141554q (1988).
35. A. Sadakane, T. Iwachido, K. Toei, Bull Chem. Soc. Japan **48**, 60 (1975).
36. K. Kimura, T. Maeda, T. Shono, Anal. Lett. Part A **11**, 821 (1978).
37. K. Kimura, T. Tsuchida, T. Maeda, T. Shono, Talanta **27**, 801 (1980).
38. J. Strzelbicki, R.A. Bartsch, Anal. Chem. **53**, 1894 (1981).
39. Yu. G. Mamedova, A.L. Shabanov, G.A. Babav, A.M. Baba-Zade, Zh. Anal Khim. **38**, 1578 (1983) A.A **46**, 7B24 (1984).
40. B.P. Bubnis, G.E. Pacey, Talanta **31**, 1149 (1984).
41. Y. Katayama, K. Nita, M. Uedo, H. Nakamura, M. Takagi, K. Ueno, Anal. Chim Acta. **173**, 193 (1985).
42. S.W. Kang, C.M. Park, W.F. Koo, K.J. Kim, S.M. Lee, C.H. Chang, Taehen Hwahakhoe Chi **32**, 443 (1988) C.A. **110**, 77407a (1989).
43. Y. Kudo, Y. Takeda, H. Mastsuda, Anal. Sci. **6**, 211 (1993).
44. K. Nakashima, S. Nakatsuji, S. Akiyama, T. Kaneda, S. Misumi, Chem. Letter. **111**, 1781 (1982).
45. Y.P. Wu. G.E. Pacey, Anal. Chim. Acta. **162**, 285 (1984).
46. T. Shono, K. Kimura, M. Tanaka, S. Kitazawa, Japan Kokai Tokkyo Koho C.A. **107**, 189997x (1987).

47. K. Kimura, S. Kanokogi, M. Yokoyama, Analyt. Sci. **12**, 399 (1996).
48. I.M. Kolthoff, M.K. Chantooni, J. Chem. Engg. Data **42**, 49 (1997).
49. M. Jawaid, F. Ingman, Talanta **25**, 91 (1978).
50. G.E. Pacey, Y.P. Wu, Talanta **31**, 165 (1978).
51. G. Xi, X. Yong, Fenxi Huaxue **13**, 452 (1985) A.A. **48**, 3B44 (1986).
52. S. Motomizu, M. Onoda, Anal. Chim. Acta. **214**, 289 (1988).
53. R. Escobar, C. Lamoneda, F.J. Barragan, A. Guiraum, Analysis **16**, 189 (1988) A.A. **52**, 6B20 (1989).
54. N.A. Vetyutneva, Farm Zh. 53 (1989) A.A. **52**, 2B16 (1990).
55. Sh. K. Norov, A. Yu. Tsivadze, I. Ya. Kachkurova, M.T. Gulamora, Zh. Neorg Khim **36**, 433 (1991) C.A. **114**, 254822x (1991).
56. T. Iwachido, A Sadakane, K. Toei, Bull. Chem. Soc. Japan **51**, 629 (1978).
57. A. Yu Nazarenko, I.V. Pyatnitskii, T.A. Stolyarchuk, Zh. Anal. Khim. **36**, 1719 (1981) A.A. **42**, 6B32 (1982).
58. N.P. Aleksyuk, A. Yu. Nazarenko, I.V. Pyatnitskii, Zh Anal. Khim. **37**, 2147 (1982) A.A. **45**, 3B25 (1983).
59. R. Huang, D. Li, M. Xie, J. Xu, L. Zhou, Fenxi Huaxue **10**, 423 (1982). A.A. **44**, 4B20 (1983).
60. T. Iwachido, M. Minami, H. Naito, K. Toei, Bull. Chem. Soc. Japan. **55**, 2378 (1982).
61. A. Yu Nazarenko, Ukr. Khim. Zh **49**, 279 (1983) A.A. **45**, 3B8 (1983).
62. F. Zhong Ti. Liao, Fenxi Huaxue **12**, 513 (1984) A.A. **47**, 2B26 (1985).
63. A.P. Arias, G.D. Blanco, A. Sanz. Medel, Mikrochem. J. **30**, 58 (1984) A.A **47**, 5B21 (1985).
64. F. Zhong. Ti. Liao, Huaxue Shiji 8 257 (1986) C.A. **106**, 77835r (1981).
65. S. Motomizu, M. Onodo, M. Oshima, T. Iwachido, Analyst. **113**, 743 (1988).
66. R. Escobar, E. Lamoneda, F. Depablos, A. Guiraum, Analyst **114**, 533 (1989).
67. X. Hong, Z. Bei, M. Zhu, Lihua Jianuan Huaxue Fence **26**, 110 (1990) A.A. **53**, 2D28 (1991).
68. N.P. Maksyutina, N.A. Vetyutheva, F.A. Mytchenko, Farm Zh (Kiev) **56**, (1990) A.A. **46**, 5B32 (1984).
69. K. Nakashima, Y. Yamawaki, S. Nakatsuji, S. Akiyarma, T. Kaneda, S. Misumi, Chem. Letter 1415 (1983) A.A. **46**, 5B32 (1984).
70. Y. Katayama, R. Fukuda, T. Iwasaki, K. Nita, M. Takagi, Anal. Chim. Acta. **204**, 113 (1988).
71. S. Motomizu, M. Oshima, N. Yoneda, T. Iwachido, Anal. Sci. **6**, 215-22 (1990) A.A. **53**, 9D32 (1991).
72. H.Z. Li, Y.X. Bai, Y.D. fang, Fenxi Huaxue **22**, 1283 (1994) A.A **57**, 6D35 (1995).
73. B.S. Mohite, S.M. Khopkar, Anal. Chim. Acta. **206**, 363 (1988).
74. N.V. Deorkar, S.M. Khopkar, Bull. Chem. Soc. Japan **64**, 1962).
75. K. Nakagawa, K.S. Okada, Y. Inoue, A. Tai, T. Hakushi, Anal. Chem. **60**, 2527 (1988).
76. V.K. Manchanda, C.A. Chang, Anal. Chem. **58**, 2269 (1986).
77. M.I. Saleh, A. Salhin, B. Saad, Analyst. **120**, 2861(1986).
78. Y.K. Agarwal, P. Srivastav, Talanta, **44**, 1307 (1997).
79. N.V. Deorkar, S.M. Khopkar, Analyst **114**, 105 (1989).
80. N.V. Deorkar, S.M. Khopkar, J. Radioanal and Nucl. Chem. Articles **134**, 433 (1989).
81. N.V. Deorkar, S.M. Khopkar, J. Radional. and Nucl. Chem. Articles **130**, 433 (1989).

82. B.S. Mohite, J.M. Patil, D.N. Zambare, J. Radioanal. and Nucl. Chem. **170**, 215 (1993).
83. N.V. Deorkar, S.M. Khopkar, Anal. Chim. Acta. **245**, 27 (1991).
84. R.P. Namdeo, S.M. Khopkar, Ind. J. Chem. **34A**, 840 (1995).
85. G.D. Blanco, J.S. Arribas, A. Sanz. Medel, Talanta 29 761 (1982).
86. N.V. Deorkar, S.M. Khopkar, Analyst **116**, 961 (1991).
87. B.S. Mohite, J.M. Patil, J. Radioanal and Nucl. Chem. **150**, 207 (1991).
88. S.S. Rane, S.M. Khopkar, Ind. J. Chem. Tech. **3**, 363 (1990).
89. S. Abe, K. Fujii, T. Sone, Anal. Chim. Acta. **293**, 325 (1994).
90. H. Koshima, H. Onishi, Anal. Chim. Acta. **232**, 287 (1990).
91. I. Ando, D. Ishimura, K. Ujimoto, H. Kurihara, Inorg. Chem. **35**, 3504 (1996).
92. M.K. Beklemishev, N.M. Kuzmin, Y.V. Zolotov, Zhur analit Khim **44**, 356 (1989) A.A. **51**, 11B88 (1989).
93. M. Yoshio, M. Ugamura, H. Noguchi, M. Nagamstsu, Analyt Letter Part A **11**, 281 (1978).
94. N. Alizadeh, M. Shamispur, Talanta **40**, 503 (1993).
95. V.E. Poladyan, L.M. Burteneko, L.M. Avlaosovich, A.M. Andrianov, Zh. Neorg. khim **32**, 737 (1987) CA **107**, 13552j (1987).
96. N. Eiko, O Kimiyo H. Namiki, Buneski Kagaku **31**, 602 (1982) A.A **44**, 4B25 (1983).
97. Y.A. Zolotov, V.P. Ionov, N.V. Nizieva, A.A. Formanovskii, Dokl. Akad. Nauk SSSR **277**, 1145 (1984) A.A. **47**, 5A16 (1985).
98. N.V. Isakova, Y.A. Zolotov, V.P. Ionov, Zhur. Anal. khim **14**, 859 (1989) [AA **52**, 4B14 (1990)].
99. L.P. Poddubnykh, Y.A. Zolotov, N.M. Kuzmin, S.G. Dimitrinko, Zhur Anal. khim. **43**, 255 (1988) CA **107**, 47419c 1988.
100. L.P. Poddubnykh, S.G. Dimtrienko, M.M. Kuzmin, Y.A. Zolotov, Zavod Lab **53**, 1 (1987) AA **49**, 12B156 1987.
101. N.V. Isakava, Y.A. Zolotov, V.P. Ionav, Zhur Anal. khim **44**, 1045 (1989) CA **112**, 478939 (1990).
102. R.G. Vibhute, S.M. Khopkar, Mikro Chim, Acta **106**, 261 (1992).
103. Z. Zhou, X. Zhang, Huaxue Xuaxue Xuebao **46**, 496 (1988), A.A. **51**, 3B86 (1989).
104. R.G. Vibhute, S.M. Khopkar, Chemia analytza (Warsaw) **37**, 15 (1992).
105. R.G.Vibhute, S.M. Khopkar, Anal. Chim. Acta. **222**, 215 (1989).
106. A. Kumar, S.M. Khopkar, Chem. and Env. Res. **4**, 145 (1995).
107. R.G. Vibhute, S.M. Khopkar, Ind. J. Chem. **28A**, 1080 (1989).
108. N. Kaneko, N. Shinohara, H. Nezu, Bunseki Kagaku **42**, 299 (1993) A.A. **55**, 10E7 1993.
109. I.V. Pyatnitskii, N.P. Aleksyuk, A.Y. Nazarenko, Zh. Anlit khim **38**, 2176 (1983) A.A. **46**, 11B10 (1984).
110. A.Y. Nazarenko, V.I. Leonenko, Visn Kiev Uni. (Ser) khim **66**, (1985) A.A **47**, 12B125 (1985).
111. V.V Sukhan, O.I. Kroniskovskii, A.Y. Nazarenko, Zhur anlit khim B 1953 (1988) A.A. B 7B5 1989.
112. V.V Sukhan, A.Y. Nazarenko, Izv. Vysstn Uchebn Zavad khim khim Technol. **31**, 39 (1988) A.A. **51**, 7B72, 1989.
113. R.G. Vibhute, S.M. Khopkar, J. Ind. Chem. Soc. **66**, 720 (1989).
114. E.A. Novikov, L.K. Shpigun, Y.A. Zolotov, Zhur anal Khim **44**, 422 (1989) AA. **51**, 11B46 1989.
115. R.G. Vibute, S.M. Khopkar, Talanta **36**, 957 (1989).

116. R.G. Vibhute, S.M. Khopkar, Bull. Bis. Inst. **55**, 5 (1988).
117. K. Nakashima, N. Mochizuki, S. Nakatsuji, S. Akiyama, T. Kaneda, S. Misumi, Chem. Pharm. Bull. **32**, 2468 (1984) A.A. **46**, 12C20 1984.
118. W. Szczepaniak, B. Juskowiak, Chem. Analyt. **32**, 787 (1987).
119. A.B. Arias, G.D. Blanco, A. Sanz-Medel, Microchem J. **32**, 296 (1985) A.A. **48**, 9B35 (1986).
120. E. Chapoteau, B.P. Czech, C.R. Gebauer, K.W. Leong, A. Kumar, Eur. Pat. Appl. 287 326 (1988) CA **111**, 93494n (1989).
121. B.P. Czech, E. Chapoteau, C.R. Gebauer, A. Kumar, W. Zazulak, Anal. Chim. Acta. **241**, 127 (1990).
122. J. Ghasemi, M. Shamsipur, J. Coord. Chem. **31**, 262 (1994).
123. C. Kalidas, H. Schneider, Indian. J. Chem. **34A**, 126 (1995).
124. D. Espersen, A. Jensen, Anal. Chim. Acta. **108**, 241 (1979).
125. J. Ghasemi, M. Shamsipur, Polyhedron **15**, 923 (1996).
126. B. Juskowiak, Chem. Anal. **37**, 479 (1992) A.A. 55 8D66 (1993).
127. S.M. Khopkar, M.N. Gandhi, Crown ethers and cryptands in solvent extraction (unpublished monograph).
128. A.J. A. Drumhiller, J.L. Laing, R.W. Taylor, Anal. Chim. Acta. **132**, 315 (1984).
129. V.J. Mathew, S.M. Khopkar, J. Radioanal. and Nucl. Chem. Letters **201**, 281 (1995).
130. V.J. Mathew, S.M. Khopkar, Chem. Anal. (warsaw) **42**, 651 (1997).
131. Mayuri Gandhi, Solvent extraction of metal pollutants from aquatic environment with cryptands. Ph. D. thesis I.I.T. Bombay (1992).
132. B. Juskowiak, W,.Szezepaniak, Chem. Anal. (warsaw) **32**, 121 (1987).
133. W. Szezepaniak, B. Juskowiak, Anal. Chim. Acta. **140**, 261 (1982).
134. W. Szezepaniak, B. Juskowiak, Mikrochim Acta. **II**, 237 (1987).
135. H. Kawamoto, W. Ishida, K. Tsunodo, H. Akaiwa, Buneki Kagaku **41**, 281 (1992) A.A. **55**, 2D93 (1993).
136. K. Sasaki, G.E. Pacey, Anal. Chim. Acta. **174**, 141 (1985).
137. Y. Katayama, R. Fukuda, K. Hiwatari, M. Takagi, K. Hokuku, Asahi Kogyo Gijutsu Shoreikai **48**, 193 (1986) C.A. **107**, 32225w (1987).
138. K. Wilcock, G.E. Pacey, Talanta **38**, 1991 1315
139. Y. Katayama, R. Fukuda, T. Iwasaki, K. Nita, M. Takagi, Anal. Chim. Acta. **204**, 113 (1988).
140. M. Shiga, H. Nishida, H. Nakamura, M. Takagi, K. Ueno, Bunseki Kagaku **32**, 293 (1983) A.A. **46**, 5B59 (1984).
141. W. Szezepaniak, W. Ciszewska, B. Juskowiak, Chem. Anal. (warsaw) **37**, 465 (1992).
142. A. Yu. Nazarenko, T.A. Bykh, Zh. Anal. Khim **38**, 1946 (1983) A.A. **46**, 9B27 (1984).
143. E.I. Morosanova, O.A. Kosyreva, N.M. Kuzmin, Yu. A. Zolotov, Zh. Anal. Khim **43**, 1614 (1988) A.A. **51**, 3B48 (1989).
144. A. Yu. Nazarenko, V.V. Sukhan, E.D. Velidchenko, Izr. Vyssh Vehebn Zaved Khim Tekhnol **32**, 57 (1989), A.A. **52**, 8B23 (1990).
145. W. Szezepaniak, B. Juskowiak, W. Ciszewska, Anal. Chim. Acta. **156**, 235 (1984).
146. B. Vaidya, M.D. Porter, M.D. Utterback, R.A. Bartsch, Anal. Chem. **69**, 2688 (1997).
147. K. Saito, Y. Masuda, E. Sekido, Bull. Chem. Soc. Japan **57**, 189 (1989).
148. T. Maeda, K. Kimura, T. Shono, Chem. Lett. 275 (1982) A.A. **43**, 3A10 (1982).
149. E. Sekido, K. Chayama, M. Muroi, Talanta **32**, 797 (1985).
150. M. Muroi, A. Hamaguchi, E. Sekido, Anal. Sci. **2**, 351 (1986) C.A. **106**, 26978g (1987).

151. B. Saad, S.M. Sultan, Talanta **42**, 1349 (1995).
152. M. McCarrick, S.J. Harris, D. Diamond, Analyst. **118**, 1127 (1993).
153. F. Amaud. Neu, R. Arnecke, V. Baohmer, S. Fanni, J.L. M. Gordon, Marie-Jose, schwing, Weill and W. Vogt, J. Chem. Soc. Perkin Trans. **2**, 1855 (1990).
154. P.D. Beer, M.G.B. Drew, A Grieve, M.I. Ogden J. Chem. Soc. Dalton Trans. 3455 (1995).
155. I. Yoshida, N. Yamamoto, F. Sagara, K. Ueno, D. Ishii, S. Shinkai, Chem. Soc. Lett **12**, 2105 (1991) A.A. **54**, 12D53 (1992).
156. D. Malkhede, S.M. Khopker, J. Radio analytical and nuclear chemistry Letters **241**, 179 (1999).
157. Y. K. Agarwal, M. Sanyal, J. Radioanal. and Nucl. Chem. **198**, 349 (1995).
158. D. Malkhede, P.M. Dhadke, S.M. Khopkar, Ind. J. Chemistry, 38A, 1079 (1999).
159. D. Malkhede, P.M. Dhadke, S.M. Khopkar, Analytical Science 5, 1 (1999).
160. R.M. Khandwe, S.M. Khopkar, Talanta **46**, 515 (1998).
161. A. Gupta, S.M. Khopkar, Talanta **42**, 1493 (1995).
162. V.J. Mathew, S.M. Khopkar, Talanta **45**, 1699 (1997).
163. D. Malkhede, P.M. Dhadke, S.M. Khopkar, Proceedings of International Conference in Solvent Extraction 1999 held at Barcelona, Spain (1999) July.
164. S. Namura, H. Taniguchi, S. Tamura, Chem. Letters 1125 (1989) A.A. **52**, 6A8 (1990).
165. D. Malkhede, Solvent extraction separations of selected elements with hexaacetyl calix (6) arene, Ph. D thesis University of Bombay, Bombay (2001).
166. D. Malkhede, M.A. Phadke, P.M. Dhadke, S.M. Khopkar, Candian Journal of Analytical Sciances 43, 143 (1999).
167. D. Malkhede, P.M. Dhadke, S.M. Khopkar, Ind. J. Chemical Technology **9**, 7 (2000).
168. R.M. Williams, J.W. Verhoeren, Recuiel **111**, 531 (1992).
169. H.S. Ajgaonkar, S.M. Khopkar, Chemia analytiza (warsaw) **44**, 61 (1999).
170. M.N. Gandhi, S.M. Khopkar, Ind J. Chemistry **30A**, 706 (1991).

8
Extractive Emission Spectroscopy with Crown Cryptands and Calixarenes

8.1 Introduction

Crown ethers, cryptands and calixarenes have been extensively utilised for the solvent extraction separation of various elements. After extraction, metals were stripped into the aqueous phase and determined by suitable physical mehod of analysis. Such physical methods included not only absorption spectrophotometry but also emission spectroscopy. Such emission methods involved either molecular emission or atomic emission spectroscopy. For the analysis of alkaline earths or alkali metals flame emission spectroscopy was commonly used.

As discussed in previous chapter extractive photometric methods had a distinct advantage from the view point of speed of analysis as it was possible to directly determine the metal colorimetrically in the organic phase. The color development was facilitated by use of suitable chromogenic counter anion or chromogenic ligand. Such chromogenic ligand were used after extraction either in the organic phase or in the aqueous phase. A large number of chromoionophores were synthesised to facilitate such extractive determinations. Such compounds included thiacrown ethers, azacrown ethers or lariat ethers with side arm containing effective chromophoric group.

However direct applications of the techniques mainly involving either atomic emission or molecular emission for the organic phase was not always fissible. Although several attempts were made to explore suitable counter anions like erythrosine, eosine, rosebengal extra which were capable of exhibiting fluoroscence, it was not always practicable to make use of such anions. On account of the existance of high ionic concentration they created new problems of setting up of competitive equilibria between them and chromogenic ligand, but in case of atomic emission analysis best results were obtained by stripping the metal from the organic phase and resorting to direct AES or AAS analysis.

In order to circumvent these problems, attempts were made by large number of researchers to synthesise new crown compounds containing fluoroionophore. Many such fluorogenic crown ethers were synthesised. They found extensive applications in fluoroscence measurement. In fact fluorimetry permitted analysis up to the level of 10^{-9}, i.e. nanogram concentration. In radioimmunoassay in medical therapy one was confronted with the analysis of metal at very low ion concentration. This problem was sorted out by commercial applications of fluoroimmunoassay. Amongst array of metals the $4f$ inner transition lanthanide series of elements showed a great promise. Just as europium complex was used commonly as internal standard in electron spin resonance spectroscopy; similarly europium complexes with fluorogenic crown ethers made a miracle. Other metals in the series like terbium, dysoprosium and samarium showed same promise. In fact not only extractive fluorimetry was greatly facilitated but also the mechanism of complexation, transfer of electrons from ligand to metals, shielding of metal complex from adverse influence of s- and p-orbitals and shift in band spectra to region was very well understood. Such knowledge provided a definite channel for designing new fluoroionophore for analysis.

The analytical chemistry had witnessed great progress due to discovery of new organic ligands with better selectivity or sensitivity. Such selectivity was accomplished by effective substitution in sterically sensitive position.

The common example was that of 1-10-phenanthroline(I), 4-7 diphenyl 1-10 phenanthroline(II) and 2-9 dimethyl-4, 7 diphenyl 1-10 phenanthroline (III). First two ligands were good for extraction of iron (III) but third ligand was good for extraction and complexing only copper (III). This was facilitated by substitution of methyl group in 2-9 position which hintered complexation with iron (III) in octahedral environment but it facilitated the complexation of copper (II) in tetrahedral environment. This phenomena is termed as steric hindrance for selectivity.

This idea gave rise to intensive investigation of specially designed switch on-crown ethers. These compounds had the property to change their basic behaviour by effective substitution or introduction of side arm called 'antena'. These synthetic modes gave rise to photo responsive crown ethers. There were others involving pH sensors, temperature sensors or redox reactions sensors compounds, but from analytical view point photo responsive sensors had great utility. They consisted of simple compounds, bis crown ethers or crown compounds capped with cations or anions. Some progress was made in chemistry of cryptands by the introduction of bipyridyl or quinolyl group. Unfortunately no such revolution had been so far made in chemistry of calixarenes.

Although not much investigations on sensitivity, selectivity and specificity of photogenic crown ethers had been made in greater details; one is expected to witness such investigations in future in few decades. Therefore attempts have been made in this chapter to discuss fluoroionophore, ring substituted fluorogenic crown ethers, applications of fluorscence, atomic emission and atomic absorption after or during extraction of metals with the crown ethers, cryptands and calixarenes.

At the end of the chapter some glimpses are provided for switch on crown ethers. It was possible to demonstrate the effective substitution or structural modification in ligands which in turn influenced analytical selectivity for such analysis.

8.2 Fluoroionophores

Fluoroscence analysis was preferred to colorimetric analysis as former was less influenced by impurities and sensitivity was relatively higer. The fluoroionophores are shown in structure (IV) and (V). They were obtained by Mannich reaction [1].

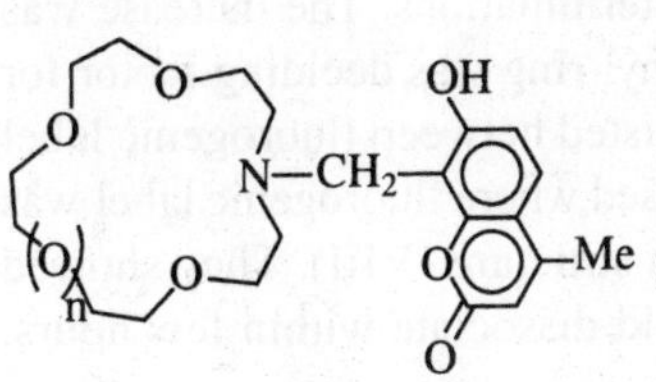

(IV) 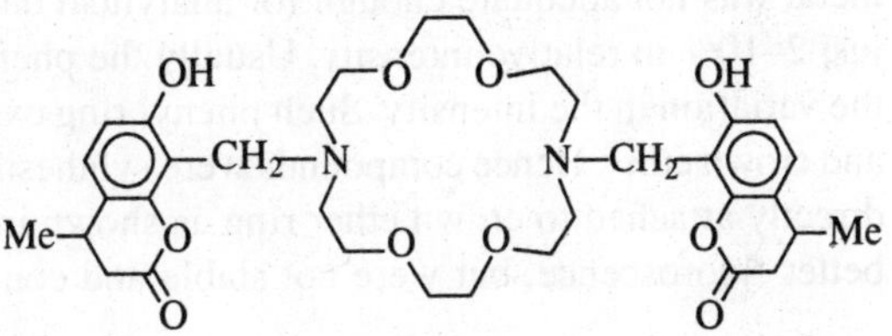   (V)

These compounds can be used for fluorimetric determination of lithium and calcium at nanogram concentrations. Some ligands were not good due to acidic nature and possessing weakness to form salt with amines. This was because the salt extraction depended upon nature of solvent used. By process of ionisation a red or a bathochromic shift was observed, while complexation with metal a blue or hypsochromic shift was noted due to strong H-bonding between the phenolate oxygen and NH (+) hydrogen. Blue shift was indication of stable complex formation. Other fluoroionophores are shown in structures (VI) and (VII).

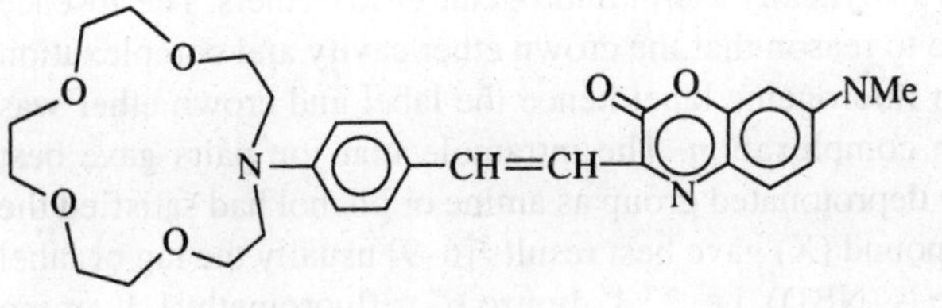

(VI) 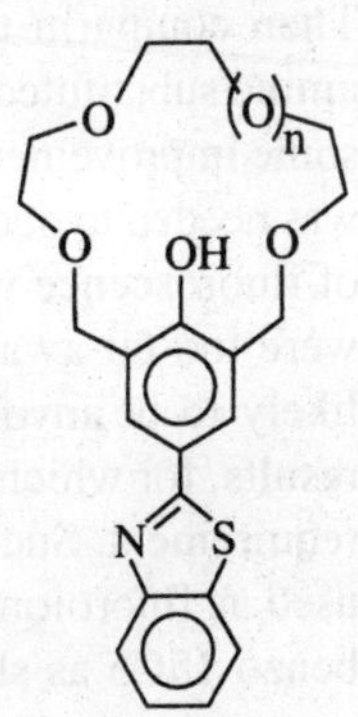   (VII)

The fluoroionophore (VI) involved the principle for guest selective colour complexation where heteroatom, viz. N-binds guest while in fluoroionophores (VII) the two heteroatoms, viz. N and S were attached to hydrogen attached guest cation. It was always better to tailor made a fluoroionophore to complex with metal cation which acted as guest molecule. Various process for such synthesis were suggested by many workers.

8.3 Ring Substituted Fluorogenic Crown Ethers

In previous chapter we had seen how to modify crown ethers by incorporation of chromophoric group. They gave red shift in absorption spectrum on complexation. The shift was as large as 150 nm, due to the intramolecular ion pair formation between complexed metal like that of s-block elements and the anionic functional group of the singly deprotonated chromophoric crown ether. It gave stable complex with good selectivity [2, 3]. On same lines, efforts were made to synthesise the fluorogenic compounds similar to chromogenic crown ethers. Large number of compounds with fluorogenic labels were synthesised [4]. Unfortunately change in intensity of the fluoroscence on complexation with metal was not adequate enough for analytical determinations. The increase was just 2–10% in relative intensity. Usually the phenyl ring was deciding factor for the variation in the intensity. Such phenyl ring existed between fluorogenic label and crown ether. Hence compounds were synthesised where fluorogenic label was directly attached to crown ether ring as shown in structure (VIII). They showed better fluorescence, but were not stable and could dissociate within few hours.

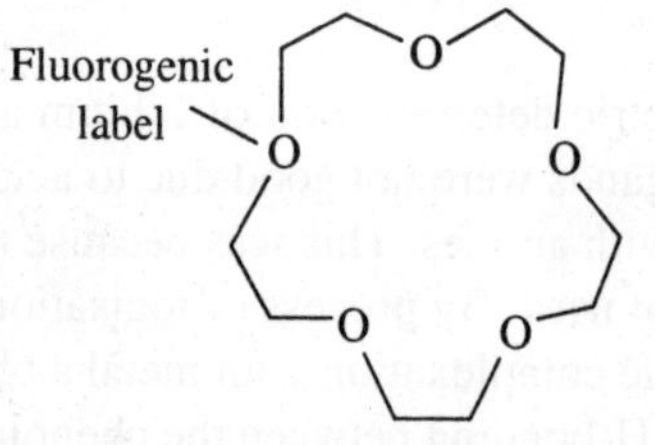

Aminomethyl 15C5
(VIII)

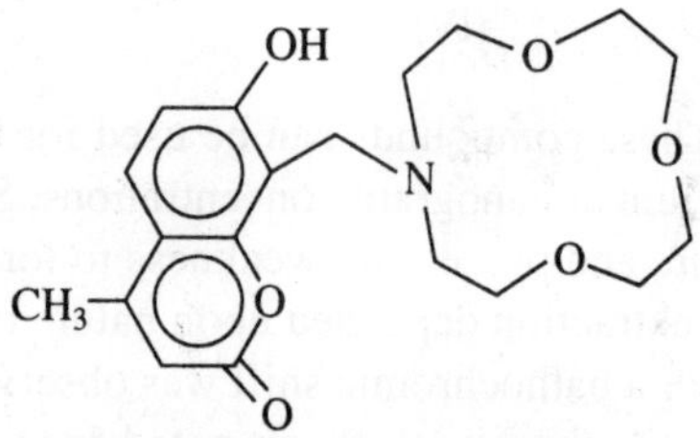

Coumarin azo 12C4
(IX)

Then coumarin type [5] fluorogenic label attached through -CH_2 group to an amine substituted crown ether were prepared (XI) as shown below. They showed some improvement in fluoroscent characteristics on examination. A new approach was needed to obtain the analytically useful fluorogenic crown ethers. The absence of fluoroscence was due to reason that the crown ether cavity and complexation were too far away from fluorogenic label hence the label and crown ether was likely to be involved in complexation. The intramolecular ion pairs gave best results, for which easily deprotonated group as amine or phenol had satisfied the requirement. Such compound (X) gave best results [6–9] usually the tag or label used as fluoroionophore is (NBD), i.e. 2″ 4′ dinitro 6″ trifluoromethyl 4′ amino benzo 15C5 as shown in structure (XI).

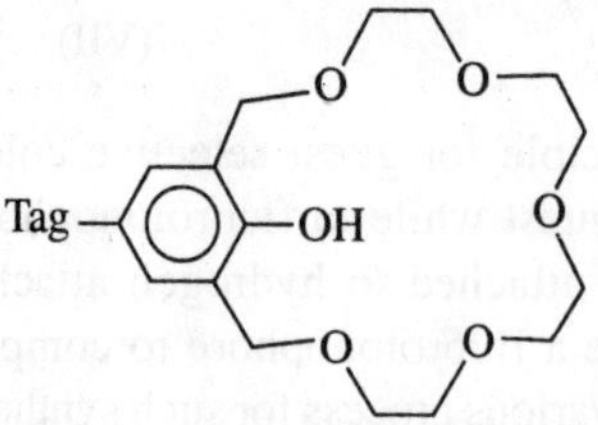

Interannular 17C%)
(X)

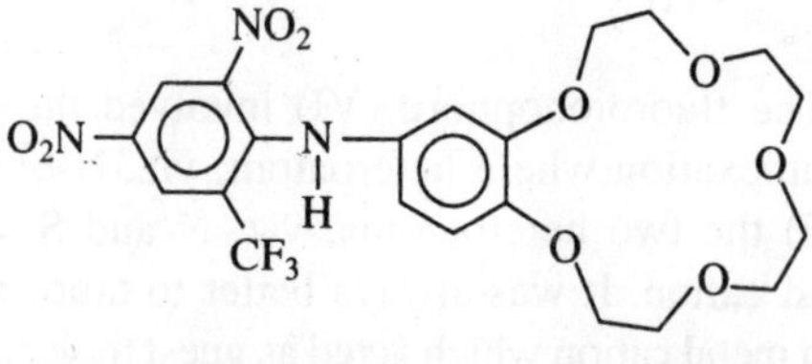

(XI)

Such tag or label was attached to the compound shown in structure (VIII) or structure (X). In order to study the effect of the substitution in the sterically sensitive positions two compounds 4-bromoanisole-18C5 and 4-bromophenol 18C5 were prepared in picrate ion pair complexes. The results for partition studies showed that phenolic compound was more soluble than anisole crown ether (e.g. 1.1×10^{-2} M for anisole while 6.8×10^{-3} M for phenolic derivative) gave partition coefficient as 3.25×10^{-4} with anisole 8.29×10^{-4} with phenolic form [10] at pH 3.0. The crown ether with a deprotonated phenol group acted as counter ion in ion pair complexation and inhibited or prevented the ion pair formation involving picrate anion. When these compounds were tested as the extractant for alkali metals like say rubidium the anisole derivative was more efficient than phenol (log D for anisole 1.84 while log D was 1.44 for phenol analogue). The former was very selective for potassium. (log D 2.45 Vs log D 1.63) At pH 3.8 for anisole and phenol form resp. Thus tagging this kind of compound generated a ligand with required effective selectivity for potassium. This concluded that fluorogenic crown ethers with a phenol group in the crown ring complex were best for alkali metals; due to tagging or labelling. It was interesting to study the spectra, a deprotonated compound showed blue shift of 30 nm whenever protonated from 330 to 350 nm [11–19].

This was due to blue shift in absorption spectra resulting in enhancement of molar absorptivity. The introduction of anionic moiety in crown ether ring increased the fluoroscence on complexation, because the phenolic group was deprotonated and formed intramolecular ion pair. The crown ring itself was restricted in movement as oxygen atom in ring interacted with alkali metals. New compound gave large fluoroscence intensity if it was present singly and it was not in the mixture. The study of fluoroscent intensity with pH showed, it did not change with pH of solution thus crown ring was attached to alkali metal. No quenching effect was observed. Thus the fluorescence intensity and selectivity was altered on complexation. This selectivity could not be changed by size change in cavity. A balance between size of cavity in free ligand and cavity made after complexation was essential.

8.4 Luminescence Characterisation of Lanthanides by Crown Ethers

Amongst transition and intertransition elements the lanthanides and specially europium, terbium and dysoprosium had been fully explored for complexation with fluorogenic crown ethers. The main reason was that these compounds were used in fluoroimmunoassay in biomedical investigations in place of radioimmunoassay. The significant reason was that these techniques were applicable at less than microgram concentration, i.e. nanogram concentration of metals. For detection of small amount of ion these methods were best suited for the purpose. Crown ethers as such were extensively utilised for the solvent extraction of lanthanides [20–23]. The crown ethers were also used as the synergistic extractants. Lanthanides were generally determined after extraction by AAS, AES or radiochemical methods [24–34]. In spite of that these methods were still not very popular with chemist because these methods had low sensitivity, high interference. On the contrary indirect analysis with fluorogenic counter part increased the sensitivity [35–38] but showed loss in specificity. The methods involving fluoroscence for determination of lanthanides was ideal. Amongst lanthanides trivalent samarium, europium, terbium and dysprosium were known to exhibit luminescence in solution. The luminescence quantum yield (ϕ) needed special technique like time resolved luminescence. The ϕ value was increased in highly absorbing compound. This was because energy donor and transfer to lanthanides itself acted as acceptors. Thus the ϕ value could be increased. The energy transfer occurred between aromatic hydrocarbons and lanthanide ions. One observed luminescence as narrow lines [39–42]. The distance between donor and acceptor could be as large as 100°A [43–44]. Further the donor was not necessarily chelated to metal ion but only was near the counter anion.

In solvent extraction, lanthanides complex with crown ether and the cation crown ether complexes were extracted in the organic solvents by ion pair formation with appropriate counter anion. Picrate was best as counter anion. It was used all along on account of its lipophilicity. Thus it was possible to use counter anion to transfer excitation energy to lanthanide ion to increase the luminosity of radiation. Most effective counter anion for lanthanides was benzoate [45]. It could readily transfer energy to the lanthanide and enhance intensity of fluoroscence to facilitate extraction as well as direct determination. Crown ethers generally used for the purpose were 12C4, 15C5, 18C6 with benzoate as counter ion. The main cause of enhancement in fluoroscence intensity was the lack of quenching by -OH group in the organic phase. Benzoate gave good results because its absorption spectra did not overlap with the absorption spectra of lanthanides. The triplet energy of benzoate anion at 27,000 cm^{-1} [46] was much higher than the extracted state energy levels of lanthanide ions. It was having fairly good solubility in organic solvents. The pka value of benzoate was low with good value for water soluble dissociated form and fine lipophilicity. The enhancement in the emission intensity was due to energy transfer from benzoate counter anion to lanthanide ion. Usually energy transfer from naphthalene, 1-naphthaldehyde, benzil, benzophenone, benzaldehyde, tryphane and tyrosine to terbium or europium was reported [45, 46]. In such cases lanthanide was directly complexed with aromatic compounds. It was thus confirmed that donor aromatic compound counter anion (cf benzoate) of extracted complex transferred energy to the lanthanide ion acceptor (terbium, dysoprosium and europium) even at lower concentration. The lanthanides were least soluble in organic solvent, and that too only in the presence of the benzoate anion luminescence intensity was increased as compared to same energy in the aqueous solution, because of the quenching by high frequency vibrations, i.e. over tones of the OH groups of water. Excellent results were obtained with ethylacetate as the solvent [47]. For europium (III) energy transfer was 12.6% with life time of 110. Thus upto 67 times enhancement in luminescence intensity of the lanthanide ions was accomplished by simple extraction followed by measurement of fluoroscent intensity. The best part of the story was that crown ethers provided some degree of selectivity in extraction methods. The sensitivity was much higher and was further increased by proper selection of counter anion.

Thus compartmentalised approach of crown ether, benzoate and lanthanide furnished promising results. Luminescence in aqueous phase was much less than in the organic phase. Four times enhancement was observed even in chloroform. Indirect excitation with benzoate ion enhance the intensity by energy transfer process. Such kind of method was extended to the other lanthanide cations in subsequent work.

8.5 Lanthanide Complexes as Supramolecular Photochemical Devices

An important arena of supramolecular chemistry was the encapsulation of lanthanides in order to form complexes. This also involved the investigation on luminescence characteristic of metal ion with appropriate organic ligand [48]. This was related to the long life time and line like emission bands rendering ions luminescent. Such property was due to excited emitted state and ground state had same number of f-electrons and f-orbitals were well shielded behind s- and p-orbitals. The inter electronic repulsion, spin-spin coupling in coordination sphere and ligand field was decided by electronic f^n configuration. Usually the transition between states of f^n configuration was strictly forbidden. The coordination environment influenced the luminescence intensity as well as its life time via solvent molecules attached to metal ions and low lying short lived excited states. Non radiative loss was properly controlled by selecting suitable coordination environment. The luminescence intensity was related also to efficiency of absorption of light. The supramolecular chemistry permitted to design lanthanide complexes with them for the purpose of light absorption and luminescence emission experiments.

Lanthanide ions showed low absorption coefficient [48] in UV and visible region, but one could construct complexes which would have better absorption in ligand centered or change transfer bands. Ordinarily the lanthanides could not form stable complexes due to its peculiar electronic configuration. But cryptands were exception to this rule. They had spheroidal cavities and binding sites having oxygen or nitrogen as donor atoms. They were capable of encapsulating ions and protecting them from surrounding interactions [49–51]. It was possible to have intense luminescence effect due to 'antenna' effect [52] in the ligand. The antenna effect was specified as the conversion of light process by absorption of energy and transfer emission cycle involving absorbing ligand and emitting metal ion. Factors contributing luminescent intensities were ligand absorption intensity. How efficient was process to transfer energy from ligand to metal and the overall luminescence due to metal was not known.

The complexation of europium (III) with cryptand-2, 2, 2, gave a spectra ($\varepsilon = 100$) with weak bands [53–55] which originated due to ligand to metal charge transfer transition (abbreviated as LMCT). The band of higher energy at $\lambda = 248$ nm was due to amine nitrogen while other oxygen gave band at $\lambda = 298$ nm. There was remarkable increase in life time of the luminescence on using heavy water instead of water on account of coupling with high energy O-H oscillators. In Eu (cryptand-2, 2, 2) complex europium was well shielded from interaction of solvent molecules [56]. Usually europium aqueous ion coordinated around 9–10 water molecules. However their replacement with fluoride anions by ion pairing changed the intensity of luminescence. The lifetimes were much smaller for [Eu (cry-2, 2, 1)]$^+$ complex due to replacement of few coordinated water molecules. A fraction of decrease in radative decay constant was observed due to coordination with fluoride anions. This was perhaps due to blue shift of f and Tb^{3+} excited states [57].

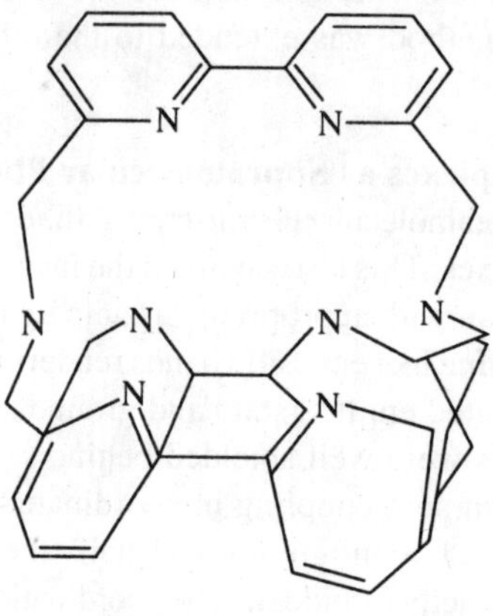

(XII)

The bpy, bpy bpy cryptates as shown in structure (XII) showed intense absorption bands in UV due to π-π^* transitions in bpy unit. There was red shift and decrease in energy. Further attempts were made to improve luminescence of lanthanide bpy, bpy, bpy, cryptates by replacing one bpy unit with bpy O_2 or biq O_2 unit. The cation like say europium was nicely protected from interaction with water; with improved light conversion efficiency over earlier Eu (III) cryptates.

Further 2, 2' bipyridine chromophoric group were used as building block to get branched macrocyclic ligand (XIII).

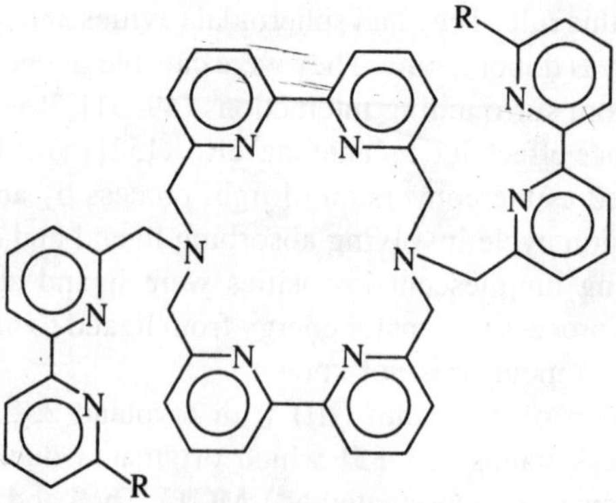

R = H, CH_3

(XIII)

In such of the complexes the ligand shielded the metal ions towards solvent molecule reaction in comparison to cryptates. In addition 2, 2' bipyridine chromophoric unit was used to prepare tripode and tetrapode [58] complex with europium. They were extensively studied by many workers.

However most fascinating results were obtained with the supramolecular compounds like calix(n)arene with the ligand was p-t-butyl calix(4)arene tetracetamide (XIV).

[Chemical structure diagram labeled (XIV) showing a calixarene unit with t-Bu, CH₂, O, CO, NEt₂ groups, with subscript 4]

It had encapsulated trivalent europium and terbium ions [59] to form complexes which were soluble in water. They had intense luminescence characters [60]; with only single water molecule coordinated to the trivalent metal ions. The absorption peaks were at 273 and 282 nm. Within europium and terbium the latter compound (XIV) had showed long luminescence life time irrespective of the temperature. However for europium complex low value was due to existence of ligand to metal charge transfer state in the complex structure.

Thus in summary it was seen that

1. bpy absorption led to red shift and shift was large for branched macrocylic ligand but for cryptates shift was low.
2. The introduction of heterocyclic N-oxide groups in ligands containing bpy unit gave remarkable change in absorption spectra. The absorption spectra of complex of calixarene (XIV) had weaker intensity.
3. For both europium and terbium complexes with ligand (XIV) value of K_{nR}^{x} (T) was small as these ligand did not promote excited state.
4. Cryptands shielded metal from water interaction less efficiently than branched ligands.
5. N-oxide group led to decrease in coordinated water molecule. Water coordination was obstructed by steric hindrance and negative charge of oxygen atom.
6. Europium and terbium complexes were best with luminescence labels in fluoroimmunoassay [51]

8.6 Analytical Application in Fluoroscence Spectroscopy

After discussion of fluoroionophores, or fluorogenic crown ethers, lanthanide complexes with crown ethers and the calixarenes it is worthwhile to discuss applications of these macrocyclic and supramolecule compounds in extractive fluorimetry for quantitative chemical analysis.

Crown ethers were used for complexation of metals like alkali metal, few lanthanides and silver lead and thallium [62–76]. Mostly aza crown compounds were used for the complexation of the alkali metals. Diaza 18C6 or its derivatives were used for the complexation of lanthanides specially europium and terbium. The aza analogue of DB18C6 or 18C6 was used for the determination of thallium and lead. The fluoroscent derivative of 14C4 was best for analysis of lithium [63]. With eosin as the counter anion [65] potassium was extracted and fluorimetrically analysed with 18C6 in ethylene chloride. A novel method for analysis of lead from water consisted of its extraction with 18C6 with eosin as counter anion permitting separation and determination of complexes in one single operation [76].

Blanco and his group [76-84] described the applications of cryptands with appropriate counter ion for extractive and fluorimetric determination of metals like alkali metals [77-79], zinc [80], cadmium [81], lead [82-93] and mercury [84]. In most of these cases concentration of ligand used was decimolar. Either chloroform or dichloroethane were used as the diluents for cryptands. Cryptand-221 or Cryptand-2, 2, 2 was employed as the extractant. They have invariably used eosin as the counter anion which was also fluoroionophore. All these metals were directly extracted and determined in the organic phase by fluoroscence measurements. The concentration of the eosin employed was same as that of cryptands ranging from 0.1-0.5 M in chloroform. The pH of extraction varied from pH 7.0-8.5 and it was never less than pH 7.0. These methods proved useful as they permitted separation of these metals from real samples like sugars, coal, soft drinks etc. [7-84].

Since the discovery of calixarene was of recent origin calix(4)arene specially was used for extraction and fluorimetric determination of sodium [85-86], europium [87-88] and lead [89-91]. As shown earlier t-butyl calix(4)arene diamide gave excellent fluorescence with sodium. Calix (4) arene triacids were used for extraction of europium in methanol media [87]. The calix (4) arene incorporating 2, 2' bipyridylazine form was used for extractive fluorimetric analysis of terbium [88]. Metals like lead, silver, sodium showed intense fluorescence with calixarene ionophores but overall response for other transition metals was very poor.

The various conditions such as pH, kind of extractant used, solvent, counter anion employed wave length for excitation as well as for analysis of various metals are summarised in Table 8.6.1 through Table 8.6.3 for crown ethers, cryptands and calixarenes respectively.

8.7 Atomic Emission Spectroscopic Analysis with Crown Compounds

In the earlier section we had considered in great details the applications of molecular luminescence spectroscopy with crown ethers, cryptand or calixarenes for the analysis of several alkali metals and lanthanides. For that purpose many time specific fluorogenic crown ethers were designed and great deal of attendation was paid towards transfer of energy between ligand like crown ether cryptand and metals like lanthanide. The entire concept was based on encapsulation of metal ion and shielding it from coordination environment influenced by s- or p-orbitals.

However in case of atomic emission spectroscopy there was no special need for tailoring specially designed crown compounds. Most of the process involved solvent extraction separation of these compounds followed by flame emission spectroscopy of the s-block metals. Only in few instances metals belonging to d- or f-block series were analysed by emission spectroscopy. For such purpose crown ethers gave favourable results but not cryptands or calixarenes. Since such analysis involved alkali metals or alkaline earth elements, flame emission spectroscopy was economically best suited for the job, only in exceptional circumstances specialised methods like inductively coupled plasma atomic emission spectroscopy methods were utilised.

Table 8.6.1 Applications of crown ethers in fluorimetry

Element	Optimum Conditions of Extractive Fluorimetry
Alkali metals	4-bromocresol 18C5 or 4-bromophenol 18C5 was used for the extraction, but addition of fluorogenic group decreased the selectivity of the ligand [61].
Li (I)	0.1 M of 16 (benzothiazothiazol 2-yl) 3, 6, 9, 12 tetraoxabicyclo (12, 3, 1) octa [18] 14, 16 trien-18ol in benzene formed complex and it was determined at 375 mm [62]. Fluoroscent derivative of 14C4 in methylene chloride was complexed and determined at 261 nm but sodium showed interference [63].
Na(I)	N-(9-anthylmethy1) monoazo-18C6 and 15C5 analogue with N-(anthylmonoazo) 18C6 complexed and extracted lead complex acted as proton scavanger [64].
K(I)	18C6 in methylene chloride in the presence of eosin extracted it and determined at 322 nm but lead, amonia interfered [65] 4-acryloy1 amido B18C6 extracted it and determined at 360 nm [66] N-(4 methyllium beli ferone 8y1 methy1) monoazo 15C5 and 18C6 extracted it in chloroform. Potassium from serum was analysed [67] 15C5 and dibromic acid glucose was also extracted [68] as sandwich complex.
Eu (III)	At pH ~ 6.5, 0.2M polymer bound diaza 18C6 with 2-thenoy1-trifluoroacetone was extracted and stripped with water and analysed at 612 nm [69] 18C6 or 15C5 in aqueous organic solvent acted as synergist along with benzoate as the counter anion. The extraction was possible in the presence of terbium and dysoprosium [68] poly (2 methacryloylox methy1 15C5) with AlBN as an initiator complexed it and was determined at 435 nm [70]. Aza crown ether modified with cyclodextrin was used to get swing in structure [71].
Tb (III)	1, 4, 7, 10 tetraza cyclodecane containing naphthy1 chromophore was studied in water and acetonitrile [72].
Cu (II)	N, N' bis (1-pyrene carbony1 7,16 diaza 18C6 complexed shared spectroscopic changes [73].
Ag (I)	At pH 8.6, 2, 3, 5, 6, 8, 9, 11, 12 octahydrobenzol 13-dioxa - 4, 7, 10 tri thia cyclo pentadecane formed complex which was determined at 558 nm. Only mercury (I) showed strong interference [74].
Tl (I)	At pH 11-12, 0.1M aza analogue of DB18C6 in chloroform extracted it with diphenylamine as the counter anion and was analysed by x-ray fluoroscence [75].
Pb (II)	18C6 in the presence of eosin extracted and determined it at 549 nm lead from waste water was thus analysed [76].

8.8 Applications of Emission Spectroscopy for Analysis

Most of the alkali metals [92–100] were extracted by crown ether and determined directly in the organic phase by flame emission spectroscopy. The crown ethers used were 21C7, DC18C6, DB18C6, 18C6 etc. The solvents used were one which were suitable for voltalisation in flame photometry. They included toluene, ethylenechloride, methylene, chloride, kerosene [97] benzene. The concentration of the extractant employed was in the range of 0.01-0.5M in appropriate diluent. The counter anions preferred during extractions were noninterfering in final

8.6.2 Application of cryptands in fluorimetry

Element	Optimum Conditions of Extractive Fluorimetry
Ca (II)	0.1M of cryptand 221 in dichloromethane extracted it with 0.15M eosin as ion pair complex. It was analysed at 552 nm calcium from sugar [77].
Sr (II)	0.17M of cyrptand-222 in chloroform extracted it with 0.67M eosin as counter anion and determined at 552 nm [78]. About 0.1M cryptand-222 in chloroform extracted it at pH 8.5 with 0.1M eosin and determined at 532 nm [79]. In such process strontium was first extracted in dichloroethane [79]
Zn (II)	0.1M cryptand-221, in dichloroethane in the presence of 0.1M eosin extracted and determined at 537 nm zinc from coal was analysed [80].
Cd (II)	0.2M cryptand -221 in ethylene chloride extracted it with 0.2M eosin and was analysed at 555 nm but lead, copper and mercury showed strong interference [81].
Pb (II)	0.15M of cryptand-222 in chloroform extracted it with 0.35M eosin and was determined at 552 nm. Lead from soft drinks was analysed [82]. A novel method for sequential separation of lead and cadmium at pH 7.8 was developed. Lead was extracted with cryptand-222 in chloroform with eosin and was determined at 552 nm, while cadmium was extracted with cryptand-221 under identical conditions [83].
Hg (II)	At pH 8.5, 0.1M cryptand-222 in chloroform extracted lead while cryptand-222 in dichloroethane extracted mercury [84] calcium, strontium, iron, aluminium showed interference.

Table 8.6.3 Application of calix (n)arenes in fluorimetry

Element	Optimum Conditions of Extractive Fluorimetry
Na (I)	p-tertiary butyl calix(4)arene diamide derivative was used [85] to get fluorescence. Tris (6, 6′ oligoethylene glycol- 3, 3′ bipyridazine was also used [86] with ruthenium chloride to form complex.
Eu (III)	Calix(4)arene triacids were explored in methanol to form complex [87] calix(4)arene receptor incorporating 2, 2′ (dipyridene-b-yl) methyl and 9-methyl 1-10 (phenanthioline-2yl) was also complexed to exhibit fluoroscence [88] lead calixarene ionophore were used [89]. Metals studied along with lead were silver and sodium.
Pb (II)	It showed intense fluoroscence intensity with some selectivity for sodium ions with no response for transition metals [90, 91, 92].

determination. They consisted of bromide, picric acid latter being most favoured counteranion. In few instances synergistic extractions involving TBP [97] TTA [101] hydroxamic acid [102] were employed. In most of the instances best method was to strip the metal from the organic phase to the aqueous phase followed by its direct flame photometric analysis. Therefore for better voltality dilute hydrochloric acid was preferred as the stripping agent. Mohite and Khopkar [24] [93–96] and [98] made significant contribution in solvent extraction and flame photometric determination of alkali and alkaline earths with crown ethers

[93]. The proposed methods permitted sequential separations of calcium, strontium and barium [98]. The pH employed for the extraction although varied in acid region, it was mainly confined to pH region of 2-4.5. Only strontium had a very broad range of extraction [24]. These methods permitted analysis of metals from real samples. Cryptand-2, 2, 2 was used for potassium while calix (4) arene (N-p chlorophenyl) derivative was used for extraction of thorium.

The optimum conditions for extractions of different metals mainly with crown ethers, cryptands and calixarenes are best summarised in Tables 8.7.1 to 8.7.3.

Table 8.7.1 Application of crown ethers in emission spectroscopy

Element	Optimum Conditions for Extractive Emission Spectroscopy
Alkali metals	They were extracted with 21C7 on dedceylnaphthalene sulphonic acid in toluene from nitric acid media [92] DC18C6 was used for the extraction of alkali metal picrates [93].
K(I)	Poly DB18C6 extracted it from 0.5-8M bromide solution and it was stripped with water containing 0.1-8M acetic or sulphuric acid and analysed at 767 nm [95]. Also extracted at pH 2.0 with 0.01 of 18C6 in dichloroethane or dichloromethane in the presence of 4-10 M $\times$ 10^{-3} picric acid and stripped with 6M hydrochloric acid. Potassium from milk soft soap was analysed [93] 0.1M of reagent in tetrachloroethane extracted it and stripped with mineral acid [95].
Rb(I)	At pH 3.7 it was extracted with 0.01M DC18C6 in dichloromethane with 0.01M picric acid and was stripped with 2M nitric acid and was analysed at 780 nm. It was analysed from mineral deposits [96].
Cs(I)	Was extracted with 0.01M DB18C6 in kerosene with tributylphosphate as synergist and was subsequently stripped with 8M nitric acid [97].
Ca(II)	At pH 3-9 it was extracted with 0.01M 18C6 in dichloroethane with 0.01M picric acid and was stripped with 1M hydrochloric acid and determined at 423 nm. The sequential separation of calcium, strontium and barium was also possible [98]. It was analysed from plant milk seeds.
Sr(II)	Was extracted with 0.1M DC18C6 in dichloromethane from 1.5-2M nitric acid and was back washed with water when best results were obtained with 1.5-2M nitric acid as the strippant [99]. It was also extracted at pH 3-9 with 0.01M 18C6 in chloroform ethanol mixture from 0.01M picric acid and stripped with 1M nitric acid and determined at 460 nm [24]. Strontium in geological samples was analysed. At pH 3.0 it was extracted with 18C6 polymer bound in ethanol nitric acid mixture and was stripped with water at 60°C [100].
Eu(III)	At pH 7.5, 0.18 mM of 1-7 diaza 4, 10, 13 trioxycyclopentadecane N N' diacetic acid a new macrocyclic ionophore in the presence of 2-thenoyltrifluoroacetone in benzene extracted it and was stripped with 1M hydrochloric acid and determined at 433 nm. Similarly yttrium and lucetium were also extracted with 1-10 diaza 4, 7, 13, 16 tetraoxacyclo-octadecane N N' diacetic acid [27].
V(IV)	Was extracted with 0.1% 4, 13, diaza 7, 10 dioxahexadecane 1, 16 dionic acid dichloride in acetone with crown hydroxamic acid in chloroform. Also extracted and determined at 395 nm and analysed by inductively coupled plasma atomic emission spectroscopy [101].

Table 8.7.2 Application of cryptands in emission spectroscopy

Element	Optimum Conditions for Extraction Emission Spectroscopy
K(I)	0.1M cryptand -222 extracted it from chloroform, toluene or nitrobenzene. Lithium chloride was used as the salting out agent. Various alkali metals were thus separated [102].

Table 8.7.3 Application calixarene in emission spectroscopy

Element	Optimum Conditions for Extraction Emission Spectroscopy
Th (IV)	At pH 6.0 about 0.2% of 25, 26, 27, 28, tetrahydroxy 5, 11, 17, 23 tetrakis (N-p chlorophenyl) calix (4) arene in ethylacetate extracted it and was stripped with nitric acid and was determined at 450 nm and analysed with inductively coupled plasma atomic emission spectroscopy (ICP-AES) at 284 nm [103].

8.9 Atomic Absorption Spectroscopy with Crown Compounds

As a matter of fact one may wonder why it was necessary to resort to separations when it was possible to determine metal by direct atomic absorption spectroscopy? Further how does AAS belongs to the category of emission methods? The answer for both is simple. Atomic absorption spectroscopy was not the answer to all the problems of analysis. In many instances AAS failed for instances metals like uranium had appreciable detection limit (e.g. 138 μg/ml). Many metals like tungsten, titanium formed carbide in graphite furnace. Metals like arsenic selenium, tellurium, antimony needed hydride generation while mercury needed the cold vapour method for analysis. In order to circumvent these problems, one had to use separation technique involving solvent extraction. Further AAS was grouped with AES from the view point of principle of absorption or emission. However flame emission spectroscopy and atomic absorption spectroscopy had too many things in common like choice of solvent, spectral interferences, method of evaluation, sample processing etc. hence it is clubed with AES in this chapter for purpose of discussion.

Amongst crown compounds, crown ethers were extensively used not merely for transition metals but also for several s-block metals which were generally analysed by flame photometry. Such metals included caesium, strontium, potassium etc. The cryptands in comparison were used for both analysis of transition metals like manganese, copper, cadmium, nickel but also for main group metals like lead and thallium with AAS. Unfortunately calix(n) arene in combination with AAS was less frequently used for only isolated case of the analysis of silver with suitable derivative of calix (4) arene.

8.10 Analytical Application of AAS with Crown Ethers and Cryptand Extractions

The alkali and alkaline earths were analysed by AAS after extractions with crown ethers with DC18C6, DB24C8 with picrate or (bis 2-ethylhexylphosphoric acid) (HDEHP) in toluene, dichloromethane or triethanolamine [105–108]. In few instances, caesium [105] was stripped to the organic phase and determined subsequently by AAS. In transition metals manganese cobalt, palladium, copper, silver, cadmium and mercury were extracted [108–121].

On average pH for quantitative extraction was pH 5-7.0. Variety of crown ethers such as B15C5, 18C6, DB18C6, DC18C6, DB24C8 were used. In few instances synergic extraction involving TOPO [105] TTA [108, 119] and PMBP [108] were carried out. The thioderivatives were less commonly used [115, 121] for metals like silver and copper. Azacrown ether like ADB 18C5 in chloroform was used with less advantage from the view point of spectrophotometry in the visible region. In all such determination the concentration of the reagent employed was very low, viz. 0.1-001M of the reagent in suitable diluent like dichloroethane, chloroform, methylene chloride, hexanone [21] or dichlorobenzene [119]. In several cases stripping of the metal to the aqueous phase was accomplished with dilute hydrochloric, nitric or perchloric acid [106] salts like potassium sulphate were rarely used [122]. Only in exceptional case organic phase was aspirated in the flame. In comparison cryptands had definite advantage over crown ether when used with atomic absorption spectroscopy. Cryptands formed relatively strong complexes with three dimensional structures. The reagent concentration employed was very small. As low as 0.0025M of it could be used [122]. The only limitation was the use of strong colour forming counter anions such as rose bengal extra, erythrosine, eosin, or tropaelin-000. This inhibited the natural process of direct spectrophotometry in the organic phase. Hence in most of the cases the elements were stripped into the aqueous phase and then determined by atomic absorption method.

The significant contribution was made by Gandhi and Khopkar [123–128] in the extractive atomic absorption spectroscopic determination of metal pollutants from marine aquatic environment. The important work covered metals like nickel [125], copper [124], thallium [126] and lead [127]. The work on manganese [123] and cadmium was also interesting. The cryptands used were cryptand-222, cryptand-221 and cryptand-222B. The counter ions used consisted of the erythrosine [123, 124], rose bengal extra [125] or eosin [127]. Very low concentration of counter anion was used. As low as 0.0003M of the cryptand could accomplish quantitative extraction of say copper [123]. The diluents used for cryptands consisted of chloroform, dichloromethane, toluene but best results were obtained with chloroform. The pH for quantitative extraction varied from pH 5.5 to 6.8. The mineral acids used for stripping of metal consisted of hydrochloric acid [123], perchloric acid [124], sulphuric acid [125]. It was noteworthy that acid concentration employed was little higher say 1-5M of mineral acid because the complexes formed with cryptands were relatively strong. Copper was separated from aerosols, effluents and sediments while nickel was also isolated from environmental samples. Thallium (I) could be easily separated from lead, zinc, chromium, vanadium and caesium in mixtures. Lead was separated from tin, arsenic, mercury, indium in mixtures.

The various methods are summarised in Table 8.10.1 to 8.10.3 involving for extraction by crown ethers and cryptand followed by subsequent analysis by AAS.

Table 8.10.1 Application of crown ethers in atomic absorption spectroscopy

Element	Optimum Conditions of Analysis
Alkali metals	Were extracted with DC18C6 on polyurethane foam in the presence of picric acid [104].
K(I)	DC18C6 extracted it in the presence of bis (2-ethylhexyl) phosphoric acid in toluene from nitrate solutions. Silver, lead and thallium (I) were coextracted [105].
Cs(I)	At pH 3.0, 0.03M DB24C8 in dichloromethane and methanol extracted it from 0.1M picric acid and was stripped with 3M perchloric acid [106]. It was separated from zirconium, thorium, hafnium in fission products [106].
Sr (II)	At pH 8.5, 0.5M of 2(sym dibenzo 16C4) carboxylate derivative and 4 (symn dibenzo 16C5) oxyhexanoic acid in triethanolamine were extracted from nitric acid media [107].
Mn (II)	At pH 6.0-6.8, 0.1M DB18C6 in dichlorobenzene with 0.1M of 2-thenoyl-trifluoroacetone as the synergist was extracted and determined by atomic absorption spectroscopy at 280 nm [108]. Manganese from natural water was analysed. [108].
Co (II)	At pH 3-5 it was extracted with B15C5, 18C6, DC18C6, DB18C6, DB24C8 with PMBP as the synergistic agent in dichloromethane and was separated from caesium. The best extraction of course was with 18C6 [109] in dichloroethane.
Pd (II)	Extracted with aza DBI8C5 in chloroform from 0.1M hydrochloric acid and it was determined at 247 nm. Many noble metals showed interference [110].
Cu(II)	Extracted at pH 5.0 with 0.5M of 6, 8, 15, 17 tetramethyldibenzo 5, 9, 14, 18 tetraazocyclo tetradecane [111]. Good extraction was possible at pH 4.0-5.0 [112]. At pH 5-8 macrocyclic fomazans extracted it in methylisobutylketone or methylethylketone from picric acid [110, 113]. Also at pH 4-5 cyclotetrathioether extracted it in dichloromethane from the picric acid [114]. Cyclic and acyclic tetrathioether in 1, 2 dichloroethane extracted it with hexanitrodiphenyl amine [115].
Ag(I)	At pH 4.5, 0.8M of DB18C6 and trioctylphosphine oxide (TOPO) in chloroform extracted it from picric acid and was stripped with 3M nitric acid. Fourteen macrocycles including tetrazocyclohexadecane in chloroform and hexanone were also used [117]. An adduct was formed [116]. DB18C6 in chloroform dichloroethane extracted it from picrate, chlorate or diphenyl aminate media [118]. It was extracted at pH 5.4 with 0.5 mM thiacrown in dichloromethane from picrate media [121] but manganese, nickel, zinc, cobalt were not extracted with it from picrate media.
Cd(II)	At pH 7.4-7.9 it was extracted with 0.01M DB18C6 in dichlorobenzene in the presence of 0.1M thenoyltrifluoroacetone and was stripped with 0.1M hydrochloric acid and was determined at 228.8 nm [119].
Hg(II)	Bisthiacrown, viz. (p-nitrophenylazo) reagent extacted it and was stripped with nitric acid [120] but copper cadmium were coextracted with mercury. At pH 5.0, 0.5M DB18C6 in hexanone extracted them and were subsequently stripped with ammonium acetate and potassium sulphate mixture and determined at 767 nm [122].

Table 8.10.2 Applications of cryptands in atomic absorption spectroscopy

Elements	Optimum Conditions of Analysis
Mn(II)	At pH 5-8, 0.0025M cryptand -222 in chloroform extracted in the presences of 0.0015M erythrosine as the counter anion. It was stripped with 5M hydrochloric acid and was determined at 279.5 mn. Manganese from industrial effluents was analysed [123].
Cu(II)	At pH5.5, 0.0003M cryptand-222 in dichloromethane extracted it with 0.0003M erythrosine B as the counter anion. It was subsequently stripped with 0.5M perchloric acid and was analysed at 324.7 nm. It was possible to separate it from iron, cobalt, manganese and nickel in waste water [124].
Cd(II)	At pH 6.5 extracted with 0.00025M cryptand-221 in chloroform with 0.001M rose bengal extra as the counter ion. It was stripped back to the aqueous phase with 1M sulphuric acid and was determined at 276.8 nm. Cadmium from industrial effluents was also analysed [125].
Ni(II)	At pH 7.5, 0.001M cryptand-222 in dichloromethane with 0.0001M erythrosine extracted it and was latter stripped with 1M hydrochloric acid and analysed at 352.4 nm. Nickel from waste water analysed (127)
Tl(I)	At pH 6.5, 0.001M cryptand -222 in chloroform extracted it with 0.001M erythrosine as the counter anion and was then stripped with 0.1M sulphuric acid and it was directly determined at 276.8 nm. It was analysed from polluted water samples [126].
Pb(II)	Microgram concentration of lead was extracted at pH 5.5 in presence of 0.0005M eosin with 0.00047M of cryptand-222B in toluene. It was stripped with 1M hydrochloric acid and was analysed at 217 nm. Lead from samples of sediments and aerosols was analysed [128].

Table 8.10.3 Application of calixarene in atomic absorption spectroscopy

Element	Optimum Conditions of Analysis
Ag	Calix (4) arene derivative in chloroform was used for the extraction along with palladium, platinum in mixture for fortyeight hours at 30°C [129].

8.11 Recent Advances in Switched on Crown Ethers

These are useful in the spectroscopy. Crown ethers and cryptands readily form complex with cations. Attempts were made to change cation binding capacity of these compounds, by switching mechanism. They could also be used to identify cations. In order to do this four modes of switching on crown ethers were discovered. They consisted of following crown ethers such as pH, thermal or redox switching and photoresponsive crown ethers.

8.11.1 pH Responsive Crown Ethers

The pH switching crown ether was a effective way by which compounds charge state was changed. This was done by lowering pH. The ionisation control or pH switching was done by means of compounds which could charge state as modified. (XV) was obtained by lowering pH.

232 Analytical Chemistry of Macrocyclic and Supramolecular Compounds

The four N-atoms were strong donors and if pH was decreased these N-atoms were protonated because an electron pair formed bond to H-making it occupied and not to allow to bind with cation and N-atom became positively charged which in turn repeated positively charged cation (structure XV). At low pH it was poor ligand however at higher pH it was capable of complexing cations. It had thus low-high (binding strength) switch dependent on pH.

(XV)   (XVI)

The compound (XVI), viz. calix (8) arene had 1-8 phenolic hydroxyl groups which could be lost. At high pH it was ionised with anions acting as strong donor while at low pH protonated form took place and cation was liberated. Therefore they acted as liquid membrane. Same proton switch transfer could be effected in cryptands (see reaction XVII). The process of complexation changed properties of donor, e.g. picric acid and sodium picrate had different UV spectra and calcium picrate had altogether different spectra from these two. The spectral change could be carried out by introducing side arm. The pH switching rendered side arm a powerful donor group.

8.11.2 Thermal Switching
It has many potential uses. A solid polymeric support a liquid crystal and DA18C6 substituted by fluoro or hydrocarbon chain gave an ideal membrane system. This membrane exhibited high transport capacity with rise in temperature which could be much above transition temperature of the liquid crystals.

8.11.3 Redox Switching
This involved redox chemistry much of it will be discussed in next chapter. It was based on the idea that a compound from one form could be transferred into another form by process of electron transfer. The addition of electrons (i.e. reduction) made the system electron rich and good binder for cation while removal of electron (i.e. oxidation) made it deficient of electrons and bad cation binder; with no change in electron density. This was verfied by cyclic voltammetry

where a solution having reducible substance contacted electrode with varying potential from high to low and vice versa. These ideas were of use for cation transport across membrane.

8.12 Photoresponsive Switching Crown Ethers

So far we have dwelt on phenomena of emission by atoms as well as molecules and absorption by atoms in the presence of normal crown ethers cryptands or calix(n)arenes. It was interesting to note that their characterisation under the influence of photochemical reactions. Such systems were encountered in nature; with light important for life process. Here a photo antenna to capture a photon was linked with a functional group to interpret subsequent transformations. Such transformations were associated with photo induced structural changes acting as photo antennae artificial photo responsive substances could act as photo antenna. One must have high reversibility and large electostatic changes. If one of the photo antennas like azobenzene, dimerised anthracene or spiropyranemicrocyanine interconversion was combined with a crown ether its properties could be controlled by on-off light switch.

Most important group was azobenzene bridged crown ethers and cyclophane type crown ethers. Biscrown ethers or those capped with anionic or cationic groups could also act as photoresponsive crown ethers. Photoisomerisation had undergone geometrical change as indicated in structures (XVII) and (XVIII).

(XVII) UV light / Visible light *-Photoisomerisation (XVIII)

The compoud was converted to compound (XVII). In UV light and reverts to (XVIII) again in visible radiation [130, 131]. Compound (XVIII) had bound sodium while compound (XVII) bound selectivity potassium ion. The N_2O_4 crown ring was enlarged to accommodate potassium during photo isomerisation [132].

In cyclophane type crown ethers photodimerisation of anthracene acts as photochemical switch for photoresponsive crown ethers. The compound (XIX) was synthesised where two anthracene molecules were incorporated in ring [133].

The intermolecular photodimerisation proceeded in presence of sodium ion or template cation. However synthesis of compound (XIX) was not simple. Large structural change was witnessed simultaneously with photochemical reversibility. Large change with no steric strain was ideal for molecular design of such crown ethers. Thus one can conclude that metal affinity and selectivity

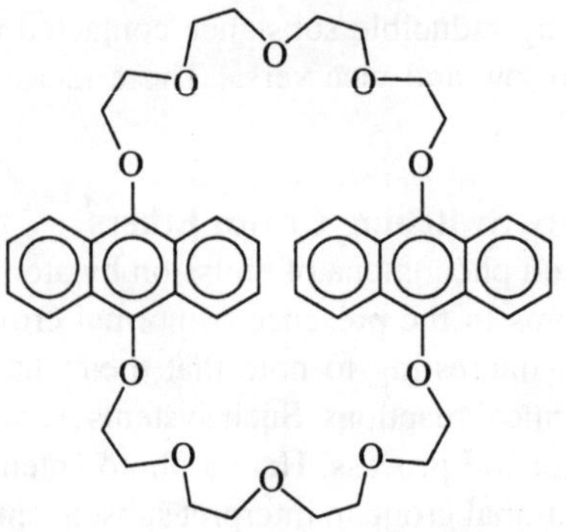

(XIX)

for photoresponsive crown ethers was controlled by light which in turn changed the cavity shape via change in geometry of photo functional entity. The other kind of photoresponsive crown ether which complexed readily with divalent metal ions like alkaline earth were also considered.

Bis crown ethers were used to entrap metal cations which had large ionic size than cavity of the crown. If fittment was exact, 1:1 complex was formed but one with large size it formed 1:2 the sandwich type complex, with lower metal selectivity. A photo induced change in the distance between two crown rings was noticed with change in ion binding ability. Shinkai [134–137] prepared a butterfly crown ether (XIX), in the presence of alkali metal photo stationary state concentration was increased when cation formed was 1:2 complex. A metal cation formed bridge between two crown rings. It readily extracted sodium. This effect was used in ion transport accross membrane. The photoresponsive change was reflected with electrical conductance change [138]. It was more in UV than in visible light. It was function of mobility of charged particles in solution so increase in conductances led to molecular size diminishment. The distance in two ionophoric group was possible with poly functional groups other than azobenzene by designing 'molecular tweezers'. The next category was the crown ether with capped anionic or cationic group produced photoresponsive compounds. The overall ion affinity was controlled by change in spatial configuration of anionic groups. A compound which could respond to photo radiation by liberating an anion cap, directed to photocontrol of ion binding ability [139]. Such compounds is represented in structure (XX).

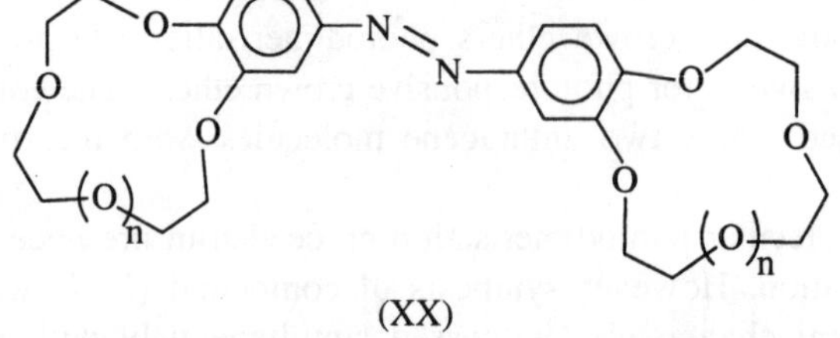

(XX)

Thus switched on crown ethers were important in cation binding process. Crown compounds act as recognisation centres. These photochemical switching mode involve use of antenna to capture light and to display structural changes. In few compounds colour and isomerisation property was of significance. The dual action of photocontrol and colorimetric display of complexation was presenting a challenge in cation binding and in measurements [140–142].

Conclusion

In conclusion one can infer that use of suitable counter anion like eosin or erythrosine will not lead to the development of fluoroscence but the specially designed fluoroionophore or fluorogenic crown ethers and switch on photoresponsive crown ethers could do this job. Therefore they were synthesised. The basic idea of complexation of metals with crown ethers cryptand or calixarene was supplemented with rapid determination of these metals after stripping into the aqueous phase with fluorimetry, flame photometry or atomic absorption spectroscopy.

There was spur of activity in development of fluoroionophores and ring substituted fluorogenic crown ether. The exact mechanism of transfer of energy from ligand to metal and shielding of metal in the case of inner transition metals from influence of s- and p-orbitals was well understood. The greatest applications of luminescence method was extended to analytical chemistry of lanthanides. The effective role of counter anion like benzoate was well stressed. The luminescent complexes of supramolecular compounds like calix (4) arene tetracetamide was investigated. The bipyridyl and similar analogues of cryptands appeared to be promising for complexation of inner transition elements. The analytical applications involving extractions with these compounds followed by analyses by emission technique was well illustrated in different sections.

The photoresponsive crown ether have great future in analytical chemistry. Apart from host guest concept widely used, switch on kind compounds have great significance to improve selectivity of analysis during extraction. They have bright future in fluoroimmunoassay investigations. The introduction of suitable antennae in side arm can make radical change in stereoselectivity of complexation and hence extraction. We expect to witness spectacular progress in next few decades in future.

References

1. H. Nishida, Y. Katayama, H. Katsuki, N. Nakamura, M. Takagi, K. Ueno, Chem. Letters 1853 (1982).
2. J.P. Dix, F. Vogtle, Chem. Ber. **113**, 457 (1980).
3. M. Takagi, H. Nakamura, K. Ueno, Anal. Letter **10**, 1115 (1977).
4. J.L. Stegar, Ph.D. thesis Miami University (1984).
5. M. Hiraoka, Crown ether and analogus compounds Elsevier Pub. Co. (1992) p 321.
6. K. Nakashima, S. Nakatsuji, T. Kaneda, S. Misumi, Chem. Letter 7311 (1982).
7. T. Kaneda, S. Nakatsuji, S. Akiama, S. Misumi, Tetraheron Letter **22**, 449 (1981).
8. K. Nakashima, S. Yamawaki, S. Akiyama, T. Kaneda, S. Misumi, Chem. Letters 1415 (1983).

9. S. Kitazawa, K. Kimura, T. Shono, Bull. Chem. Soc. Japan **56**, 3253 (1983).
10. H. Forrest, G.E. Pacey, Talanta **36**, 335 (1989).
11. G.E Pacey, B.P. Bubnis, Anal. Letter **13**, 1085 (1980).
12. G.E. Pacey, Y.P. Wu, B.P. Bubnis, Analyst **106**, 636 (1980).
13. Y.P. Wu, B.P. Bubnis, G.E. Pacey, Syn. Comm. **11**, 323 (1981).
14. B.P. Bubnis, J.L. Steger, Y.P. Wu, L.A. Meyers, G.E. Pacey, Anal. Chim. Acta. **139**, 307 (1982).
15. Y.P. Wu, G.E. Pacey, Talanta **31**, 165 (1984).
16. G.E. Pacey, B.P. Bubnis, Tetrahedron Letter **25**, 1107 (1984).
17. Y.P. Wu, G.E. Pacey, Anal. Chim. Acta. **162**, 285 (1984).
18. B.P. Bubnis, G.E. Pacey, Talanta **31**, 1149 (1984).
19. K. Sasoki, G.E. Pacey, Anal. Chim. Acta. **174**, 141 (1985).
20. C. Cossy, A.E. Merbach, Pure and applied Chem. **60**, 1785 (1986).
21. L.F. Lindoy, Chemistry of macrocyclic ligand complexes Cambridge Press (1989).
22. K.L. Cheng, K. Ueno, T. Imamura, Handbook of organic analytical chemistry CRC Press Baton Rouge (1982).
23. W.L. Hinze, D.N. Armstrong (Ed), Ordered media in Chemical separation, American Chemical Society (1987).
24. B.S. Mohite, S.M. Khopkar, Anal. Chem. **59**, 1200 (1987).
25. L.M. Tsay, J.S. Shih, S.C. Wu, Analyst **108**, 1108 (1983).
26. W. Wenji, C. Bozhong, J. Zhong-Kao, W.J. Ailling, J. Radio and Chem. **76**, 49 (1983).
27. V.K. Manchanda, C.A. Chang, Anal. Chem. **59**, 813 (1987).
28. Y. Hasegawa, S. Haruna, Solvent extn. and ion exchange **2**, 451 (1984).
29. Y. Hasegawa, M. Masuda, K. Hirose, Y. Fukuhura, Solvent extn. and ion exchange **5**, 255 (1987).
30. H.F. Aly S.M. Khalifa, J.D. Navratill, M.T. Saba, Solvent extn. and ion exchange **3**, 623 (1985).
31. D.D. Ensor, G.R. McDonald, C.G. Pippin, Anal. Chem. **58**, 1814 (1986).
32. J. Tang, C.M. Wai, Anal. Chem. **58**, 3235 (1986).
33. J.C. Bunzil, Handbook of Physics and Chemistry of rare earths Ed. K.A. Gschneidner, L. Eryring. North Holland Amsterdam Vol. 9 (1987).
34. K.Nakagawa, S. Okada, Y. Inoue, A. Tai, T. Hakushi, Anal. Chem. **60**, 2527 (1988).
35. G. Stein, E. Wurzberg, J. Chem. Phy. **62**, 208 (1975).
36. E. Soini, T. Lovgren, CRC critical Reviews in Anal. Chem. **18**, 105 (1987).
37. M. Morin, R. Bador, H. Dechaud, Anal. Chim. Acta. **219**, 67(1989).
38. R.P. Fisher, J.D. Winefordner, Anal. Chem. **43**, 454 (1971).
39. H. Shou, J. Ye, Q. Yu, J. Lumin, **42**, 29 (1987).
40. A. Heller, E. Wasserman, J. Chem. Phy. **42**, 949 (1965).
41. J.R. Escabiperet, F. Nome, J.H. Fendler, J. Am. Chem. Soc. **99**, 7749 (1977).
42. G.C. Correll, R.N. Cheser, F. Nome, J. H. Fendler, J. Am. Chem. Soc. **100**, 1254 (1978).
43. P.W. Schuller, Biochemical Fluoroscene Concept Ed. R.F. Chen H. Edelhoch Marcel Dekker New York (1975) Vol. 1 Chapter-5.
44. T. Forster, Modern quantum Chemistry-Ed. O. Sinanaglu Academic New York (1965) part III.
45. C.D. Tran, W. Zhang, Anal. Chem. **62**, 835 (1990).
46. H. Baba, M. Kitamura, J. Mol. Spectro **41**, 302 (1972).
47. W.D. Horrocks, M. Albin, Progress Inorg. Chem. Ed. S.J.-Lippord, Wiley New York p.1 Vol. 31 (1984).

48. J.M. Lehn, Angew Chem. **29**, 1304 (1990).
49. B. Dietrch, J.M. Lehn, J.P. Sauvage, Tetrahedron 29, 1647 (1973).
50. J.M. Lehn, J.P. Sauvage, J. Am. Chem. Soc. **97**, 6700 (1972).
51. B. Alpha, J.M. Lehn, G. Mathis, Angew Chem. **26**, 266 (1987).
52. J.M. Lehn, V. Balzani, Ed. Supramolecular Photochemistry Reidel Dortrech 29 (1987).
53. N. Sabbatini, S. Dellonte, M. Ciano, A. Bonazzi, V. Balzani, Chem. Phy. Letters **107**, 212 (1984).
54. G. Blasse, M. Buys, N. Sabbatini, Chem. Phy. Letters **124**, 538 (1986).
55. N. Sabbatini, S. Dellonte, G. Blasse, Chem. Phy. Letters **129**, 541 (1986).
56. P.J. Breen, W. Dew. Horrock, Inorg. Chem. **22**, 536 (1983).
57. N. Sabbatini, S. Perathoner, G. Lattanzi, S. Dellonte, V. Balzani, J. Phy. Chem. **91**, 6136 (1987).
58. R. Ziessel, J.M. Lehn, Hel. Chim. Acta. **73**, 1149 (1990).
59. G. Calestani, F. Ugozzoli, A. Aruduini, E. Ghidini, R. Ungaro, J. Chem. Soc. Chem. Comm. 344 (1985).
60. N. Sabbatini, M. Guardigli, A Mecati, V. Balazani, R. Ungaro, E. Ghidini, A. Casnati, A. Pochini, J. Chem. Soc. Chem. Comm. 878 (1990).
61. N. Sabbatini, M. Guardigli, J.M. Lehn, Coordination Chem. Review **123**, 201 (1993).
62. K. Nakashima, S. Nakatsuji, S. Akiyama, I. Tanigawa, T. Kaneda, S. Misumi, Talanta **31**, 749 (1984).
63. K. Kimura, S. Iketani, T. Shono, Anal. Chim. Acta, **203**, 85 (1987).
64. A. Prasanna De Silva, A. Saliya De Silva, J. Chem. Soc., Chem. Commun. 1709 (1986).
65. A. Sanz-Medel, G.D. Blanco, A.V.R. Garcia. Talanta **8**, 425 91981).
66. M.E. Diaza Garcia, F. Alava Moreno, A. Sanz–Medel, Milkcochim. Acta. **113** 211 (1994).
67. H. Nishida, Y. Katayama, H. Katusuki, H. Nakamura, M. Takagi, K. Ueno, Chem. Letter 1853 (1982) A.A. **44**, 6B32 (1983).
68. T.D. James, S. Shinkai, J. Chem. Soc. Chem. Commun. 1483 (1995).
69. S.V. Bel'tyukova, T.B. Kravchenko, A. Kh. Zitsmanis, G.M. Balamtsarshvili, A.S. Roska, E.V. Malinka, Zhur. Anal. Khim. **45**, 1906 (1990), A.A. 53, 5D54 (1991).
70. G. Adachi, H. Nakamura, T. Mishima, J. Shiokawa, Chem. Express **2**, 341 (1987), C.A. **107**, 155103n (1987).
71. Z. Pikromenou, D.G. Noccra, Inorg. Chem. **31**, 531 (1992).
72. A. Beeby, D. Parkar, J.A. Gareth Williams, J. Chem. Soc. Perkin Trans. 2, 1565 (1996).
73. Greg. E. Collins, Ling-Siu-Choi, J. Chem. Soc., Chem. Commun. 1135 (1997).
74. M-Oue K. Kimura, T. Shono, Analyst. **113**, 551 (1988).
75. M.K. Beklemishev, L.I. Gorodihova, N.I. Shevtsov, L.M. Kardivarenko, N.M. Kuzmin. Zhur. Anal. Khim. **44**, 1058 (1989) A.A. **52**, 5B33 (1990).
76. A. Sanz. Medel, G.D. Blanco, E. Fuente, S.J. Arribas, Talanta, 31, 515 (1984).
77. G.D. Blanco, E. Andres, Garcia. E., A.E. Fuente, Abrodo.P. Arias, Milkrochim Acta, **3**, 259 (1990).
78. G.D. Blanco, A.E. Fuente, Abrodo. P. Arias, Mikrochim Acta III, 59 (1989).
79. G.E. Andres. A.E. Fuente, G.D. Blanco, Anal. Letter **25**, 339 (1992).
80. E.A. Gracia, G.D. Blanco, Mikrochim Acta. **124**, 179 (1996).
81. G.D. Blanco, A.E. Fuente, G.E. Andres, Abrodo.P. Arias, Talanta, **36**, 1237 (1989).
82. G.D. Blanco, A.E. Fuente, A. Sanz-Model, Talanta **32**, 915 (1985).

83. G.D. Blanco, G.E. Andres, Analyst. **115**, 89 (1990).
84. G.E. Andres, A.E. Fuente, G.D. Blanco, Anal. Letter **27**, 775 (1994).
85. H. Murakami, S. Shinkai, J. Chem. Soc. Chem. Commun. 1233 (1993).
86. H. Durr, R. Schwartz, I. Willner, E. Joselevich, Y. Eichen, J. Chem. Commun. 1338 (1992).
87. D.M. Rudkevich, W. Verboom, E.V. Tol, C.J.V. Stavern, F.H. Kaspersen, J.W. Verhoron, D.N. Reinhoudt J. Chem. Soc. Perkin. Trans. **2**, 131 (1995).
88. A. Casnati, C. Fischer, M. Guardigli, A. Isernia, I. Manet, N. Sabbatini, R. Ungaro, J. Chem. Soc. Perkin Trans. 2, 395 (1995).
89. B.M. Lynch, M.M. Ryan, B.S. Greaven, G. Barrettt, M.A. Mckervey, S.J. Harris, Anal. Proc. **30**, 150 (1993) AA **55** 8D43 (1993).
90. P. Linnane, J.D. James, S. Shinkai, J. Chem. Soc. Chem. Commun. 1997 (1995).
91. E.V. Dienst, B.H.M. Snellink, I.V. Piekartz, J.F.J. Engbersen, D.N. Reinhoudt, J. Chem. Soc. Chem. Commun. 1151 (1995).
92. W.J. McDowell, G.N. Case, J.A. McDonough, R.A. Bartsch. Anal. Chem. **64**, 3013 (1992).
93. B.S. Mohite, Ph.D. Thesis Solvent extraction of alkali and alkaline earth elements with crown ethers, I.I.T. Bombay (1986).
94. B.S. Mohite, S.M. Khopkar, Ind. J. Chem. **22A**, 962 (1983).
95. B.S. Mohite, D.N. Zambare, B.E. Mahadik, Anal. Chem. **66**, 4097 (1994),
96. B.S. Mohite, S.M. Khopkar, Talanta, **37**, 565 (1985).
97. I.H. Gerow, J.E. Smith, Jr. and D.M. Wijun, Sep. Sci. Technol. 16, 519 (1981).
98. B.S. Mohite, S.M. Khopkar, Analyst, 112, 191 (1987); Anal. Chem. **59**, 1200 (1987).
99. E.A. Filippov, V.V. Yakshin, V.M. Abshkin, V.G. Fomenkov, I.S. Serebryakov, Radiokhimiya, **24**, 214 (1982) A.A. **44**, 2B34 (1983).
100. G. Zirnhelt, M.J.F. Leroy, T.P. Brunette, Y. Frere, P. Gramain, Sep. Sci. Technol. **28**, 2419 (1993).
101. P. Srivastav, Y.K. Agrawal. Analysis **24**, 13 (1996).
102. V.V. Yakshin, A.T. Fedorova, B.N. Laskorin, Zhur. Anal. Khim. **40**, 45 (1995) A.A. **48**, 1B 37 (1986).
103. Y.K. Agrawal, M. Sanyal. Analyst, **120**, 2759 (1995).
104. A.S. Khan, W.G. Baldwin, A. Chow, Cand. J. Chem. **59**. 1490 (1981).
105. G.A. Clark, R.M. Izatt, J.J. Christensen, Sep. Sci. Technol. **18**, 1473 (1983).
106. R.G. Vibhute, S.M. Khopkar, J. Radioanal. Nucl. Chem. Articles **152**, 487 (1991).
107. H. Bukowsky, E. Uhlemann, M.D. Gloek, H. Mosler, Z. Chem. **30**, 73 (1990) A.A. **53** 5D32 (1991).
108. M. Billah, T. Honjo, K. Terada, Anal. Sci. **9**, 251 (1993).
109. B. Rusdiarso, A. Messoudi, J.P. Brunette, Talanta **40**, 805 (1993).
110. M.K. Beklemishev, S.E. Panfilova, A.B. Vatynskii, N.M. Kuzmin, Y.A. Zolotov, Vestn. Most. Uni. Ser-2 Khim, 30, 168 (1989) A.A. **52**, 1B80 (1990).
111. Y.A. Zolotov, N.V. Nizeva,V.P. Ionov, D.M. Kumina, O.V. Ivanov, Mikrochim. Acta. I (5-6), 381 (1983).
112. N.V. Nizeva, V.P. Ionov, I.V. PLentev, D.M. Kumina, V.M. Ostrovskays, I.A. Dyakonova, Y.A. Zolotov, Dokl. Akad. Nauk. SSSR, **274**, 611 (1984) A.A. **46**, 9B23 (1984).
113. N.V. Isakova, Y.A. Zolotov, V.P. Ionov, Zhur. Anal. Khim. **44**, 1045 (1989) A.A. **52**, 5B25 (1990).
114. K. Saito, S. Murakmi, A. Muromatsu, E. Sekido, Anal. Chim. Acta. **237**, 245 (1990).
115. K. Chayam, E. Sekido, Anal. Chim. Acta. **248**, 511 (1991).

116. Y. Hasegawa, K. Suzuki, T. Sekine, Chem. Lett. **8**, 1075 (1991) A.A. **42**, 2B44 (1992).
117. Y.A. Zolotov., V.P. Ionov, V.A. Bodnya, G.A. Larikova, W.Y. Nizeva, G.E. Vlasova, E.V. Rybakova, Zh. Anal. Khim. **37**, 1543 (1982) A.A **44**, 6A5 (1983).
118. Y.A. Zolotov, E.I. Morasanova, S.G. Dmitrienko, A.A. Form, anovsky, G.V. Ivanov Mikrochim Acta III (5-6), 399 (1984).
119. M. Billah, T. Honjo, Z. Anal. Chem. **357**, 61 (1997).
120. M. Muroi, E. Sekido, Anal. Sci. **9**, 691 (1993).
121. E. Sekido, H. Kawahara, K. Sujik, Bull. Chem. Soc. Japan. **61**, 1587 (1988).
122. E. Nakamura, I. Fujisawa, H. Namiki, Bunselki Kagaku, **32**, 332 (1983) A.A. **45**, 6H56 (1983).
123. M.N. Gandhi and S.M. Khopkar, Anal. Sci. **8**, 233 (1992).
124. M.N. Gandhi and S.M. Khopkar, Mikrochim. Acta. **111**, 93 (1993).
125. M.N. Gandhi and S.M. Khopkar, Chemia. Analityczna, **37**, 437 (1992).
126. M.N. Gandhi and S.M. Khopkar, Anal Chim Acta. **270**, 87 (1992).
127. M.N. Gandhi and S.M. Khopkar, Chemical and Environmental Research **1**, 389 (1993).
128. M.N. Gandhi and S.M. Khopkar, Ind. J. Chem. **30A**, 706 (1991).
129. K. Ohto, H. Yamaga, E. Murakami, K. Inoue, Talanta **44**, 1123 (1997).
130. S. Shinkai, T. Ogawa, T. Nakaji, Y. Kusano, O. Manabe, Tetrahedron Letter **45**, 69 (1979).
131. S. Shinkai, T. Nakaji, Y. Nishida, T. Ogawa, O. Manabe, J. Am. Chem. Soc. **102**, 5860 (1980).
132. H.L. Ammon, S.K. Bhattacharjee, S. Shinkai, Y. Honda, J. Am. Chem. Soc. **106**, 262 (1984).
133. H. Bouas, Laurent, A. Castellan, M. Daney, J.P. Desvergne, G. Guinand, P. Marsan, M.H. Riffand, J. Am. Chem. Soc. **108**, 315 (1986).
134. S. Shinkai, T. Ogawa, Y. Kusano, O. Manabe, Chem. Letter 283 (1980).
135. S. Shinkai, T. Nakaji, T. Ogawa, K. Shigomatsu, O. Manabe, J. Am. Chem. Soc. **103**, 111 (1981).
136. S. Shinkai, T. Ogawa, Y. Kusano, O. Manable, K. Kikukawa, T. Goto, T. Masuda, J. Am. Chem. Soc. **104**, 1960 (1982).
137. S. Shinkai, K. Shigmatsu, Y. Kusano, O. Manabe, J. Chem. Soc. Perkin Trans. I 3279 (1981).
138. S. Shinkai, Crown ethers and analogous compounds Ed. M. Hiraoka Elsevier Publishers p. 335 (1992).
139. S. Shinkai T. Minami, Y. Kuano, O. Manabe, J. Am. Chem. Soc. **104**, 1967 (1982).
140. S. Shinkai, Cation binding by macrocycles Ed. Y. Inoue and G.W. Gokel, Marcel Dekker p. 397 (1990).
141. S. Shinkai, O. Manabe, Topic in current chemistry **121**, 67 (1984).
142. G.W. Gokel, Crown ethers and cryptands – Royal chemical society London p. 145 (1991).

9
Electroanalytical Methods with Crown Compounds

9.1 Introduction

Electroanalytical methods like spectral methods play a vital role in the study of the coordination complex of metals with crown ethers, cryptands and calixarenes. Amongst electroanalytical methods, one has to consider important techniques like potentiometry, conductometry, voltammetry inclusive of polarography and cyclic voltammetry and coulometric methods of analysis.

The investigations involving variation of pH of the solution or release of protons during complexation reactions was the early applications of the potentiometric methods in the analysis of metal complexes with crown compound. The use of ion selective eletrode was the striking example in potentiometric applications for charaterisation as well as for quantitative determination of metals as crown complexes. Most of these potentiometric techniques involved use of saturated calomel electrode (SCE) as the reference electrode. The principal use of potentiometry was to evaluate stability constants of the complexes of metals with crown ethers or cryptands. Apart from this, the technique was used for the determination of formation constants of metal complexes involving crown compounds. In comparison the technique was not used for study of calixarene complexes.

The conductometry methods were more popular in comparison to potentiometric methods. These techniques not only facilitated the evaluation of stability constants and the formation constants of metal ligand complexes but also helped to evaluate the vital information on ionic mobilities of ions. Such data had great significance in study of membrane transport phenomena. Most of these studies were confined to monomeric (1:1) complexes. Only in rare instances, the technique was extended to polynuclear complexes. A very interesting relationship between size of metal and cavity of the crown ether and stability of complexes was established. A definite knowledge of exact relationship between stability constants and diluents was also established.

Polarography and cyclic voltammetry techniques were most popular for the studying the nature of complex. It was the knowledge of half wave potential ($E^{1/2}$) which led to not only identification of metal encountered in complexation mechanism but also the evaluation of their stability and the formation constant. Polarography was more agressively used for the study of metal complexes with crown ethers and cryptands while cyclic voltammetry was predominantly utilised to study the complexes of metals with calix (n) arenes. These technique proved to be the boon for study and syntheses of switch on/off type of the supramolecular compounds. These methods also proved to be useful to ascertain individual concentration of the crown ethers or cryptands in solutions.

Finally coulometry which involved direct applications of the Faradays law of electrolysis was used with certain limitations. The coordination complexes of mercury with cryptands, i.e. macrobicyclic polyether were best investigated by this method. The extent of utility of coulometry useful for study of coordination complexes of metals with crown compounds was unknown. In fact it was an ideal method of quantification; since usually the interest of chemist also centred around analysis. It was generally thought that the electroanalytical methods would best fit the requirements of analysis. As such it was seen that the methods in demand were potentiometry, conductometry and cyclic voltammetry. Therefore an endeavour is made in this chapter to consider each technique individually.

9.2 Potentiometry

The term macrocyclic ligand covered the natural and synthetic macromono or macropolycylic compounds where in the heteroatom comprised of one-third or one-fourth of total number of atoms forming ring. The minimum number of heteroatoms was three to six (I). In sixteen membered ring. Macromonocyclic compounds were called coronand's irrespective of heteroatoms being oxygen, sulphur or nitrogen. The replacement of one of the oxygen atom in crown ether molecule by sulphur or nitrogen or inclusion of benzo or cyclo group (II) connecting pair of heteroatoms was indicated through prefix, e.g. DB18C6. The macrobicyclic (III) or macropolycyclic (IV) compounds were called 'cryptands'. The open chain compounds were called as 'podands'.

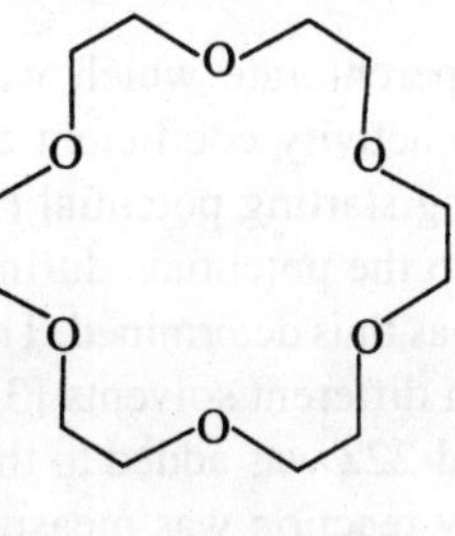

(I) 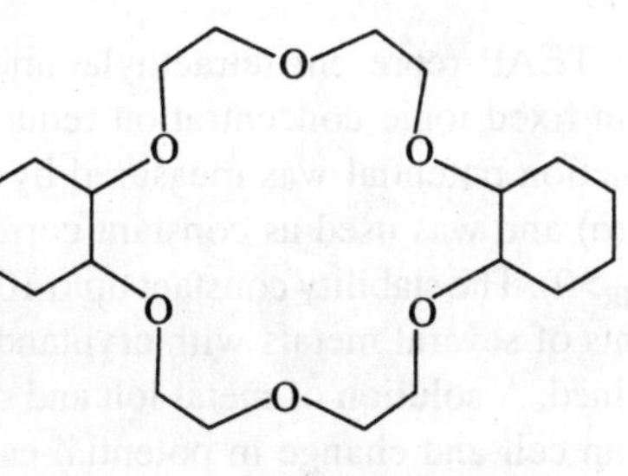 (II)

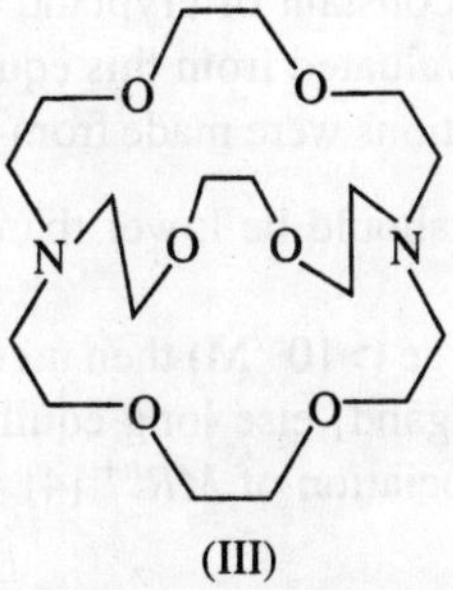

(III) 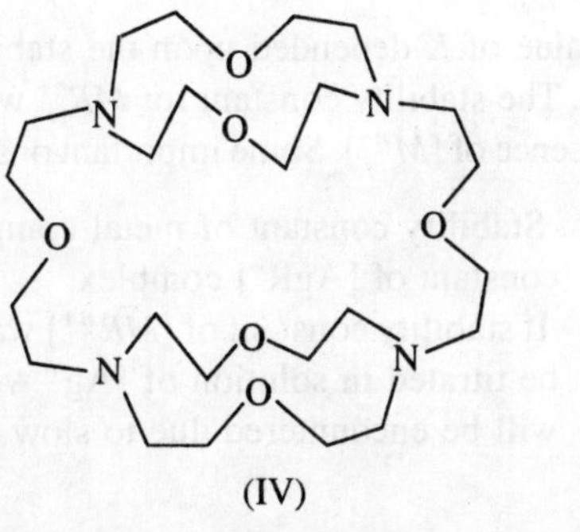 (IV)

The thermodynamic stability constant K was defined by the equation

$$M^{n+} + R = MR^{n+}, K = \frac{[MR^{n+}]}{[M^{n+}][R]} = \frac{[MR^{n+}]\gamma\, MR^{n+}}{[M^{n+}][R]\gamma\, M^{n+} + \gamma R} \tag{1}$$

where γ-represents activity coefficient of species. In potentiometric methods the stability constant of complex was easily calculated if concentration of the metal ion and that of the ligand was known along with free metal ion concentration at equilibrium. The method was valid for low concentration of metal ions. However one needed stable electrode which could display a reversible reaction with metal ion.

The potential E of cation selective electrode against SCE arose from redox equilibrium or from concentration gradient of ions on membrane, e.g. Ag/Ag electrode. Its potential in Nernst' terms was

$$E_{Ag} = E^{\circ}_{Ag} + \left(\frac{RT}{nF}\right) \ln a_{Ag^+} \quad \text{If } n = 1 \tag{2}$$

Then $\quad E^{\circ}_{Ag} = \frac{RT}{F} \ln^a [Ag^+] \tag{3}$

The accurate potential measurement was possible even with low $[Ag^+]$ concentration. Silver was studied with azacrown ethers or cryptands. In general cryptand 2, 2, 2 in propylene carbonate was used [1] with the Kolthoff type reference cell [2] as shown in Fig. 1.

| Ag | Ag^+, Cryptand 2 2 2
PC, 0.05M TEAP | PC
0.05M TEAP | Ag^+
PC, 0.05M TEAP | Ag |

Fig. 1 Kolthoff Type cell

Here TEAP represent tetraethylammonium perchlorate which was used to maintain fixed ionic concentration requiring no activity coefficient correction. The junction potential was measured by knowing starting potential (when C_{HR} was zero) and was used as constant correction to the potentials during titration (i.e. $C_{HR}>0$). The stability constant upto 10^{-16}M was thus determined. The stability constants of several metals with cryptand-222 in different solvents [3] was thus determined. A solution of metal ion and cryptand-222 was added to the solution of Ag^+ in cell and change in potential caused by reaction was measured as:

$$Ag^+ + MR^n = Ag\,R^+ + M^{n+} \tag{4}$$

The value of K depended upon the stability constant of cryptand complex and $[M^{n+}]$. The stability constant for MR^{n+} was evaluated from this equation even in the absence of $[M^{n+}]$. Some important observations were made from these studies:

(a) Stability constant of metal complex should be lower than the stability constant of $[AgR^+]$ complex.
(b) If stability constant of $[MR^{n+}]$ was large ($>10^{-7}$M) then metal ion should be titrated in solution of $[Ag^+$ with ligand$]$ else long equilibration time will be encountered due to slow dissociation of MR^{n+} [4].

(c) The titration in absence of inert salt ensures constant ionic concentration but might need liquid junction potential correction [5].
(d) The consequences of ion pair formation constant could be ignored for s-block metals even in solution of methyl alcohol or any similar nonaqueous solvents.

The glass electrode was also used. The stability constant for ammonia, silver and alkali metals was evaluated by it [6] in protic solvents with mixture of dipolar aprotic solvents. Glass electrode was cation selective but not cation specific. The response was good in the range of $10^{-6}-10^{-7}$ M. The glass electrode was extensively used for ascertaining the stability constant of complexes of azacrown ethers or cryptands because such complexation was pH dependent. The usual procedure considered of the titration of free ligand with acid alone and titrating again in the presence of metal ions [7]. A protonated form of metal ion complex $(RMH)^{2+}$ could also be present.

9.3 Analytical Applications of Potentiometry

The accuracy of stability constant was influenced by condition of half cell or salt bridge. The uncertainity same time was associated with liquid function potential which must remain constant during titration. As a precautionary measure titration was commensed when the drift in liquid function potential was decreased to 2 mV per week [4]. It was possible to carry out titration without liquid junction but suitable electrode working in the combination with single solution was required. A typical cell satisfying this need was developed [8] in study of complexes of barium with 18C6.

This technique was extended for analysis of metals consisting of alkali metals and [9–25] alkaline earins and lanthanides [21–28].

The purity of crown ether was ascertained by nonaqueous titration in nitromethane or methylene chloride [9] with perchloric acid solvents like acetonitrile (CH_3CN) was used [12]. The titrations were carried out not only to evaluate the value of stability constant but also to understand the structures of protonated liquids [13]. Perchloric acid aniline perchlorate or alkali thiacyanate were commonly used as the titrants. The extracted constant were evaluated in order to understand transfer of complex cation through membrane [14]. The stability of complexes with 18C6 in propylene carbonate were investigated. The mechanism of extraction of BOB 14C4AA complex with sodium and potassium was ascertained by potentiometric methods [17]. The existance of sandwich type complex for 18C6 with potassium was confirmed by potentiometric technique [19]. The technique of linear regression method was resorted to evaluate the stability constant of complex of B15C4, 15C5, B18C6, 18C6 with sodium [20].

Very few transition metals were studied by this method. Copper was an exception to this generalisation. Mononuclear and binuclear complexes of copper were thus investigated [21]. The nature of 1:1 complex of metals with several metals including those belonging to s-block were studied [22, 23]. Lithium was potentiometrically titrated with cryptand-222 and cryptand-211 in triethanolamine [24]. Lithium was readily complexed due to the process of encapsulation [25]. Cryptand-222 in different organic solvents was complexed with barium and stability constant was evaluated in acetonitrile [27]. A definite relationship between ionic charge and the donicity was established in dimethylsulphoxide complex of europium [26].

The various method involving potentiometric techniques for the evaluation of stability constant, nature of complex formed, mechanism of formation of complex is summarised in Table 9.3.1.

Table 9.3.1 **Analytical applications of potentiometry with crown ethers**

Element	Conditions of Analysis
Alkali metals	When diethyldibenzo 30C10 or 24C8 or DB18C6 was complexed with alkali metals and if extraction constants were measured potentiometrically, it showed that the selectivity depended upon the transfer of complex cation through membrane [14]. With 18C6 and 12C4 the complexes of alkali metals and ammonium ion in various solvents were thermodynamically studied [15].
$H^+(I)$ $D^+(1)$	The study of the complexation of proton and deutron ions with crown ethers like 18C6, B18C6, DB18C6 in propylene carbonate showed that 18C6 complex was most stable while others formed less stable complex (log K ~ 3.32 vs 6.35 for 18C6) [16].
Li(I)	The two phase potentiometric pH titration of bis-t octylbenzo 14 crown 4 acetic acid BOB14C4AA complexes of lithium and sodium showed that the ion exchange extraction mechanism was best for extraction of sodium at pH 1.4-1.8 and was best for lithium at pH 1.7-2.3 from chloride solution [17].
K(I)	18C6 potassium ferricynide complex in acetonitrile was used for the titration of potassium nitrite oxalate or thiocyanate, when alkaline earths showed no interference [18]. When the complex of potassium with 18C6 in methanol water was titrated, it showed formation of ML and ML_2 type sandwich type complexes [19].
Na(I)	The macrocyclic polyethers formed complexes of sodium with crown ethers such as B15C5, 15C5, B18C6 18C6. When titrated they gave stability constant by linear regression technique [20].
Cu(II)	6-amino 5, 7 dioxo-1, 4, 8, 11 tetra-azocyclo tetradecane formed binuclear yellow complex and mononuclear pink complex in alkaline media [21].
Amines	Aniline and DB18C6 were titrated with 0.1M hydrochloric acid in nitromethane gave two inflexation point (o-aniline gives single point) addition of lithium formed crown complex which in turn facilitated ascertaining amine concentration [22].

Crown ethers were determined by potentiometric analysis using nonaqueous solvents like methylene chloride or nitromethane [9]. The mixtures of DB24C8, 18C6, DB18C6 and 15C5 were titrated in nitromethane with perchloric acid [10]. Crown ethers were titrated in nitromethane in solution of aniline perchlorate (0.05M) with perchloric acid [11] 4, 4'(5') di-t-butyldibenzo 18C6, 4, 13 dithiabenzo 18C6 and dibenzo 30C10 were titrated in acetonitrile with thiocyanate of sodium or potassium or anilium perchlorate. For 4, 13 dithiabenzo 18C6 best titrant was sodium thiocyanate. Protonation of diazocrown and cryptand in water was done not only to know stability constant but also to understand structure of mono and bis protonated ligands [13].

9.4 Conductometric Studies of Complexes with Crown Ethers

These studies provided definite information on the existance of the complexation phenomena between macrocyclic or macrobicyclic compounds and the metal. It also threw light on the stability constant of such complexes. It also gave information on ionic mobility which favoured transport mechanism. The mobility was measured as velocity per unit field strength in term of ion dissociation constant (K_D) and the distances as real way to ion pair (a°).

Such stability constant were determined by potentiometry as discussed earlier or by voltammetry, spectrophotometry, NMR or thermal mehods. The determination in nonideal solutions was some what cumbersome. Conductometry however was exception to rule as it provided reliable information even at very low concentrations. The structures of crown metal complex was determined interms of mobility, K_D and a° A complex formation constant was also evaluted. The discussion was valid for monomeric and neutral complexes [29] only.

9.4.1 Evaluation of Formation Constant from the Conductance Data

During the measurement of conductance a correction for viscosity changes and the association between anion cation was neglected. The usual experimental set up was used for such determination. Usually molar conductivity was evaluated as the function of ionic strength of the electrolyte concentration. The magnitude of limiting molar conductivity and extent of association of cation and anion depended upon the inter ionic forces of mobility of ions. In order to get conductance parameter a solution of crown compound was kept in cell and its resistance was measured. Then step by step electrolyte concentration was increased such that total concentration of crown compound was five times as large as that of concentration electrolyte. Once again resistance was measured and conductance data was analysed to get $\Lambda^0,\ |\ a°\ |\ |\ K_A\ |$, where Λ^0 was molar electrolytic conductance, K_A was ionpair association constant and a° was activity of ion in question.

9.4.2 Stability of Crown Complexes with Metals

Number of factors usually influenced the stabilities of complexes. They included size of cation and cavity size of crown ether, charge of cation, kind of donor atom, electron density of crown compound, type of substituent ligands, flexibility of the ligand ring, and physical characteristics of the solvents. It will be clear from some of the cited examples [30]. The complexation between DB18C6 and Na^+ was entropy and enthalpy driven. Here entropy gain was larger, dimethyl formamide molecule participated in solvation cell but such participation was controlled by counter anion. The donor number of a solvent was responsible for stability sequence of same crown metal ion complexes among the solvents [31]. The substituent effect on the stability of complex ion was great in comparison to the effect of ring size [32]. If one used acids like picric, benzoic and toluene sulphonic acid as counter anion during complexation [33] the affinity for complexation with crown ether showed an increase in proportion to their pK value in the aqueous solution. The stabilities of complexed ions were affected by steric hindrance and quality of diluent used [34]. Other factors influencing stability included size of cation cavity, number of donor atoms like oxygen, ring flexibility and capacity of cation to solvate [35]. It was observed that more stable complex was formed when the ratio of cation and cavity size was close to unity [36].

The observations with cryptands like cryptand-2, 2, 2 were little different. For instance the stabilities of their complexes with heavy alkali metals was greater as bicyclic ligand formed much stronger complexes than crown ethers [37]. Selectivity was dependent on nature of diluent used. Since aliphatic ether oxygen atom was more basic than aromatic ether oxygen atom, interaction of former was stronger than that of latter. The conductivity studies were utilised to understand the role of interactive forces on cation macrocyclic ligand complexation process. The electrostatic ion dipole forces, dielectric constant and dipole moment of ligand were important parameters in phenomena of complexation [38] however Takeda [39–43] showed that there was no relationship between stabilities of crown complexes and dielectric constants of the diluents.

9.4.3 Mobilities of the Complexes for Transport

The mobilities of 18C6 and cryptand-222, were not same in acetonitrile or methanol inspite of the fact that size and cavity size of both were more or less same so also mobilities of DB18C6 and DB24C8 was identical as equivalent (λ^0) conductivity was same for their complexes with alkali metals. The best fitting complex like one with 18C6 and Na^+ had large mobility. Thus metals which had an optimum sizes for crown ether cavities were best shielded by crown ether from solvent molecule while for larger or smaller cation [42] in relation to hole of crown ether the interaction among ions trapped and solvent molecule reduced the ionic mobilities. The mobility sequence of crown metal complexes was decided by size sequence of complex. Large complexes were generally less mobile [43]. The consideration of mobility was prime importance in study of membrane transport. For the aprotic solvents (e.g. acetonitrile, propylene carbonate, dimethyl formamide and dimethyl sulphoxide) the donicity was very crucial while considering mobility of metal complexes. The mobilities of 18C6, B18C6, DB18C6 complexes with potassium in protic solvent was more than in aprotic solvent due to presences of H-bond in solvents but H-bond between protic solvent molecule and ether oxygen atoms of the crown ethers had diminished the mobility of crown complexes.

The solvation properties were also studied by conductance measurement, as this information was of significance in understanding the mechanism of extraction by the process of solvation.

9.4.4 Electrical Conductivity

The electrical conductivity of solutions having fixed concentration of completely dissociated metal salt was reduced on account of formation of complex with crown ethers or cryptands. Such variation in magnitude of the electrical conductance was used to ascertain not only value of the stability of the complex but also composition of the solution. The latter were ascertained by examining sharp break in graph of equivalent conductance against concentration of ratio, viz. (C_R/C_M^{N+}). The breaks were sharp for stable complexes [44]. The sensitivity of ionic interaction was dependent upon concentration. The ion pair formation constant was assumed to be not significant but total concentration of cations solvated and complex ions was held constant. At coordination number, a sharp break was visualised. For 1:1 complex method was good. Thus for stable complexes the value of stability constants evaluation by the conductance method was dependable as it relied on lower limit of metal cation concentrations ($\sim 10^{-5}$ M) but for complexes with lower stability higher concentration was required to form sufficient amount of complex. Such method was used for determining stability constant of complexes of crown ether with alkali metals in acetonitrile [45].

9.5 Analytical Applications of Conductometry with Crown Compounds

The work on complexation was mainly confined to evaluation of (a) stability constant of ligands (b) formation constants of the metal crown or cryptand complexes and (c) the study of ionic mobility from view point of membrane transport phenomena.

The metals which were largely studied were alkali metals, alkaline earths [54] silver and mercury [55]. The crown compounds frequently employed consisted of 16C5 18C6, 15C5, 21C7, B15C5, DB18C6, DC18C6, DB15C5 and cryptand-222. In comparison to crown ethers, the cryptands and calixarene were less frequently investigated. The studies were invariably carried out in the water soluble solvents like methanol, propylene carbonate, or in immiscible organic solvents like chloroform, acetonitrile, nitrobenzene and dimethyl formamide.

The decrease in extractibility of alkali metals when 16C5 was substituted with side arms was significant as shown by conductance measurements [46]. The direct current conductometry indicated that polycrown ether was potential compound for solid state electrolytes [47] by study of mobility of ions. Ion pairing was neglected in the high permittivity solvents like propylene carbonate. This observation was significant when alkali metal complexes with 18C6 were investigated [48]. The study of ion-ion and ion solvent interaction of lithium with picrate or perchlorate as the counter anion with 15C5 in nitromethane gave interesting results [49]. Lithium picrate was nonconducting material but 15C5 complexation showed sudden rise in the conductivity. Sometimes conductance studies were also used to understand cation binding properties of polybis (benzo 21C7) derivatives [50]. The concentration of crownethers was also ascertained by conductance measurements in solvents like chloroform which were nonpolar and had low dielectric constants [51]. A remarkable decrease in molar conductance of alkali metals in acetonitrile was noticed by addition of B15C5 as it conformed complexation mechanism [52]. The complexation between potassium and 18C6 in methanol dimethyl formamide solution was studied at various temperature [53]. The propylene carbonate and acetonitrile were most favoured solvents for perchlorate as common counter anion. Thus alkaline earths were complexed with 18C6, DB18C6 or DC18C6 to show formation of 1:1 complex with highest mobility for the magnesium [54]. The thiaza derivatives of DB15C5 were used to complex silver and mercury in ethylenechloride, dimethylsulphoxide and the measurement of electrical conductance of such complexes confirmed the complexation and their stability [55]. Iodine crown ethers complexes were also investigated [56].

The various factors influencing complexation were investigated by conductance measurements. Such details are presented in Table 9.5.1.

248 *Analytical Chemistry of Macrocyclic and Supramolecular Compounds*

Table 9.5.1 Analytical application of conductometry with crown compounds

Element	Optimum Conditions of Measurement
Alkai metals	1, 4, 7, 10, 13, pentaoxacyclohexadecane 16C5 on complexation with the alkali metals showed drop in extractibility and selectivity in methanol solutions showing side arm substitution had lariat effect [46].
Li(I)	Poly (16 crown 5-methacrylate) formed complex with lithium perchlorate which was amorphous and polymeric. It behaved like ionic conductor [47]. The complexes of 18C6 with lithium and other alkali metals when studied conductometrically in propylene carbonate showed that ionpairing was significant and high permitting solvents like propylene carbonate correlated the cation crown molecular conformation [48]. Ion-ion and ion solvent interactions of lithium picrate or perchlorate in nitromethanol with 15C5 showed nonconductivity with lithium picrate but not lithium perchlorate; 1:1 complex was formed there was sudden rise in conductance due to ionpair complexation [49].
Na(I)	Conductance studies of cation binding properties of poly and bis (benzo 21 crown-7) derivative showed caesium was best extracted while sodium was least extracted with dimer of the compound [50] but guanidium cations had no strong interaction. Sodium picrate in chloroform was used to conductometrically determine the crown ether contents. An equation to correlate conductance and concentration was obtained [51]. In acetonitrile the binding of B15C5 with alkali metals when studied showed decrease in molar conductivity due to complexation. The complex being 1:1 stoichiometry [52].
K(I)	With dimethylformamide and methanol or acetonitrile or propylene carbonate the complexation of potassium with 18C6 was conductometrically studied at varying temperature when it was seen that conductance varied with complexation [53].
Mg(II)	15C5, DB18C6, DC18C6 were complexed with alkaline earths and silver in propylene carbonate and studied by conductance measurement when 1:1 complex was noted with greater mobility for Mg-18C6 complex [54].
Hg (II)	Thiaza derivatives of DB15C5 viz (6, 6, 14, 15 dibenzo 10 thia 1, 4 dioxa 7, 13 diazacyclopenta decane) and analogues compounds were complexed with mercury and silver in dichloroethane, dimethylsulphoxide and nitrobenzene with picrates anion. The conductivity was studied to ascertain existance of 1:1 complex [55].
I^{3-}	The complexation of iodine with 18C6, 7, 16 diaza 18C6, cryptand -222 in chloroform when studied showed gradual release of tri-iodide ion from complex the cryptand-222 complex were more stable [56, 57].

9.6 Polarography

This was unique technique which was carried out on dropping mercury electrode (DME). The initial signal was slow on variation of potential, due to the existance of residual current. The final signal in form of diffusion current (id) was plotted against potential (E) with saturated calomel electrode (SCE). The diffusion current (id) was directly proportional to the concentration of electrochemically reducible species. This was expressed by Nernst Equation

$$E = E_{1/2} - \frac{RT}{nF} \ln\left(\frac{i_d}{i_d - i}\right) \tag{6}$$

Here $E_{1/2}$ was half wave potential, it was potential at which the current was equal to id/2 (see Fig. 2). It was similar to formal potential of redox system provided the diffusion coefficients of oxidised and reduced species were equal [58, 59]. The complexation reaction influenced the polarogram. Such effect depended upon the magnitude of stability constant [59]. For moderate complex $E_{1/2}$ was shifted to more negative potentials on successive addition of ligand though small effect was noticed on diffusion potential (RT/nF) In. The $\Delta E_{1/2}$ changed in the half wave potential was defined as

$$\Delta E_{1/2} = (E_{1/2}) M^{n+} - (E_{1/2}) MR^{n+} = RT (\ln K_s + \ln [R]) \tag{7}$$

A large excess of ligand over metalion $[R]$ was equated to total ligand concentration $C_R = [R] + [MR^n]$ this was true for weak complexes if $[R]$ was large. A stability constant was obtained as intercept of graph of $\Delta E_{1/2}$ versus nC_R. This facilitated the evaluation of overall stability constant (βn) and coordination number (n) from the slope and intersection respectively.

In an eventuality the stabilities of the complex was very high, dissociation rate constants was low and further reduction process was noted at $(E_{1/2}) MR^{n+}$ irrespective of free metal ion reduction $(E_{1/2}) M^{n+}$. The difussion currents of the two kinds of reducible ions, viz. free as well as complexed was determined separately and was proportional to the concentration then only (n) coordination number was found out. For stable complex in absence of free metal ions and free ligands concentration stability constant was not obtained. In such case one of them which was in excess concentration was removed quantitatively.

Thus there were wide range of methods to evaluate stability constant of metal complexes with either crown ether or cryptands. No much information was available on complexes of calixarenes. The methods as potentiometry, polarography or cyclic voltammetry, and conductometry were based on chemical or physical factors typical of equilibria or individual metal entity. In electrochemical measurements only it was viable to evaluate stability constant for complexes with $K_s > 10^{-6}$M. Since they had great sensitivity potentiometric methods for many ions (e.g. Ag^+, H^+) and capacity to incorporate complexation of other ions via competitive equilibria made them popular with the research workers.

Recently a beginning has been made to extend electrochemical methods for the study of complexes of the supramolecules like calix(n)arene and calixresorcinarene. Among these calix(n)arenes studied, calix(4)arene had been extensively investigated specially by voltammetric methods involving cyclic techniques. Therefore the role of cyclic voltammetry in the characterisation of crown metal complexes is now considered.

9.7 Cyclic Voltammetry

Cyclic voltammetry was one of the few technique used for verifying electron transfer behaviour specially where chemical reactions were coupled to the electron transfer [60]. It was useful for study of reducible crown compounds and finding out effect of s-block metal ions on electrochemical behaviour of these reducible ligands. In cyclic voltammetry, solution containing a reducible material was contacted an electrode, the potential of which was changed [60]. The term cyclic referes to potential which was altered from high to low and then back to the higher starting point or vice versa [61]. It was hence called cyclic voltammetry. Thus electrode potential was scanned back and forth with relatively high rate in comparison to polarography at stationary electrode (e.g. DME). Each of the oxidation reduction procedure gave two peaks one for reduction and other for oxidation as Epc and Epa respectively. These potentials differed by 57/n mV at room temperature. This Epc or Epa equalled $E_{1/2}$ for reversible process and was equal to formal standard potential of the system when diffusion coefficients of the oxidised or reduced forms were more or less same [62] (see Figure 3).

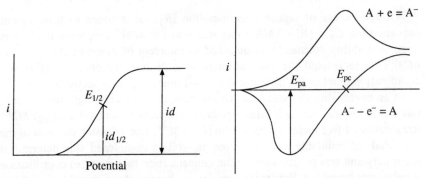

Fig. 2   Polarogram   Fig. 3   Cyclic voltammetry

Unlike polarography, cyclic voltammetry was used for evaluation of stability constants, though less accurate values were obtained than those obtained by polarography. In metal complexes with ligand both coordination number and stability constant was evaluated. In this technique separate peaks were obtained for free and complexed metal ions if metal concentration was higher than that of the ligand.

9.7.1   Cyclic Voltammetry of Lariat Ethers and Cryptands

It was used to assess redox switching ability of lariat ethers bearing side arms based on the nitrobenzene group. The electrochemical reduction of nitrobenzene side arm increased the electron density and donor ability of nitrobenzene moiety, so that over all increase in thermodynamic stability of sodium complex took place. Cyclic voltammetry of azobenzene showed only one electron reduction of the azobenzene moiety. This selectivity loss observed with nitrobenzene based lariat ethers complexes was not of general observation. On the contrary selectivity was retained upon electrochemical switching to a higher binding state if three dimension cavity was present in neutral ligand. The digital simulation results of cyclic voltammetry led to following observations.

(a) For high binding complexes with $K_L > 10^{-4}$, ligands give rise to simultaneous generation of waves for complexed and uncomplexed ligands and observed potential could give binding enhancement factor.

(b) For intermediate binding complexes with $K_L = 10^{-3}$ would exhibit two wave and enhancement factor was easily evaluated.
(c) Finally in low binding complexes with $K_L < 1$ one could get wave which was shifted anodically as the concentration was increased.

Several examples illustrated above observation were cited in many publications [63–67]

9.8 Analytical Applications of Voltammetry with Crown Compounds

Amongst voltammetry methods polarography and cyclic voltammetry had extensive applications in the characterisation and determination of complexes of metals with crown ethers and cryptands. In contrast very little information on complexes of calixarene and calixresorcinarene was available in the literature using these techniques.

Amongst metals studied the mention must be made of alkali metals [68–73] lanthanides [74–75] typical transition elements like silver, cobalt, ruthenium, copper, mercury and few main group metals like lead indium.

With crown ethers as the complexing ligands lithium, potassium, lanthanides, silver, lead and cobalt, ruthenium, palladium, and silver were studied. Such studies included also aza crown ethers. The most commonly used crown ethers consisted of B15C5, 18C5, 15C5, DB24C8 and diaza crown ethers. The cyclic voltammetry studies of complexes of lithium with crown ether attached to anthriquinone unit showed two electron reduction [71]. Chronopotentiometry was used in conjunction with cyclic voltammetry [68]. The electrochemical separation of ionic compounds with electroconductive stationary phase coated with crown ether was carried out successfully [70]. The transfer of potassium from water to ethylenechloride by DB18C6 was followed by cyclic voltammetry and chronoamperiometry [72]. Ion transfer was mainly due to diffusion of species. Polyferrocene bis B15C5 receptor as redox responsive molecular switch was developed by Beer [73]. The existance of complexation of lanthanide with 18C6 and 15C5 was investigated by cyclic voltammetry [74]. New europium complexes were also investigated [75]. Anodic stripping voltammetry was carried out with Nafion crown ether film electrode. The studies were restricted to silver with DC18C6 on Nafion 117 electrode [76]. Preconcentration of lead was done at crown and cryptand containing chemically modified electrodes [77]. Most guest complexes of lead showed modification in electrochemical profiles [78] where reduction potential was dependent on complex stabilities. Crown ethers like DB18C6 and aza analogue were used in stripping voltammetry of palladium with graphite electrode. Palladium could be quantitatively analysed in crown ethers solution in aqueous perchloric acid [82]. Similarly stripping voltammetry of silver was carried out with thia crown compounds by selectively accumulating it at electrode and gradual stripping [83]. Cryptands containing ferrocene unit were studied by the technique of cyclic voltammetry [85]. Cage like azo macrocyclic compounds of cryptands with few metals were also throughly investigated [87]. No doubt some work on cryptands was certainly interesting.

Since last two years some work on electrochemical behaviour of various calixarenes [88–95] was reported. Such work was mainly focussed on cation binding abilities of supramolecules. Calix(4)arene diguinone derivative was thus studied. The electrochemical detection of conformational change on cation binding of switched calix(4)arene was investigated with potassium as guest ion. The results showed conformational change of calix(4)arene from cone to partial cone configuration [89]. Calixarene were also used as the modifiers for carbon paste electrode, for detection of lead, copper and mercury at microgram concentration [90]. In fact most of the work was concerned with either cyclic voltammetry or anodic striping voltammetry leading to the quantitative analysis. However applications of the polarography was very much restricted to study of metals like lithium [92], mercury [94], indium [95]. In dimethyl sulphoxide complexes of thorium with various crown ethers like 12C4, 15C5, 18C6 and DC18C6 were polarographically studied to understand reaction mechanism of thorium crown ether complexes [93]. It was first time, technique of the differential pulse polarography was used to study complexes of mercury with crown ethers in various solvents [94] with DC18C6 > 18C6 > 15C5 > DB18C6 > DB24C8 > B15C5 > 12C4. It was observed that there was inverse relationship between complex and donor number of solvent. The coordination complexes of indium with 18C6, DC18C6, 15C5, B15C5 were studied with AC and DC polarography [95]. Novel amperometric detectors coated with calixarene were then developed [96].

The summary of various determinations involving use of cyclic voltammetry, anodic stripping voltammetry and AC as well as DC-polarography is presented in Table 9.8.1.

Table 9.8.1 Analytical applications of voltammetry with crown ethers

Element	Conditions for Analysis
Li(I)	With Ag-AgCl electrode and Pt-electrode and picrate in 0.01M lithium chloride nitrobenzene containing tetrabutylammoniumtetraphenylborate (0.05M) facilitated transfer of cations in the presence of crown ethers [68]. The electrochemical behaviour of anthriquinone substituted carbon pivot lariat ethers and podands containing alkali metals showed nucleophilic substitution due to electronic and steric factors [69]. The ion transfer voltammetry of lithium, sodium on crown ethers coated column at 0.35 V potential indicated desorption of these ions but not of potassium, due to preferential association of potassium ions with crown ether during extraction [70]. Cyclic voltammetry studies of alkali metals complexed with crown compound annexured with anthriquinone unit showed two one electron reductions and producing anion radicals to attract cations. Lithium was complexed highest while caesium was least attracted [71].
K(I)	The ion transfer of potassium from water to dichloroethane by DB18C6 when studied by cyclic voltammetry indicated that ion transfer process was governed by diffusion of participating species and not merely by charge transfer [72]. Electrochemical study of polyferrocene bis B15C5 complexed with potassium (1:1) showed that receptor underwent anodic perturbation of redox-ferrocene couple [73].

(Contd)

Element	Conditions for Analysis
Ln(III)	The voltammetric investigation of lanthanide complexes with 18C5 and 15C5 indicated two reduction waves corresponding to two species [74].
Eu(III)	The N, N'N'N" tetrakis (2quinolinemethyl)-1oxide-1, 4, 8, 11 tetrazo cyclotetradecane with europium formed stable complex (1:1), cyclic voltammetry showed reversibly in two oxidation state, with formation of nonluminescent complex [75].
Ag(I)	The aniodic stripping voltammetry on Nafion-117 and DC18C6 crown ether film electrode was used for quantitative analysis, in 10M nitric acid. Silver was deposited with no interferences from lead, antimony, bismuth, selenium or copper but potassium interfered so also gold and mercury [76].
Pb(II)	18C6, DB18C6, poly (dibenzo 18C6), DB24C8 and cryptand-222 were mixed with carbon paste to form chemically modified electrode. This showed complexing capabilities during preconcentration process for lead, and mercury, millimicron of lead was analysed by pulse voltammetry [77]. The electrochemical profiles of lead ion in crown type host guest complex indicated a correlationship of reduction potential and complex stability [78]. Adsorptive stripping analysis and monolayer properties of crown ethers was studied, for trace measurements of 13-membered crown ethers [79].
Co(II)	Aza 12C4, 15C5, 18C6 and diaza 18C6 reacted with 2-Indanone to form enamine 2-indenyl crown ethers which with Co(THF)$_3$Br$_2$ gave cobaltenium crown ether with four oxidation states in cyclic voltammetry [80].
Ru(III)	The B15C5 vinyl bipyridyl ruthenium complexes when studied by cyclic voltammetry in acetonitrile showed that B15C5 unit were sensitive to supporting electrolyte and triscomplex of ruthenium was reductively electropolymerised and showed transparant electrochromicity [81].
Pb(II)	DB18C6 azo analogue in 0.1M perchloric acid with graphite electrode was used for determination by stripping voltammetry [82].
Ag(I)	A carbon paste electrode modified with thiacrown ethers was used to study accumultation of silver in stripping voltammetry in sodium perchlorate [83].
Ferrocene	1, 1 'ferrocenediyl (bis methylene bispyridinium) chloride tosylate with diaza 15C5 gave 2, 2' addition product, cavity could accommodate two metals [84].

Table 9.8.2 **Analytical application of voltammetry with cryptands**

Element	Conditions for Analysis
Alkali metals	Cryptands containing ferrocene units were studied by cyclic voltammetry a shift in $\Delta E_{1/2}$ was observed due to complexation with metals [85]. The complexation with cryptand containing ferrocene was studied by cyclic voltammetry to get various physical parameters such as K_2 and ionic radius/charge ratio [86].
Cu(II)	Mono and polynuclear cryptand complexes of cage like azamacrocyclic compounds were studied by cyclic voltammetry, which showed lower valent copper complex was stabilised due to formation of binuclear complex [87].

254 *Analytical Chemistry of Macrocyclic and Supramolecular Compounds*

Table 9.8.3 Analytical application of voltammetry for calixarenes complexes

Element	Condition for Analysis
NH_4^+	Calix(4)arene diquinone was studied for its redox and cation binding properties in acetonitrile, new peaks were noted due to the electrochemical reduction and receptor molecule, ammonia could form H-bond with reagent [88].
K(I)	4-tert butyl calix (4) arene bearing 1-8 bis (ethoxyethylene) anthriquinone bridge with alternate phenolic oxygen showed enhanced binding with the selectivity dependent on phenolic oxygen [89] for potassium but not for sodium.
Pb(II)	Carbon paste electrode containing calixarene as the modifier when used showed decrease in concentration of metal ions due to complexation [90]. Calix (n)arene and amines interaction studies by electrochemical devices indicated proton transfer reaction resulting new electrolyte [91]. Influence of 18C6 on polarographic behaviour of cations showed complex with crown was readily reduced. For lithium triple complex was formed [92].
Th(IV)	The complexes with 12C4, 15C5, 18C6 and DC18C6 were measured for conductance in dimethylsulphoxide and redox reaction was studied by polarography, complex was 1:1 involving 1-electron reduction and could be hydrolysed [93].
Hg(II)	Complexed with cyclic crown ethers in nitrobenzene, dimethylformamide or acetonitrile were studied to understand stoichiometry and stability [94] of complex.
In(III)	The complexes with 18C6, DC18C6, 15C5, B15C5 were studied in aqueous solution by polarography and formation constant was evaluated [95, 96].

9.9 Coulometry and Its Analytical Applications

In comparison to potentiometry and conductometry, coulometry was rarely used to study coordination complexes of crown ethers, cryptand and calixarenes. As a matter of fact this technique was mainly used for quantitative analysis by extending Faradays laws of electrolysis confirming that the amount of material generated either at cathode or anode was directly proportional to the quantum of electricity passed through the system. This technique was very rarely used for structure elucidation. However it was ideal technique for transition metal complexes involving mainly cryptands.

The common was use of coulometry to study complexes of mercury with various cryptands. This technique [57] involved electrogeneration of mercury

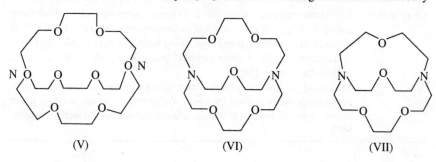

(V)　　　　　　　　　(VI)　　　　　　　　　(VII)

ions by utilising cryptands. The coulometric titration with electrogeneration of mercury (II) and formation of 1:1 complex of mecury (II) along with cryptand-222 (V) cryptand-221 (VI) and cryptand-211 (VII) was thoroughly investigated.

Mercury (II) was generated with mercury pool cathode and platinum as anode. Coulometric titrations were carried out potentiometrically with SCE as reference electrode at pH 6.0-7.0 tetramethyl ammonium acetate was used as the supporting electrolyte. For cryptand-222 (V) method could operate at microgram concentration cryptand-211 (VII) was useless as it could not form stable complex while third ligand cryptand-221 (VI) gave high results.

9.10 Electrochemistry of Supramolecular Compounds

The main objective of electrochemistry is to evaluate redox potentials and to correlate those with molecular properties of these compound [97]. It was powerful mode to monitor supramolecular events as self aggregation on solid electrode material, create electronic and structural modifications. Redox modifications could be used to increase or decrease the binding affinity of ligands for the guest molecule or ions. Such ligands were designated as redox switches [98–101]. Electrochemistry led to significant structural changes in supramolecular system. The nuclearity of the complex depended on redox state of chelated ions. By controlling measured redox state the ions one could switch between di, tri, tetra or polynuclear complexes. It was used to control binding state, self-assembly and production of new compound. Redox switching was thus interesting area of research. The oxidised and reduced states of switchable molecule did show varying affinity for second species. It determined thermodynamic stability of complex between receptor and guest species. The lower peak currents were noted for complexed species in comparison to free species. This was valid for small guest and large host. The ferrocene based system was developed after 1980. It underwent monoelectronic reversible oxidation to ferrocenium at suitable potential, was insensitive to the protonic solvents and molecular oxygen. Lot of work was reported on self-assembly of porphorin and binding properties of supramolecular [102] compounds.

The calixarenes were a special class of cyclophanes capable of binding both neutral and charged guest. The studies were considered under group where redox unit was integral part of the σ-bonded framework and the complexes of electroactive calixarenes had electroactive guest species. They had further groups as on having redox active unit and others which had electroactive pendant group covalently attached to ethers. The electroactive calixarene dealt with type number and location of the active group in the macrocycle. The over all size was of crucial importance. In small ligands like calix(4)arene distance and rotational flexibility influenced the redox active unit. The pendant electroactive groups were affected less by ring. Calix(4)arene was first synthesised [103–104]. These compounds had two electron reduction for each quinone of which one was reversible and other was nonreversible [105–106]. As ring was enlarged the structure led to behaviour analogues to noninteracting identical redox unit [107]. The binding of calixarene with metals like alkali have great effect on the electrochemistry [108–110]. Thus electron reduction potential of these compounds shifted from 90–600 mV in presence of bound ion. The binding ion in turn could change conformations of calixarenes [111–112].

The calixarene with nitrogroups in para position of the phenol rings were also studied [113] with different pattern of electrochemistry. The cyclic voltammetry of a calixarene bearing four nitrogen aromatic groups had two electron reduction waves, each wave corresponding to the simultaneous reduction of diagonally opposite pairs of nitroaromatic units. The attachment of pendant electroactive groups to the upper, lower or methylene side arm positions was also studied [114]. In complex between sulphonic acid calix(6) arenes, in presence of calixarene had an impact on electrochemical behaviour of incorporated guest [115–120].

Thus electrochemical methods could help for characterisation or synthesis of supramolecular compounds. The detection of dilute analytical solution could be accomplished by pulsing method. They could produce electronic change in the host guest molecules which influenced supramolecular aggregation.

Conclusion

We conclude from the forgoing discussion, that the electroanalytical methods were used for study of stability constant, formation constant and mobility of ions on complexation with crown compounds. Potentiometry was best suited for the development of ion selctive electrodes and study of stability constant of complexes with great accuracy. While conductometric methods were quite useful for evaluation of formation constant from conductance data. Large number of experiments were performed to ascertain the stability constants of complexes involving cryptands. The mobility consideration was of atmost importance during study of membrane transport phenomena. This was specially true for crown ether complexes with alkali metals. AC and DC polarography had been extensively used to calculate $E_{1/2}$, i.e. half wave potential which not only characterised the metal but it provided reliable method for quantitative analysis by measurement of the diffusion current with varying concentration. Cyclic voltammetry took a major share to understand electron transfer mechanism. The stability constants determined by cyclic voltammetry were less accurate. Cyclic voltammetry was used for study of complexes of metals with calixarenes however cryptand complexes were more vigorously pursued by this technique. Unfortunately the coulometry technique had major role to play only in study of mercury-cryptand complexes. The major player in electrochemical studies was supramolecular compounds like calixarenes and switch on compounds involving redox switching. They gave some information on structural transformation. Although the present progress was slow it is envisaged that these methods would play vital role in characterisation and the determination of complexes of metals with calixarene type of compounds in near future.

References

1. J. Gutknecht, H. Scheneider, J. Stroka, Inorg. Chem. **17**, 3326 (1979).
2. I.M. Kolthoff, M.K. Chantooni, J. Am. Chem. Soc. **87**, 4428 (1965) and Anal. Chem. **52**, 1039 (1980).
3. B.G. Cox, J. Gracia-Rosas, H. Schneider, J. Am. Chem. Soc. **103**, 1384 (1981).

4. M.K. Chantooni, I.M. Kolthoff, J. Solution Chem. **14**, 1 (1985).
5. I.M. Kolthoff, M.K. Chantooni, Proc. Nat. Acad. Sci. USA **77** (1980).
6. H.K. Frensdorff, J. Am. Chem. Soc. **93**, 600, 4684 (1971).
7. J.M. Lehn, J.P. Sauvage, J. Am. Chem. Soc. **97**, 6700 (1975).
8. A.Z. Gordon, P.A. Rock, J. Electrochem Soc. **124**, 534 (1977).
9. B. Spince, S. Eiduka, A. Vevris, A. Zicmanis, Lacv PSR Zinat Akad Vestis Kim Ser 566 (1986) A.A. **49**, 8C57 (1987).
10. A. Veveris, S.A. Steinberga, B. Spince, A. Zicmanis, Zhur anlit khim **43**, 1886 (1988) CA **111**, 49664u (1989).
11. A. Veveris, N.A. Petrovskaya, A. Zicmanis, Latr. PSR Zinat Akad. Vertis Kim. Ser. 736 (1989) CA **113**, 17271p (1990).
12. A. Vevris, N.N. Evdokimova, O.V. Ivanov, Zhur analit Khim **45** 1238 (1990) AA **53** 5D123 (1991).
13. H.J. Buschmann, C. Carvalho, E. Cleve, G. Weuz, E. Schollmeyer, J. Cord. Chem. **31**, 347 (1994).
14. Sh.K. Norov, Zhur anlit khim **42**, 429 (1987) CA **107**, 146321e (1987).
15. K. Ohtsu, Buneski Kagaku **46**, 167 (1997) AA **59**, 6030 (1997).
16. H.J. Buschmann, Polyhedron **11**, 559 (1992).
17. R.A. Sachleben, B.A. Moyer, F.I. Case, S.A. Garmon, Sepn. Sci. and Tech. **28**, 1 (1993).
18. I.N. Papadoyannis, Anal. Letters **18**, 2013 (1985).
19. B.S. Grabaric, M. Takcec, V. Merzel, I. Filipovic, Electroanalysis **3**, 647 (1991) CA **115**, 216220n (1991).
20. A. Goecmen, C. Erk, J. Anal. Chem. **342**, 471 (1993) AA **56**, 7D235 (1994).
21. M. Kodama, Bull. Chem. Soc. Japan **69**, 3187 (1996).
22. A. Veveris, S. Steinberga, B. Spince, A. Zicmanis Latr. PSR Zinat Akad. Vestis kim Ser 441 (1988) AA **51**, 3C45 (1989).
23. R. Abidi, F. Arnaud Neu, M.G.B. Drew, S. Lahely, D. Marrs, J. Nelson, Marie Jose Schwing Weill. J. Chem. Soc. Perkins Trans **2**, 2747 (1996).
24. G. Czervenka, E. Scheubeck, Z. Anal. Chem. **276**, 37 (1975).
25. A. Bencini V. Fusi, C. Giorgi, M. Micheloni, N. Nardi, B. Voltancoli, J. Chem. Soc. Perkin **2**, 2297 (1996).
26. R.S. Dhillon, S.R. Lincoln, A.K.W. Stephens, P.A. Duckworth, Inorg. Chim. Acta **215**, 79 (1994).
27. F. Angela, F. Danilde Namor D. Kowalska, J. Phy. Chem. B **101**, 1643 (1997).
28. A. Cassol, P.D. Bernando, G. Pilloni, M. Tolazzi, P.L. Zanonato, J. Chem. Soc. Dalton 2689 (1995).
29. Y. Takeda, "Cation binding by macrocycles Complexation of cationic species by crown ether" Ed. Y. Inoue, G.W. Gokel Marcel Dekker and Co. New Yord p. 133 (1990).
30. E. Shchori, J.J. Gradzindki, Z. Lüz, M. Shporer, J. Am. Chem. Soc. **99**, 489 (1977).
31. N. Matsura, K. Umemoto, Y. Takeda, A. Sasaki, Bull. Chem. Soc. Japan **49**, 1246 (1976).
32. R. Ungaro, B.E. Haj, J. Smid, J. Am. Chem. Soc. **98**, 5198 (1976).
33. N. Nae, J.J. Grodzinski, J. Am. Chem. Soc. **99**, 489 (1977).
34. L. Tusek-Bozic, P.R. Danesi, J. Inorg and Nucl. Chem. **41**, 833 (1979).
35. Y. Takeda, H. Yano, M. Ishibashi, H. Isozumi, Bull. Chem. Soc. Japan **53**, 1720 (1980).
36. H.P. Hopkins Jr, A.B. Norman, J. Phy. Chem. **84**, 309 (1980).
37. S. Kulstad, L.A. Malmesten, J. Inorg. and Nucl. Chem. **42**, 573 (1980).

38. A. Dáprano, B. Sesta, J. Phy. Chem. **88**, 5445 (1987).
39. Y. Takeda, T. Kumazawa, Bull. Chem. Soc. Japan **61**, 655 (1988).
40. Y. Takeda, Y. Ohyagi, W. Akabori, Bull. Chem. Soc. Japan **57**, 3381 (1984).
41. Y. Takeda, K. Katsuta, Y. Inoue, T. Hakushi, Bull. Chem. Soc. Japan **61**, 627 (1988).
42. Y. Takeda, Bull. Chem. Soc. Japan. **54**, 3133 (1981).
43. Y. Takeda, Bull. Chem. Soc. Japan **58**, 1259 (1985).
44. S. Kulstadt, L.A. Malmsten, J. Inorg and Nucl. Chem. **42**, 573 (1980).
45. D.F. Evans, S.L. Wellington, J.A. Nadis, E.L. Cussler, J. Solution Chem. **1**, 499 (1972).
46. Y. Inoue, M. Ouchi, K. Hasoyama, T. Hakashi, Y. Liu, Y. Takeda, J. Chem. Soc. Dalton 1291 (1991).
47. D. Peramuage, J.E. Fernandez, L.H. Garcia. Rubio, Macromolecules **22**, 2845 (1989).
48. M. Salomon J. Solution Chem. **19**, 1225 (1990).
49. A. Dáprano, B. Sesta, A. Princi, J. Electoanal. Chem. **361**, 135 (1993).
50. K. Kimura, S. Kitazawa, T. Maeda, T. Shono, Z. Anal. Chem. **313**, 132 (1982).
51. F.L. Fongs, Fenxi Huaxue **13**, 349 (1985) AA **48**, 1C45 (1986).
52. K.M. Tawarah, S.A. Mizyed, J. Incl. Pheno. **6**, 583 (1988).
53. G. Rounghi, Z. Eshaghi, E. Ghiamati, Talanta **44**, 275 (1997).
54. A.K. Srivastwa, B. Tiwari, J. Electro Anal. Chem. **325**, 301 (1992).
55. L.M. Kardivarenko, V.V. Bagreev, N.M. Kuz'min, Zhur Neorg. khim **33**, 2871 (1988) CA **110**, 83098h (1989).
56. A. Semnani, M. Shamsipur, J. Chem. Soc. Dalton 2215 (1996).
57. J.L. Leibengum, M. Roynette, M.H. Saltany, Analysis **9**, 3 77 (1981) AA **42**, 6264 (1982).
58. B.G. Cox, H. Schceider, Coordination and transport properties of macrocyclic compounds in solution Elsvier Public Ltd. (1992).
59. P.T. Kissinger, W.R. Heineman, Laboratory. Technique in Electro. Analytical Chemistry Marcel Dekker N.Y. (1980).
60. Y. Inoue, G.W. Gokel, Cation binding by macrocycles complexation of cationic species by crown ether Marcel Dekker p. 374 (1990).
61. D.J. Eatough, R.M. Izatt, J.J. Christensen, Thermo Chimica Acta. **3**, 203 (1972).
62. J..D. Lamb, R.M. Izatt, J.J. Christensen, Progress in macrocyclic Chemistry Vol. 12 p. 41 Wiley and Sons (1981).
63. D.A. Gustowaski, L. Echegoyen, D.M. Golli, A. Kaifer, R.A. Schultz, G.W. Gokel, J. Am. Chem. Soc. **106**, 1633 (1984).
64. D.A. Gustowaski, V.J. Gatto, A. Kaifer, L. Echegoyen, R.E. Godt, G.W. Gokel, J. Chem. Soc. Chem. Comm. 923 (1984).
65. L. Echegoyen, D.A. Gustowaski, V.J. Gatto, G.W. Gokel. J. Chem. Soc. Chem. Comm. 220 (1986).
66. A.J. Bard, L.R. Faulkner, Electrochemical methods fundamental and application John Wiley (1980) Chapt. 6.
67. S.W. Feldberg, Electrochemistry Ed. A.J. Bard Marcel Dekker (1969).
68. D. Homolka, Lequoc Hung, A. Hofmanova, M.W. Khalil, J. Koryta, V. Marecek. Z. Samec, S.K. Sen, P. Vanyseki, J. Weber, M. Brezina, M. Janda, T. Stibav, Anal. Chem. **52**, 1606 (1980).
69. D.A. Gustowski, M. Delgado, V.J. Gatto, L. Echegoyen, G.W. Gokel, J. Am. Chem. Soc. **108**, 7553 (1980).
70. T. Nagaka, M. Fujimoto, H. Nakao, K. Kakuno, J. Yano, K. Ogura, J. Electro. Anal. Chem. **350**, 337 (1993).

71. J.M. Caridade Costo, B. Jeyashri, D. Bethell, J. Electro Anal. Chem. **361**, 259 (1993).
72. P.D. Beattie, A. Delay, H.H. Givault, J. Electro. Anal. Chem. **380**, 167 (1995).
73. P.D. Beer, C.G. Grane, J.P. Danks, P.A. Gale, J.F. McAleev, J. Org Met. Chem. **490**, 143 (1995).
74. E. Guan, X. Gao, Chin. Sci Bull. **35**, 830 (1990) CA **113**, 161119n (1990).
75. M. Pietaraskiewick, J. Karpiuk, R. Bilewicz, S.P. Kasprzyk, J. Coord. Chem. **21**, 75 (1990).
76. S. Dong, Y. Wang, Anal. Chim, Acta **212**, 341 (1988).
77. S.V. Prabhu, R.P. Baldwin, L. Kryger, Electroanalysis **1**, 12 (1989) CA **111**, 125977r (1989).
78. H. Tuskube, K. Takagi, T. Higashiyama, T. Iwachida, N. Hayama, Bull. Chem. Soc. Japan **61**, 293 (1988).
79. E. Muszalska, R. Bilewicz, Analyst, **119**, 1235 (1994).
80. H. Pelenio, D. Burth, Organometallilcs **15**, 1151 (1996).
81. P.D. Beer, O. Kocian, R.J. Mortimer, C. Ridgway, J. Chem, Soc. Faraday Trans **89**, 333 (1993).
82. M.V. Tsymbal, I.Y. Turjan, Z.A. Temerdashev, K.Z. Brainina, Electroanalysis **6**, 113 (1994) AA **57**, 2D 115 (1995).
83. S. Tanaka, H. Yoshida, Talanta **36**, 1044 (1989).
84. H. Plenio, R. Diodone, J. Org. Metal. Chem. **492**, 73 (1995).
85. C. Dennis Hall, Sunny Y.F. Chu. J. Org. Metal. Chem. **498**, 221 (1995).
86. C.D. Hall, N.W. Sharpe, I.P. Danks, Y.P. Sang, J. Chem. Soc. Chem. Commun. 419 (1989).
87. C. Bazzicalupi, A. Bencini, A. Biachii, H. Cohen, D. Meyerstein, V. Fusi, L. Mazzanti, P. Palotti, B. Valtáncoli, G. Golu. J. Chem. Soc. Dalton 2377 (1995).
88. T.D. Chung, D. Choi, S.K. Kans, S.K. Lee, S.K. Chang, H. Kim, J. Electroanalychem **396**, 431 (1995).
89. D. Bethell, G. Dougherly, D.C. Cupertino, J. Chem. Soc. Chem. Comm. 675 (1995).
90. D.W.M. Arrigan, G. Sevhla S.J. Harris, M.A. Mckervey, Electroanalysis **6**, 97 (1994) AA **57**, 2D14 (1995).
91. A.F.D. Namor, P.M. Blackett, M.T.G. Pardo, D.A.P. Tanaka, F.J.S. Velavde, Pure and Applied Chem. **65**, 415 (1993).
92. S.G. Mairanovskii, S.H. Erbekov, R.U. Vakhobova, J. Electro Anal. Chem. **325**, 135 (1992).
93. Jung. Hak Jin, Jung Oh Jin, Suh Hyouek Choon, Taechan Harahakhoe Chi **31**, 250 (1987) CA **107**, 104813 m (1987).
94. A. Rouhollahi, M. Shamsipur, M.K. Amini, Talanta **41**, 1465 (1994).
95. H.A. Azab, A. Hassan, Chem. Scr. **28**, 205 (1988) CA **109**, 117108b (1988).
96. J. Wang, J. Liu, Anal. Chim. Acta. **294**, 201 (1994).
97. P.L. Boulas, M.G. Daifer, L. Eschegoyen, Angew. Chem. **37**, 216 (1998).
98. E.M. Seward, R.B. Hopkins, W. Sauerer, S.W. Tam, F. Diederich, J. Am. Chem. Soc. **112**, 1783 (1990).
99. P.D. Beer, Chem. Soc. Rev. **18**, 409 (1989).
100. A.E. Kaifer, L. Echegoyen, Cation binding by macrocycles Ed. Y. Inoue, G.W. Gokel, Marcel Dekker p. 373 (1990).
101. A.E. Kaifer, E. Mendoza, Comprehensive Supramolecular Chemistry Vol. 1 Ed. G.W. Gokel, Pergaman p. 714 (1996).
102. V. Kral, J.L. Sessler, H. Furuta, J. Am. Chem. Soc. **114**, 8704 (1992).
103. A.P. de Silva. C.P. McCoy, Chem. and Ind. 792 (1994).

104. S.R. Miller, D.A. Gustowski, Z.H. Chen, G.W. Gokel, L. Echegoyen, A.E. Kaifer, Anal. Chem. **60**, 2021 (1988).
105. J.Q. Chambers, The chemistry of quinoids compounds Vol. 2 Part I Ed. S. Patai Wiley Chap. 12 (1988).
106. J.B. Flanagan, S. Margel, A.J. Bard, F.C. Arison, J. Am. Chem, Soc. **100**, 4248 (1978).
107. A. Elkasmi, D. Lexa, P. Maillard, M. Momenleau, J.M. Saveant, J. Phy. Chem. **97**, 6090 (1993).
108. M. Gomez-Kaifer, P.A. Reddy, C.D. Gutsche, L. Echegoyen, J. Am. Chem. Soc. **116**, 3580 (1994).
109. P.D. Beer, Z. Chen. P.A. Gale, Tetrahedron **30**, 930 (1994).
110. T.D. Chung, D. Choi, S.K. Kang, S.K. Lee, S.K. Chang, H. Kim, J. Electroanaly Chem. **396**, 431 (1995).
111. J. Blixt, C.D. Detellier, J. Am. Chem. Soc. **117**, 8536 (1995).
112. M. Gomez-Kaifer, P.A. Reddy, C.D. Gutsche, L. Echegoyen, J. Am. Chem. Soc. **116**, 5222 (1997).
113. W. Verboom. A. Durie, R.J.M. Egberink, A. Asafri, D.N. Reinhoudt, J. Org. Chem. **57**, 1313 (1992).
114. P.D. Beer, Z. Chen, P.A. Gale, J. Electroanal Chem. **393**, 113 (1995).
115. L. Zhang, L.A. Godinez, T. Lu, G.W. Gokel, A.E. Kaifer, Angew Chem. **107**, 236 (1995).
116. A.R. Bernardo, T. Lu, E. Cordova, L. Zhang, G.W. Gokel, A.E. Kaifer, J. Chem. Soc. Chem. Comm. 829 (1994).
117. L. Zhang, A. Macias, R. Isnin, T. Lu, G.W. Gokel, A.E. Kaifer, J. Incl. Pheno. Mol. Recog. Chem. **19**, 361 (1994).
118. L. Zhang, A. Macias, T. Lu, J.I. Gorden, G.W. Gokel, A.E. Kaifer, J. Chem. Soc. Chem. Comm. 1017 (1993).
119. R.M. Williams, J.W. Verhoeven, Rec. Trav. Chem. **111**, 531 (1992).
120. J.L. Atwood, G.A. Koustantonis, C.L. Raston, Nature **368**, 229 (1994).

10
Ion Selective Electrodes and Membrane Transport with Crown Compounds

10.1 Introduction

The electroanalytical methods are very useful for the characterisation and determination of important physical parameters like stability constant, formation constant and mobility of coordination complexes. Such methods were mainly used for the elucidation of the structures of the ligands or the coordination complexes. The most commonly used techniques for characterisation consisted of either cyclic voltammetry or conductance measurement. Although potentiometry was used to some extent, unfortunately this technique was not employed for the quantitative analysis with crown ethers. However with the discovery of ion selective electrodes this gap was filled in. In fact liquid anion exchangers, i.e. high molecular weight amines were used extensively for the analysis of anions. Only with the discovery of crown ethers and to some extent cryptands the novel technique for the quantative analysis of cations with ion selective electrodes was discovered.

In initial investigations, macrocyclic crown ethers capable of forming 1:1 complexes were used as the neutral carriers. The work was mainly focussed on the analysis of alkali metals. The most favoured element was potassium. By drawing an anology with ion selective electrode containing valinomycine, all efforts, were diverted to develop ion specific electrode for potassium with better selectivity over sodium. This selectivity was ascertained in terms of the value of selectivity coefficient for the particular electrode. The novel techniques were developed to fabricate various kind of electrodes involving PVC-membrane by single solution or double solution methods. The one developed by Moody Thomas [7] was most popular with analytical chemist. The potential of the membrane electrode and selectivity coefficient gave clear idea about the standard and quality of the electrodes used. It provided means to check the Nernstian response during quantative analysis. There was a surge of activity for the development of cation specific electrodes. To some extent thia and aza crown ethers were employed specially for the analysis of heavy metals like thallium, zinc, cobalt and copper. Only in last decade calix(n)arenes and the spherands were used for the analysis of organic cations and metals. Similarly last few decades witnessed the brisk activity on attempts to develop ion selective electrodes containing bis (crown ether) derivatives. The success of working of such electrode depended upon the kind of aliphatic linkage attached in between two crown ether rings. Apart from symmetrical structures involving same crown ethers few attempts were made to use different crown ethers like 15C5 and 18C6 with proper aliphatic linkage between these two compouds. The work on development of novel specific ion selective electrodes is in active progress.

Apart from the development of ion selective electrodes, investigations were carried out on use of membrane for the transport of ions. Such studies were mainly focussed on solvent extraction separations involving semipermeable membrane. A detailed discussion on source and receiving phase, transport system and use of proton ionisable crown ethers was then considered. To begin with let us have discussion on the use of crown ethers in ion selective electrodes.

For continuous measurement of samples ion selective electrode was the best tool of analysis [1, 2]. They were used in clinical and biological analysis. Those electrodes of liquid membrane type generally employed crown ethers as neutral carriers, because they had specific ability to transport ion selectivity. B15C5, DB18C6, DC18C6, DB30C10 were used as ligands in nitrobenzene for ion selective studies of alkali metals [3]. Plasticized polyvinyl chloride was used as the membrane [4] 18C6 and 30C10 proved to best crown membranes. Plasticized polyvinyl chloride proved to be best material for liquid membrane [5]. The activities or concentrations of ionic species was analysed in solution containing various species by these methods. They were simple, handy, cheap compared to other analytical techniques or instruments. Neutral carrier type ion selective electrodes were most popular. The electrically neutral ionophore or ion carriers were used as the basic components of ion selective electrodes. Thus crown ethers which were best were neutral carriers for ion carriers. The monocyclic and bicyclic crown ethers were best for the purpose of analysis. The first fundamental properties of neutral carrier ion selective electrodes included membrane potential, electrode, selectivity, electrode fabrication and measurement of the potential of the system during quantitative analysis.

10.2 Characteristics of the Electrode

The membrane potential, selectivity coefficient, fabrication and measurement of potential were important function of ion selective electrodes.

10.2.1 Fabrication of Electrodes

The liquid membrane was used. They were prepared by coating of neutral carrier from organic solvent on microporous polymer film. Most useful polymeric membrane was plasticised poly (Vinyl Chloride) i.e. PVC. It contained PVC support and plasticizer as pthalate ester and phenylester derivatives acting as membrane solvent and small amount of the neutral carrier (a lipophillic salt could be added). The membranes were prepared from THF (or cyclohexanone) solutions containing requisite amount of membrane component. On evaporation a transparent elastic membrane of 0.1-0.2 mm thickness was formed. A disk was cut out of this membrane and was fixed on electrode end. The overall assembly is depicted in Fig. 1 to Fig. 3.

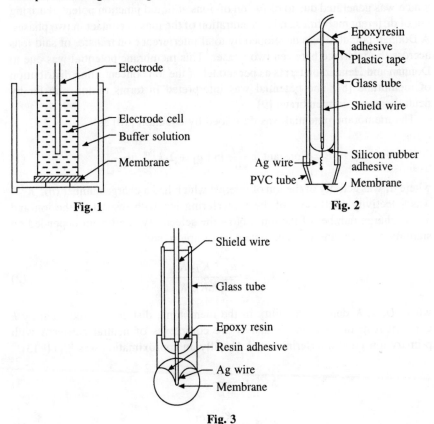

Fig. 1

Fig. 2

Fig. 3

The fabrication was quite simple. A simple rod measuring chain consisted of tube shaped plastic casting which covered reversible working reference electrode (i.e. working electrode) immersed in inner test solution which was to be analysed. The buffer and test solution were separated by the selective membrane which responded to the ion to be analysed. The membrane usually contained lipophillic podands or neutral carriers like crown ethers. In comparison the coated wire electrode was compact and simple. They were prepared by repeated dipping of metal wires of copper or silver into coating solution, composition of which was same as that of PVC membrane. The microelectrodes flow through electrodes ion-sensitive field effective transitors (ISFET) were modified version of these ion selective electrodes [6]. The electrode Fig. 1 is conventional while Fig. 2 is Moody Thomas type electrode while Fig. 3 is coated wire electrode [7] used in quantitative chemical analysis.

10.2.2 Potential and Selectivity

A membrane was thin body which separated two phases. Now if the membrane was selectively permeable for particular ion, and if it was semipermeable it was called "electrochemical membrane". A potential arising out of difference between two solutions just separated by membrane was called "membrane potential" which was generated due to diffusion of ions at.liquid junction potential arising out of different mobilities and concentration of the ions in contact in two phases. A Donnan potential was developed by total interference on transfer of said ions accross the interface between two phases. This membrane potential was due to Donnan and Henderson terms as per model of the ion concentration distribution of membrane [8]. The potential was interpreted in terms of ion transfer by neutral carriers in membrane [9].

The membrane potential was described by equation [10] as

$$E = \text{Constant} \pm \frac{RT}{nF} \ln\left[a_1 + \sum_j k_i^{Pot} \, aj\frac{ni}{nj}\right] \quad (1)$$

where a_1 was activity of the ion of interest which had a charge number (n). K_{ij}^{Pot} was selectivity coefficient of the j-interfering ion with respect to the ion and n-was charge number of the ion j. Now the selectivity coefficient depended on stability constant and dist coefficient so of anion, so

$$k_{ij}^{Pot} = \frac{\overline{u_{js}} * K_{Js} \, K_{Js}}{\overline{u_{js}} * K_{is} \, K_{is}} \quad (2)$$

where U, k, K denoted mobility in the membrane, dist coff K and stability k constant resp in solution, Js and is are complexes of neutral carrier S with primary ion i and interfering ion J. Usually an approximation was 8 [11–13]

$$K_{ij}^{Pot} = \frac{K_{js}}{K_{is}} \quad (3)$$

$$K_{ij}^{Pot} = \frac{K_{ex} \text{ for } j}{K_{ex} \text{ for } I} \quad (4)$$

Thus selectivity coefficient depended not merely on the relative stability constant but also upon distribution coefficient. The selectivity coefficient was obtained by two methods, viz. (a) separate solution and (b) mixed solution. Thus in ISE, potential between internal filling solution of fixed strength of primary cation I and a measuring solution of the primary cation and interfering cation J was given by above equation. The term k_{ij}^{Pot} was potentiometric selectivity coefficient, it specified the ability of ISE to identify the varying ions in same solution. In single or fixed interference method the potential was measured with solutions containing constant activity of interfering ion Aj and the varying activity of primary ion using ISE and reference electrode. The mixed or separate solution method involved measurement of potential of each solution provided Nernstian response was valid. The first method was most frequently employed. The neutral carriers formed 1:1 complex. The selectivity coefficient was defined as shown in above equation. It was approximately equal to the ratio of the extraction equilibrium constants for the ions I and J as shown by equation (4). It was the ratio of the complex formation constants in the aqueous solutions. Few crown ethers formed 2:1 sandwich complexes with cations which exceeded the crown ether cavity in size.

Few observations on anions study by ISE were valid for consideration. Thin membrane model [14–15] did not stipulate electrical neutrality, so hydrophobic anions were excluded from the membrane. Usually the membrane contained immobilised anions those were chemically bound to the supporting material and the mobility of anions across the membrane was very low in comparison to that of the cationic complexes. Membrane potential including anion interference was extensively studied [16]. The Nernstian response depended on the overall concentration of the netural carrier and extraction constant. The linearity of the Nernstian slope had increased by using lower concentration of the neutral carrier and high concentration of lipophillic anions in the membrane. The linearity was further improved by changing solvent the choice of which depended upon the distribution coefficient. Nitrobenzene gave best results. The change in plasticizers improved the selectivity. A good quality ISE could be prepared by selecting appropriate concentration of crown ether in membrane and accounting for solution state of the species in the membrane. Prior conditioning of the membrane with dilute solution produced faster response; or membrane component was nonpolar.

10.3 Crown Ethers as Neutral Carriers
Majority of the crown ethers were used for fabrication of ion selective electrodes for s-block elements. While those of the compounds containing nitrogen or sulphur were used for the analysis of the heavy metals like copper, mercury and thallium.

10.3.1 Crown Ethers in Electrodes
The monocyclic crown ethers were used in ion selective electrodes for alkali metals. The commonly used crown ethers were 15C5, 16C5, 18C6, 19C6, 22C7, 24C8, 30C10, 33C11 and 36C12. Many crown ethers were dissolved in nitrobenzene and were used in supported liquid membrane. Several workers [5, 17–19] prepared potassium selective electrodes against sodium, comparable to valinomycin-based electrodes. The use of high concentration of crown ether in PVC membrane had improved the selectivities of electrodes. A PVC membrane containing 12C4 and phosphotungstic acid was selective for sodium [20]. They showed that the Nernstian response in wide ion activity range and selective against other [21] interfering ions. DB18C6 was best for barium. A barium selective electrode with polyacrylamide DB18C6 conjugate membrane was best. The ISE with neutral carriers had problem in ascertaining ionic species in analysis specially in the organic solvents. This was because membrane solvent and neutral carriers dissolved in organic phase.

In such electrode crown ether as shown in structure (I) was covalently bound to hydrophilic polymer, consequently there was no loss of neutral carrier from the membrane. Such electrode was beneficial for potentiometric analysis of barium specially in organic solvents such as acetonitrile (CH_3CN).

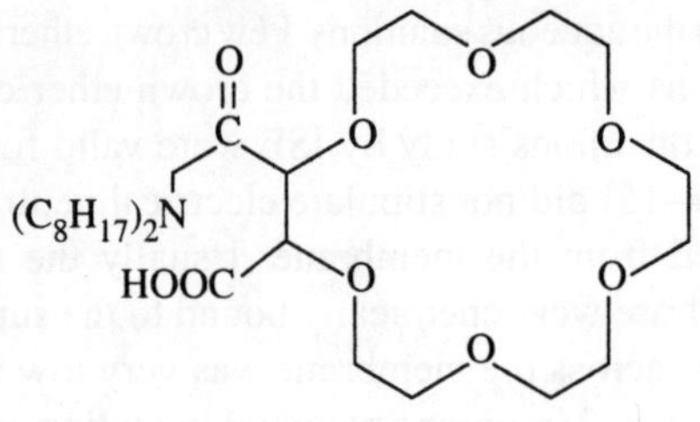

(structure I: a crown ether with pendant group containing C(=O), HOOC, and N(C8H17)2 substituents)

(I)

10.3.2 Cation Specific Electrodes

These electrodes were made so that they were specific for a particular cation. An electrode very specific for lithium was prepared in potentiometric analysis. The crown ethers having four ring oxygen atoms exhibited high selectivity for lithium. The ionophore is shown in structure (II).

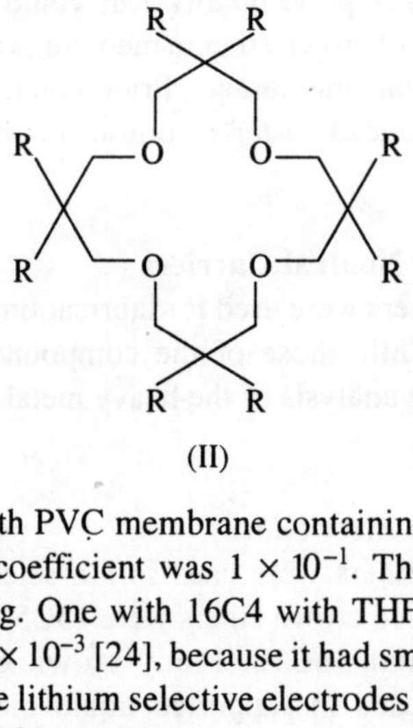

(structure II: 12-crown-4 derivative with R groups at each carbon)

(II)

This was electrodes with PVC membrane containing 12 crown-4 derivatives [22, 23]. The selectivity coefficient was 1×10^{-1}. Theoretically lithium fitted well in 15C5 or 14C4 ring. One with 16C4 with THF ring was selective with selectivity coefficient of 5×10^{-3} [24], because it had smaller cavity and presence of fused ring structure. The lithium selective electrodes based on 14C4 was quite promising for the serum lithium assay.

Several cation specific ion selective electrodes were prepared using variety of crown ethers. DB24C8 [25], dimethyl DB30C6 and many biscrown ethers [26–27] were used as the sensors for potassium. For analysis of caesium, bis (crown ether) containing 18C6 [28], bis (B15C5) [29] were best while for copper B12C4 was ideal for analysis [31]. For calcium very specific ion selective electrode were developed. The [32] organic compounds like guanidinium and alkylammonium ions, and enantiomeric recognisation of chiral amines and amino acids by optically active crown ethers was also extensively investigated [33–34]. Recently a review has appeared [30].

10.3.3 Thia and Aza Crown Ethers in Electrodes

The compounds containing sulphur and nitrogen as the donor atom were extensively used in electrodes for transition elements.

For instance thia crown ethers (III) and aza crown ethers (IV) were used as neutral carriers for nickel and copper. Even 1, 4 dithia 12C4 and 15C5 were used

270 Analytical Chemistry of Macrocyclic and Supramolecular Compounds

ether) derivatives were also quite useful. They had long term practical applications specially for analysis of sodium and potassium in the biological fluids and clinical samples.

Some bis (crown ethers) containing two different crown ethers were also used in ion selective electrodes. For instance derivative (XI) was very selective for rubidium. This was unsymmetrical bis (crown ether) contained 15C5 and 18C6 ring [46]. Unfortunately its selectivity over sodium or potassium was not very impressive. Similar methodology was employed to develope selective electrode [50] for calcium.

(XI)

In conclusion, we can say that the design of crown compounds was very important in preparation of ion selective electrode to attain high efficiency. Crown compounds should be selective for complexation and extraction equilibria. Crown compound should have flexible conformation for rapid ion exhange, to get good response on the electrode. The crown compounds should be lipophillic enough to remain in the membrane phase without dissolving out to the aqueous solution. The absence of lipophilicity led to loss of neutral carrier and destruction of the electrode. The lipophilicity could be easily increased by the introduction of long aliphatic chain in to the ring of crown ether. The crown compound should not have high molecular weight to allow high mobility. Crown ether polymers with high weight and good lipophilicity were not good for construction of ion selective electrodes. A large number of highly ion selective crown compounds were likely to be explored in future as the neutral carriers in ion selective elecrodes specially for construction of anion selective electrodes.

10.5 Analytical Applications of Crown Compounds in Ion Selective Electrodes

A large number of review articles [51, 52] have appeared on the applications of crown ethers as well as cryptands and calixarenes [54] in ion selective electrodes. The compounds with alkyl, nitro, halo substituted it in bis crown ethers were most effective in electrodes as the ionophores.

Apart from alkali metals [55–97], alkaline earths [98–102] the transition elements like iron, cobalt, nickel, copper, silver, mercury, so also maingroup

---

(III) Thia crown ethers

(IV) Aza crown ethers

for mercury and silver [21, 35]. These were highly selective electrodes. They were used for the determination of solubility product of silver chloride by titration techniques. Use of lipophillic crown ether further improved selectivity [36]. Many analogues of crown ethers have good future as neutral carriers of ion selective membrane electrodes. Spherands were promising ligands for analysis of sodium and lithium. The compound used in such electrodes is shown in structure (V).

(V)

Their decomplexation rate was too small due to high rigid structure of spherands (V). They were not best suited as neutral carriers due to slow ion exchange and reasonable response time. The selectivity coefficient was 8×10^{-3}. The response time was quite fast and acceptable.

10.3.4 Calix (n) Arenes in Ionselective Electrodes

The cyclic oligomers of phenolformaldehyde condensates were calixarene type compounds. They formed inclusion complexes with organic molecules [37]. Calixarene derivatives carrying ester and amide linkages of the phenolic oxygen complex were best suited for s-block metal [38, 39]. Calix(4)arene alkly ester derivative (VI) was used as neutral carrier for sodium selective electrodes [40, 41].

Such electrode exhibited Nernstian response for sodium activity. Calix(n)arene based electrodes were highly selective for alkali metals with the magnitude of the selectivity coefficient as 4×10^{-3}.

Thus hemispherands and calixarenes were promising as highly lipophillic neutral carriers for electrodes. In future we except brisk activity with calixarene.

(VI)

10.4 Bis (Crown Ethers) in Ion Selective Electrodes

The monocyclic crown ether were good, but had few limitations for the applications in the ion selective electrodes, but use of bis (crown ether) derivatives significantly enhanced the selectivity of the ion selective electrodes. Bis (crown ethers) were macrobicyclic polyethers having two crown ether rings with aliphatic chain, capable of forming 2:1 sandwich type of complexes which was intramolecularly bound to cation which had its atomic size larger than the cavity of the crown ether [42–44]. The two rings of crown ether exerted cooperative action to form a sandwich complex. The complexes being lipophillic readily travelled to membrane phase. They displayed high ion selectivities of membrane electrodes with crown ethers. The commonly utilised bis (crown ethers) included mainly bis (15C5), bis (12C4), bis (18C6) and few other derivatives. Thus bis (15C5) derivatives were very selective for potassium than normal monocyclic 15C5 [45]. The structure of the compound is shown in structure (VII).

(VII)

It gave value for selectivity coefficient as $K_{Na}^{Pot} = 3 \times 10^{-4}$ with NOPE used as membrane plasticizers. Rubidium did not interfere. This configuration influenced ion selectivities. Number of workers have developed ion selective electrodes for potassium with bis (15C5) derivatives [29, 46, 47]. The poly (crown ether) B15C5 were also useful for ion selective electrodes [26, 28, 48]. The compo used are shown in structures (VIII) and (IX).

(VIII) (IX)

They behaved like neutral carriers on PVC membrane in ion selective electrode but they had low sensitivity, inspite of having good selectivity. The electrod response was slow and Nernstian behaviour was bad, because of the low mobilit of poly crown ethers in membrane. The stirring of solution caused disturbanc in potential measurement if bis (B15C5) was used while electrode with bi (15C5) had no adverse effect of stirring on the magnitude of the potential. In flow injection analysis system it was quite important to take note of such variation in the potential due to stirring.

The another important bis crown ether frequently used was bis (12C4) and bis (18C6) derivatives. The former was good for sodium [49]. It formed sandwich complex due to difference in size of ion and cavity of crown ether. The chain length on it determined its selectivity bis (12C4) malonate or succinate were very good for sodium. The difference originated due to cooperative effect of two neighbouring crown ethers during formation of complex with metal ion.

$R_1 = R_2 = C_2H_5$

(X)

The compound used is shown in structure (X). In fact sodium selective electrodes containing thia crown compound was best substitute for glass electrode.

The coated wire kind of ion selective electrodes involving the bis (crown element as thallium and lead were studied [103–120] with crown ethers. So far cryptand was studied for the characterisation of zinc only while calixarenes were used for study of sodium [121–128], caesium [129–130], calcium, europium, silver and lead [130–136]. In addition to these compounds bis crown ethers which had an advantage over normal crown ethers were also used. The commonly used crown compounds consisted of 18C6, DB18C6, 14C4, 16C5, 12C4, 15C5, etc. Most of the time polyvinyl chloride was invariably used for fixing the membrane on the electrode. Aza derivatives of 12C4 was used for the analysis of alkali metals [55]. PVC plasticizer with monoazo crown ether gave excellent results, with positive response. The electrodes showed near Nernstian response for electrodes containing ionophore, NPOE, PVC and buffer. TOPO acted as synergistic agent [59] in detection of lithium. The crown ether carboxylate derivative in the presence of TOPO showed better selectivity over sodium for small cation like lithium. Nafion-901 in molten pyridinum heptachloroaluminate was used as the electrode for lithium, when cationic conductor was reversed to anionic conductor. The selectivity of electrode many times depended upon the nature of substituent like benzoylgroup on crown ether. The presence of diamide bearing derivative on ethylene chain or central carbon atom of propylene chain offered better selectivity [66] for lithium over potassium.

For analysis of sodium 12C4 on tungstophosphoric acid with PVC membrane and dipentylpthalate as the solvent was best [69]. PVC, silicon, rubber or collidion were used as membranes. For lipophillic bismonoazacrowns of 12C4, 15C5 etc response was dependent on the pH of the solution. The use of ONPOE as plasticizer was most popular, where ONPOE represented ortho nitrophenyl octylether. Alkali chloride was generally used as the internal standard. The combination of PVC and ONPOE gave best results with 16C5 for determination of sodium. With these plasticizers optically active bis (crown ethers) like 12C4 was very efficient B15C5 on polyacryl amide in acetonitrile demonstrated best Nernstian response and excellent selectivity for sodium.

For analysis of potassium, compounds analogues to valinomyan such as mono (nitrobenzene) 15C5 derivatives or bis (B15C5) linked by urethane group to ether via sulphide or alkyl chain were used for analysis of potassium in biological fluids. Naphtho crown ether on PVC based support on naphtho 16 Crown 5 or naphtho 18C6 or dinaphtho 30C10 were effective for potassium. The response however was dependent on concentration and nature of the plasticizers used. Generally tetraphenyl borate or tetrakis (4-chlorophenyl borate) were used as the lipholic salt to bestow better selectivity. Lariat ethers with side arm as the ionophores also exhibited better selectivity. The side chain improved the selectivity of other alkaline earths if they were present in the solution. In comparison to lithium, sodium, potassium no attempts were made to design ion selective electrode for rubidium although for caesium selective electrodes many attempts were made.

Like potassium, for caesium also tungtophosphate derivative on 12C4 with PVC membrane was used at pH 2.0. The 1, 4 dithia 12C4 showed excellent selectivity over transition metals like mercury or silver. Crown benzoquinones when used with membrane and ONPOE, i.e. orthonitrophenyloctyl ether also written as NPOE and potassium tetrakis (4-chlorophenyl) borate gave best selectivity for caesium over s-block metals, but many transition elements showed interference.

In comparison work on alkaline earths was very limited. Only magnesium, calcium, and barium were analysed by electrodes containing crown ethers. 13C4 was used for analysis of magnesium. The diamide derivative of diaza penta oxa cyclohexane cosane on PVC was best for calcium with PVC, and ONPOE as plasticizers when excellent selectivity was noticed over magnesium. The anthriquinone ionophores of crown ethers on PVC membrane and potassium (tetrakischlorophenyl) borate in THF gave best results for barium analysis in the real samples.

Amongst transition metals iron, cobalt, nickel, copper, silver, mercury were studied with ion selective electrodes. The thia crown ether proved to be efficient as ionophore for determination of iron (III). A PVC based membrane with suitable macrocycle as ionophore was used for potentiometric determination of cobalt, nitrogen containing donor atom like cryptand proved to be ideal for nickel analysis from real samples like choclates. Copper was analysed potentiometrically with elecrode consisting of macrocycle on PVC membrane with dibutyl or dioctyl pthalate as the plasticizer. The response time was good. Silver was analysed with poly thia ether derivative of the crown ether. The podants containing nitrogen or sulphur were equally effective. Pentathia 15C5 at pH 2.7–5.0 on tetraphenyl borate membrane with dibutylpthalate as the plasticizer was used for analysis of mercury. At the same time cadmium showed strong interference. The potentiometric titrations were carried out with EDTA for the analysis of silver or mercury employing these ion selective electrodes.

Amongst main group elements only thallium and lead were investigated. Methyl substituted derivative of DB30C10 was used as sensor in potentiometric determination of thallium. When it was noted that unsubstituted crown ethers showed better selectivity for thallium on PVC membrane over alkali metals. Another element of interest was lead, for which 18C6, B15C5, DB18C6 were used in ion selective electrodes, DC18C6 was best for lead with tetraphenyl borate in dichloromethane over alkali. It was analysed from atmosphere with 15C5 as an ionophore. Several potentiometric titrations with crown ethers like 18C6 as an ionophore were carried out with EDTA as the titrant. For heavy metals generally lipophillic acyclic dibenzoyl polyether diamides as ionophores with borate buffer showed satisfactory selectivity.

The macrobicyclic polyether, viz. cryptand 222 was used for analysis of zinc at pH 2.8-7.0 on PVC membrane with extraordinary fast (~ 10 seconds) response with long life time. It was quite effective for potentiometric titration of zinc also.

Finally amongst supramolecular compounds the calixarenes were extensively used for the analysis of sodium, caesium, calcium, europium, silver, lead and few anions. Calix(4)arene was most favoured. In comparison to tetramer work on other calix(n)arene derivative was some what lacking. As in previous work NPOE was used as the plasticizer with suitable buffer solution like borate buffer p-tertiary butyl calix(4)arene amide or ester gave good results, when dibutyl sebacate or pthalate proved to be effective plasticizer. For optical sensor chromogenic calixarenes were used as the ionophore in electrodes. Bromo derivative of tetramer was good because of the presence of $(CH_2-CH_2O)_2CH_2\ CH_2$ chain on crown group. Nernstian response was excellent for sodium, calcium and silver with 1, 3 bridged calix(4)arene. Calix(4)arene modified oligosiloxane on silicon rubber membrane was best for sodium.

Calix(6)crown was used for analysis of caesium on CHEMFET membrane—a special kind of supporting membrane. Such ionophore exhibited better selectivity over other alkali metals. Calcium was tested with calix(4)arene tetra phosphoric oxide on PVC membrane. Some ionophore was also used for the analysis of europium with ONPOE as the plasticizer.

For analysis of silver and lead thioether functionalised calix(4) arene and thioamide functionalised calix(4) arene were respectively utilised at pH 6.0 with PVC support in the presence of borate buffer. They had fast Nerstian response. As usual NPOE was used as the plasticizer. Calix(4)arene derivative with mercury plugged in it with NPOE plasticizer and PVC in THF tetrahydrodarane was best used for quantitative analysis of some anions. We expect brisk activity in future for determination of anions with calix(n)arene as the ionophores in electrodes. Various methods for analysis of metals with the crown ether, cryptands and calix(n) arene are summarised in Table 10.5.1 & 10.5.2.

10.6 Membrane Transport

Moore and Pressman [137] discovered the use of biological membrane for the transport of antibiotic valinomycin for potassium in mitochondria. The work on the exact mechanism by which these carriers facilitated ion transport was active field of study. The flow of ions across membrane occurred by process of diffusion through pores or channels or with the help of carriers. The carrier molecule was situated at the interface of aqueous solution and the membrane metal complex diffused through the membrane to other side where the cation was set free in water solution. The cycle was completed by reverse diffusion of the free carrier. This process was called "carrier relay mechanism". The neutral carriers like crown ethers allowed total charged ion to traverse through membrane and to develope decreasing potential of the membrane [138]. The important requirements of metal carriers were [139, 140].

(a) The carrier molecule must have nucleophillic group which can compete with solvent to gain cation located in hydrophoric networks.

(b) Largest number of solvent molecule should be replaced by donor group of the carrier in the coordination spheres.

(c) The hole caused by carrier group should match with atomic size of the metals to be transported and finally.

(d) The carrier molecule like crown ether should be adequately flexible to permit rapid replacement of solvent molecule at the cation.

The mobile carriers were sensitive to membrane fluidity. It was less influenced by width of membrane. It was limited by fixed diffusion rate. The carrier conduction took place with 1:1 complex at low concentration. Crown ethers acted as natural ionophores. The transporation of ions through membrane without exerting pressure

Table 10.5.1 Ion selective electrodes with crown compounds

Element	Optimum Conditions of Analysis
Alkali metals	Large number of reviews have appeared on the applications of crown ethers [51] or B15C5 ether [52] cryptands and calix(n)arenes for use in their ion selective electrodes in conductometric, chromatographic and the polarographic analysis [53]. The bis crown ethers with alkyl nitro halo substituents were effective as the ion selective electrode [54] Crown ethers like 18C6, 24C8, DB30C10 were used for the detection in form of ferrioxamine B compounds [55]. Crown ether derivatives like squardine, i.e. bis (4-monoazo 12C4) ether or bis-(4 monoazo 15C5) ether phenyl or similar derivative of 18C6 in acetonitrile containing tetra butyl ammonium perchlorate were used for analysis with applied potential of 0.8-OV at 100 mV/second [56]. Monoazacrown ether on PVC membrane electrode with plasticizer had showed good response specially for potassium [57].
Li(I)	14C4 derivative viz 3 dodecyl-3methyl, 1, 5, 8, 12 tetraoxa cyclotetradecane as an ionophore was used in polymeric membrane containing 1% ionophore 70% NPOE, 28% PVC and 0.7% K-tetra kis (4 chlorophenyl) borate for its analysis showing near Nerstian response [58] DB14C4 with TOPO (Trioctyl phosphine oxide) showed selectivity due to synergistic action [59] Dodecylmethyl 14C4 and (6 dodecyl 6-methyl) 1-1, 4, 5, 11 tetraoxacyclo tetradecane with TOPO was used for the analysis of lithium in serum but selectivity for the hydronium ion had decreased [60]. Crown ether carboxylic acid and benzyloxy methyl crown ether with 14C4, 13C4, on PVC membrane when used showed better selectivity in the presence of TOPO over sodium [61]. DB18C6 on Nafion -901 in molten pyridinum hepta-chloroaluminate was used as an electrode, a cationic conductor was changed to anionic conductor in molten salts and chloride batteries [62] 14C4 derivative carrying 2, 3, 4 substituent was used to imporve the selectivity of lithium over sodium, the selectivity was dependent on the number and kind of substituent like benzoyl group on the crown ether with diamide on crown ether was also used [63–64]. 14C4 derivative in 70% bis (1-butyl phenyl) adipate as membrane solvent, 1% of K-tetrakis4-chlorophenyl borate was used. Decalino 14C4 offered better selectivity for lithium over sodium in blood serum [65] 14C4 derivative bearing diamide group on ethylene chain or central carbon atom of propylene chain was used with better selectivity over sodium. The compound was lithium selective ionophore [66].
Na(I)	12C4 on tungstophosphoric acid with PVC membrane with dipentyl pthalate as solvent exhibited satisfactory selectivity for sodium. Other alkali metals like lithium, potassium, caesium showed interference [69] bis (monoaza 12C4) on membrane containing PVC or silicone rubber or colliodion showed good response [67] chiral bis (12C4 methyl) dialkylmalonate also was used [68] so also lipophillic bis monoazo 12C4 or 15C5 or bis 12C4 were equally effective but response depended on the pH of the solution. Bis(B12C4) derivative with orthonitrophenyl octyl ether (ONPOE) as plasticizer was responsive [70] B15C5 and naphthol 15C5 derivative was used [71] Dibenzyl 2, 2, 1 (1,4, 10, 13 tetraoxa-7, 16 diazacyclo octadecane-7, 16 diylidi carbonyl dipyrolidin-1-carboxylate) in cyclohexane with PVC membrane and nitrophenyl octyl ether as plasticizer had responded with excellent selectivity [72] DB16C5 derivative between pH 2-9 with PVC membrane and ONPOE

(Contd)

Element	Optimum Conditions of Analysis
	as the plasticizer with 0.1M sodium chloride as an internal standard was used with good selectivity over alkali and alkaline earth elements [73] DB 16C5 derivative on PVC with ONPOE as the plasticizer on membrane was used. The oxyacetic derivative and potassium (tetrakis (*p*-chlorophenyl) borate in tetrahydrofurane with sodium chloride as an internal standard exhibited good response for sodium [74] bis (12C4 methyl) malonate on PVC-ONPOE membrane or bis (12C4) methyl α, dibenzyl malonate with sodium tetra kis [3, 5 bis (trifluoro methyl) phenyl] borate in THF was also used [75] 16C5 derivative with PVC and ONPOE combination was used with merit [76]. DB16C5 derivative on PVC membrane with N, N' dialkyl substituted symm (propyl) DB16C5-yl) oxy octamide as the ionophore was best for sodium [77] New optically active bis12C4 crown on PVC membrane with ONPOE as the plasticizer was effectively used [78]. Polyacrylamide B15C5 on polyacrylamide in the acetonitrile offered good Nerstian response [79].
K(I)	Bis (B15C5) derivative containing cyclohexane or benzene ring were responsible with ciscyclohexane 1, 2 dicarboxylic acid had rapid response for potassium at low concentration [80]. Bis (crown ether) on PVC membrane with dodecyl methane bis (12C4) or bis (15C5) was used [81] eight bis crown ethers like mono (nitrobenzene 15C5) derivative and then bis (15C5) linked by urethan group to ether to sulphide or alkyl chain was used for analysis of biological samples [81] such compounds were similar to valinomycine [82], bis crown ether on PVC membrane electrode containing crown ether and 1, 3, 5 triazine ring was used 2:1 sandwich complex was formed with the metal and rubidium [83] alkylene bis (benze crown ether) film at pH 4.5-10.0 on pure graphite coated with 1, 2 dioctyl-o-phenthale and poly (chloroethylene) and THF showed efficient response [84]. Two B15C5 units linked to schiffs base and secondary amine in PVC membrane with ONPOE as plasticizer and Ag-AgCl as reference electrode exhibited excellent selectivity for potassium [85] *t*-butyl derivative of naphthol 15C5 was also used in electrode [86] bis B15C5 was also useful with bis (*o*-nitrophenyl) urathane [87]. Naphtho crown ether on PVC based support of naphthia K 15C5 or naphtho 18C6 or dinaphtho-30C10 was equally effective [88]. Their property depended on the concentration of crown and type of plasticizer used. Bis crown ether on pH membrane based on alkned alkylpolymethylene and α, α' dihydroxy polymethylene bis crown ether with 2-nitrophenyloctylether as the plasticizer gave good response [89] bis crown ether derivative was also used [90] 2-dodecyl 2-methyl propane 1-3 diyl/16 nitro B15C5-15 carbamate on 33% PVC and 65% ONPOE with tetraphenyl borate or tetrakis (4-chlorophenyl borate) as lipophillic salt when used exhibited excellent selectivity [91] bis (benzocrown ether) on PVC membrane containing bis B15C5 was useful [92] bis benzocrown ether with polymethylene bridge showed good response for rubidium and caesium [93]. B18C6 and lariat ether derivative as ionophore on PVC membrane exhibited best selectivity for potassium over alkali metals but side chain improves the selectivity of calcium and strontium. Sodium and lithium [94] crown ether modified bilayer lipid membrane or solid support with chlorides of sodium or potassium as analytical solution gave good response with silver electrode [95].

(Contd)

Element	Optimum Conditions of Analysis
Cs(I)	Crown ether at pH ~ 2.0 on PVC membrane in the presence of the tungsto-phosphate derivative on 12C4 phosphotungstic acid or 1-4-dithia 12C4 or 1, 4-dithia 15C5 showed good selectivity for caesium over sodium, mercury silver at pH > 2.0 also lanthanides could be analysed [96] crowned benzoquinones like 2-3 benzoquinones-15C5, 2-bromo-1, 4-dihydroxy B15C5 or 5-brown 2-3 benzoquinone-15C5, was used with membrane and K-tetrakis (4-chlorophenyl borate) and NPOE exhibited caesium selectivity over s-block metals. While d-block metals showed some interference [97].
Be(II)	A 9-crown-3 was used at pH 2.5-4.0 with response time of 50 seconds and with life time of 4 months. The electrode could work for the concentration of 1×10^{-6}M of the beryllium solution [11].
Mg(II)	The derivative of 13C4 a bis (13C4) in dichlorobenzene when used in membrane electrode gave Nernstian response [98].
Ca(II)	The diamide derivative of 10, 19 diaza 1, 4, 7, 13, 16 pentaoxacyclo-beneicosane on PVC membrane exhibited excellent selectivity at low concentration [99] Macrocyclic polyether dioxatetralactum as ionophore with PVC and OPOE as support exhibited excellent selectivity specially over magnesium [100]. Uncharged acyclic compounds like 3, 6 dioxa-N N N′ N′ tetraphenyl octane diamide used for extraction [101].
Ba(II)	Anthriquinone ionophore of crown ether on PVC membrane and potassium tetrakis (chlorophenyl) borate in THF gave good response similar to one given by valinomycin for potassium [102].
Fe(III)	1, 7 dithia 12C4 on PVC membrane exhibited ion selectivity over alkali metals by varying several parameters [103].
Co(III)	PVC based membrane containing macrocycle with borate buffer was used in potentiometric titration with EDTA to detect end point [104].
Ni(II)	Fourteen member macrocyclic ligands with nitrogen as donor atom at pH 3-7 on PVC support exhibited good selectivity and was used for analysis of nickel from choclate and sweets [105].
Cu(II)	Macrocyclic compound at pH 2.5 on PVC membrane with dibutyl pthalate or dioctylpthalate as plasticizer used as indicator in EDTA and potentiometric titration. The response time was as short as 30 seconds [106].
Ag(I)	Dithiamacrocycle on PVC membrane with polythiaether 6 oxa 3-9 dithiabicycle (9, 3, 1) pentadeca (1, 5) 11, 13 triene-dioctyl pthalate in THF was used [107] Podands containing nitrogen and sulphur on polymeric membrane was used but the response depended on the structure of podands [108].
Hg(II)	Thia crown ethers on PVC membrane at microgram concentration was used [109] Pentathia 15C5 at pH 2.7-5.0 on membrane containing tetraphenyl borate with dibutyl pH thalate as plasticizer when used showed good response for mercury but it showed interference by cadmium [110] while one with 1, 4, 7, 10, 13, 16 hexathiacyclo-octadecane and 1, 4, 8, 11 tetrathiacyclotetradecane was used for end point detection of potentiometric titration with EDTA [III]. Hexathia 18C6 tetrone at pH 0.5-2.0 on PVC membrane used as indicator electrode in potentiometric titration at macrogram concentration of mercury [112].
Tl(I)	4-methyl DB30C10 on PVC membrane base used as sensor in the potentiometric determination [113] unsubstituted symmetrical and

(Contd)

Element	Optimum Conditions of Analysis
	unsymmetrical crown-5 and crown-6 ether on PVC membrane showed high selectivity for thallium over alkali metals based on the knowledge of solvent extraction such electrode was developed [114].
Pb(II)	18C6, DC18C6, B15C5, was mainly used in electrode in polarographic analysis [115] DC18C6, with lead nitrate and tetraphenyl borate in dichloromethane was effective [116] over alkali metals. 15C5 on PVC based membrane was used for analysis of lead from environmental effluents [117] Dibenzopyridine 18C6 with PVC membrane gave good Nernstian response for lead and hence was used for potentiometric titration with EDTA [118] Twelve lipophillic acyclic dibenzopoly ether diamides as ionophores with PVC and borate buffer showed extreme selectivity over heavy metals when a linear response was obtained [119].
Zn(II)	Cryptand-222 at pH 2.8-7.0 on PVC membrane was used in electrode to give good response in ten seconds, stable for three months and was also used in potentiometric titration of zinc [120].

Table 10.5.2 Application of calixarenes in ion selective electrodes

Element	Optimum Conditions
Na(I)	Tetrameric calix(4)arene with potassium borate buffer with ONPOE as plasticizer was used with good selectivity both for sodium and caesium [123] p-t butyl calix(4)arene-esteramide derivative with borate buffer with tributyl sebacate or pthalate as the plasticizer was used [124] chromogenic calix(4)arene as ionophore with PVC membrane was effective as optical sensor [125] calix(4)arene bromo derivative had good response due to the presence of -$(CH_2$-$CH_2O)_2$ CH_2CH_2 linkage on crown ether [126] calix(4)arene derivative was selective for sodium, silver and calcium the Nernstian response was good for 1, 3 bridged calix(4)arenes [127] calix(4)arene modified oligosiloxane on silicon rubber membrane was used with good response with tetraalkyl ester chain [128].
Cs(I)	Calix(4)arene crown 6 derivative was used with CHEMFET membrane (i.e. PVC membrane based on ionophore plasticizer) and borate buffer with external potential of 0.5V [129] Calix(4)arene crown-6 ionophore exhibited selectivity over s-block metals [130].
Ca(II)	Calix(4)arenetetraphonic oxide on PVC membrane was used [131].
Eu(III)	Calix(4)arenetetraphosphonic oxide with ONPOE as plasticizer and p-chlorophenyl borate buffer was used in millimole concentration with lifetime of 20 days [132].
Ag(I)	Thioether functionalised calix(4)arene as ionophores at pH 2.5-6.0 with borate buffer and PVC membrane were used with response time of 10 seconds. This was also used in potentiometric titration of halides with silver nitrate as the titrant [133] calix(4)arene based ionophore with borate buffer and ONPOE or sebacate as plasticizer used with very good response time [134].
Pb(II)	Thiamide functionalised calix(4)arene as ionophore at pH 3-6 was used with ONPOE as plasticizer with K-tetrakis (4 chlorophenyl) borate [135] calix(4)arene derivative of mercury with 66% ONPOE as plasticizer and 33% PVC in THF was used for quantitative analysis of anion [136] in solution.

or potential was an interesting phenomena [141–144]. The transport of the specific ion in the liquid membrane was dependent on distribution coefficient in the membrane. The membrane potential and electroconductivity of membrane containing crown ether was thoroughly investigated [145–146] DC17C6 was used for the purpose [147–149].

10.7 Separations With Liquid Membranes

This provided a powerful tool to explore molecular recognisation skill of the crown ethers to effect separations via liquid membrane by process of solvent extraction. The process was very rapid, with neutral carriers which extracted cation and accompanying anions in order to maintain electroneutrality. The membrane systems used consisted of bulk liquid membrane (BLM), thinsheet supported membrane (T-SLM), hollow fiber supported membrane (H-SLM) and emulsion liquid membrane (ELM). The knowledge of Kex extraction equilibrium constant helped to design selective liquid membrane. The ion solvation, pairing of ion in source phase and use of proton ionisable macrocycle played a significant role in transportation.

10.7.1 Solvation of Ions

A crown compound with hydrophobic exterior and electron rich cavity was useful to bind and solvate the cation in the organic phase which was hydrophobic. The cation hydration energy influenced the extent of extraction. The cation should be large with small charge with neutral carriers like crown ether the kind of the anion which was present with cation-crown complex decided the selectivity order during extraction. Usually the selectivity order was retained in single solvent extraction and membrane kind, but at a times selectivity order was changed and extent of selectivity was changed.

10.7.2 Pairing of Ions in Source Phase

The extraction of cation (M^{n+}) and accompanying anion (X^-) by neutral carrier molecule (R) was expressed as in equation (4)

$$MX_m \, (aq) + R \, (org) = MX_m R \, (organic) \qquad (4)$$

Several metals formed stable and soluble ionpairs in water. The emulsion liquid membrane data could demonstrate selectivity of one ion over other ion by adjusting anion concentration of the source phase [150]. The knowledge of K and Kex values for equilibria as well as for anion cation interaction equilibrium constant allowed one to evaluate the selectivity parameter, useful from point of separations.

10.7.3 Proton Ionisable Crown Ethers in Extraction

To minimise the anion solvation proton ionisable crown ethers were used in the membrane transport reaction. A choice of specific crown ether could control selectivity of cation, e.g. pyridono and triazolo were generally preferred for such work. The reaction proceeded as per sequence:

$$M^+ \, (aq) + HR \, (org) = MR \, (org) + H^+ \, (aq) \qquad (5)$$

The extraction proceeded with liberation of proton in source phase. At membrane receiving phase reverse process occurred. The phenomena of coanion transport was minimised. Neutral crown ethers and proton ionisable crown ethers worked with same efficiency for separation of alkali metals by membrane transport separations. The liquid membrane transport involved extraction of cations from the source phase as shown by equation (5). Crown ether should be uniformly distributed in organic phase. In absence of proton, no transportation took place at high pH. In the presence of the acidic receiving phase transportation was much rapid. A charged based selectivity was possible with proton ionisable group in cavity of the crown ether in solvent extraction on membrane transport separation. By use of crown ether containing triazole group separation of silver [151] from other metals was carried out. The composition of membrane receiving phase controlled the driving mechanism. It was usually a difference between concentrations of transporting complex at membrane barrier with source and receiving phases. Selectivity was fissible by including singly proton ionisable moiety into cavity of crown ether.

10.8 Experimental Set up for Ion Transport Work

The reaction was carried out in the U-tube (Fig. 4). It was called 'Pressman cell'. The principle of extraction constant was generally used. Lower half was filled with chloroform and water was placed in both arms one acting as the source phase and other acting as receiving phases. Sodium was in source phase it was carried to bottom in chloroform layer with crown ether along with accompanying anion. The concentration of anion could be easily monitored. Atomic absorphen spectroscopy was used to monitor cations. In Fig. 5 use was made of beaker with glass tube half of bottom was immersed. The inner tube was filled with source or receiving phase and outer portion was filled with the complementary phase. The diameter of tube was variable. Rate of the transport of cations was easily ascertained by this assembly. The stirring efficiently prevented emulsification. Transportion occurred as shown by following cycle (Fig. 6). The driving force was concentration gradient or difference in pH of the solutions.

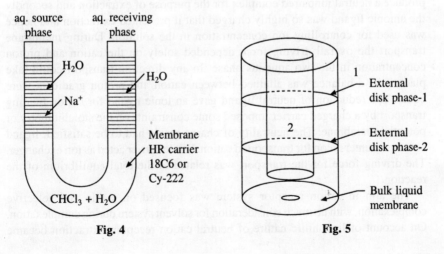

Fig. 4   Fig. 5

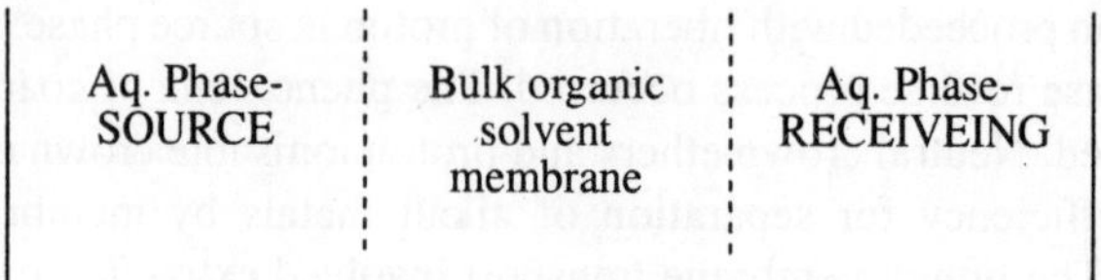

Aq. Phase-	Bulk organic	Aq. Phase-
SOURCE	solvent	RECEIVEING
	membrane	

$$M^{+n} + \text{Ligand} \rightleftharpoons MRn \rightleftharpoons \text{Ligand} + M^{+n} \quad (6)$$

Fig. 6 Cation transport

Apart from chloroform, dichloromethane and other nonpolar solvents were also used. The stiring helped mixing at phase boundries, but high rate affected transport mechanism. The binding was rapid at source phase. Cation was firmly bound to membrane and binding at the phase was fast but fragile. When cryptands were used with high pH at the receiving end promoted transport. At one end ligand was made strong binder. Crown or cryptand was added to membrane phase to form the complex with metal soluble in the organic solvent. On reacting at receiving phase, decomplexation took place as follows:

$$\underset{\substack{\text{(Source phase)}\\ [\alpha/2]}}{MX + R} \rightleftharpoons \underset{\text{(Membrane)}}{MRX} \rightleftharpoons \underset{\substack{\text{(Receiving phase)}\\ [\alpha/2]}}{MX + R} \quad (7)$$

The concentration at source was α while at receiving phase was zero in beginning. After transportation concentration was equal (viz. $\alpha/2$) at both phases. But on removing cation at receiving phase by water, transport continued and at final stage receiving phase was having concentration α while source phase concentration was almost zero. A use of switching crown ether gave good results. The cation concentration was measured in receiving phase by AAS. Even a coloured anion like picrate was measured by photometric methods.

10.9 Applications of Membrane Transport

Crown ethers were used for the transport of cations involving charged nature of the extractant. Here anionic ligand acted as a counter ion for the cation to produce a neutral ionpaired complex for the purpose of extraction and secondly the anionic ligand was so highly charged that it prevented extraction and hence was used for controlling ion concentration in the solution. During membrane transport the overall driving force depended solely on the cation and proton concentration in the two aqueous phase. In any direction transport could take place until a balance was attained between cation and proton gradients were equal. The reduction of neutral ligand gave an ionic ligand for cation binding transport by a charged carrier imposed some constraints on the stoichiometry of potential membrane. The neutrality of charge inside had to be satisfied, ligand acted as counter anion for transported cation. While carrier acted as ion exchanger. The driving force for the transport was related to the total equilibrium of the reaction.

The aim in cation receptor system was focused on stable and selective complexation, with no much consideration for solvent system during complexation. On account of lipophillic nature of neutral cation receptor, extraction became important aspect. The charged ligands due to solubility in water and capacity to bind cation provided good method of separation. As such electrostatic ion binding acted as significant force for development for newer methods of separations.

10.10 Novel Role of Crown and Podant as Ionophores

Natural ionophores formed stable complexes, and acted best host compounds. The cation transport involving neutral carriers, the rate of transport was determined by the kind of anion which accompanied cation in complexation reaction. The [152] anionic species was transported by neutral crown ethers. It was shown [153] that some ionophores behaved like anion transport carriers. However bulky residues could stop complexation reaction with guest ammonium cation of amino acid ester. Crown type nonactins had cavity for complexation and a hydrophobic, lipophillic envelope surface to promote transportation of ions. The cation recognisation phenomena was based on host guest chemistry. The studies furnished important direction for formulating new synthetic host molecules.

10.11 Transportation of Ions Across Membrane

The maintenance of the electrolyte balance needed input energy and specific ion transport reaction to decide direction and ion selectivity. Biological membrane kept electrical potential difference between two solution which originated on account of diffusion of ion from cytoplasm inline with its concentration gradient. The work required to transport ions across membrane was given by equation.

$$\Delta G = RT \ln (C_e/C_i) - zFE \text{ (i.e. Membrane potential)} \qquad (8)$$

where C and C_i are external and internal concentrations, respectively, $-zFE$ the electrical work, z the valency of ion, F the Faradays constant and E the potential inside.

The membrane permeability in terms of diffusion as related by Fick's law was given for the rate of flow across plane J as

$$J = (1/A) (dn/dt) = - D (dn/dx) \qquad (9)$$

where A is the flux, n the number of moles passing through unit area in unit time, dn/dx the concentration gradient and D the diffusion coefficient. The diffusion across membrane was dependent on concentration difference between inner C_i and outer C_e solutions. So we get Fick's law stated as

$$J \alpha (C_e - C_i) \text{ or } J = P (C_e - C_i) \qquad (10)$$

if P = permeability coefficient. The carrier mediated transport and diffusion was of significance in partitioning into membrane and diffusion of ions in membrane. The complex formation favoured partitioning. The movement of species across membrane by diffusion was passive transport. This was process which was driven by concentration gradient needing no extra source of energy. While the active transport involved movement of substance against concentration or electrochemical gradients. The bulk membrane were commonly used in chemical analysis, due to simple assembly and ease of measuring of transport process, due to diffusion controlled transport rate. The polymer supported liquid membrane

[154–155] were most useful. They carried organic carrier solution immobilised on microporous film separating two phases. The organic layer was impregnated by capillary forces and surface tension. Shinkai [156] used composite membrane of polycarbonate, N-(4-ethoxybenzylidene) 4' butylamine and crown ethers to introduce ion permeation. AAS and colorimetry were best method to monitor ion transport, but NMR methods were used to monitor ^{23}Na or ^{39}K in transportation system.

A study of [157–160] transport by crown ethers and cryptand was extensively investigated. The stability constants for complex were used to understand extraction equilibria in different systems. As a rule rate of transport increased with rise in stability of the complex, and latter decreased due to complex formation at external interface. Large monovalent anions like picrate, perchlorate permitted good transport. Within halides order varied from iodide to fluoride in decreasing order, iodide being best, due to difference in solvation energy of anions while transfering from water to aprotic solvents. The anionic ligands like carboxylic polyether or ordinary carboxylate derivatives of crown ether were used to exchange cations across the membrane. The dioxatetraza ligand formed stable complex (neutral) with copper due to ionisation of two amide groups. These had been used to transport copper against a concentration gradient [161–162]. The macrocyclic compounds could be used to get effective transfer of electrons via membrane was redox reactions at solvent water boundary layer.

A large amount of work has been published on the cation transport through membrane. Many studies were confinded to transport of biological fluids [163–165] others dealt with applications of lariat crown ethers for membrane transport [166–168]. Some studies were related to use of calixarenes for membrane transport phenomena as the biomimics [169–170]. Several applications of the crown ethers, cryptands, and calixarene were possible. So far the field is still new, very few experiments have been carried out but the technique appears to be most challenging from the view point of biological separations.

10.12 Analytical Applications of Membrane Transport with Crown Compounds

Several studies on ion transport were carried out with crown ethers, armed crown ethers, cryptands and the calixarenes [171]. Amongst the metals studied with crown ethers largely included alkali metals, alkaline earths [180–185] and light transition elements like cobalt, silver, cadmium [186–192]. The cryptands in comparison with crown ethers were used less frequently. The metals studied included alkali metals like sodium and potassium. In comparison calixarenes were used for ion transport studies of metals like copper and gold and large number of alkali and alkaline earths.

The transport studies with crown ethers for alkali metals included crown compounds like DB16C5, 18C6, DB18C6, DC18C6 and few bis (crown ether) carboxylate derivatives [185]. Porous poly propylene support was used in liquid membrane studies. Azobis B15C5 gave best separations [174]. Faster transport rate with increasing the concentration of crown ether in study of sodium with 2, 3 benzylmethyl 18C6 were used but addition of methyl group in 8, 11, 15 positions showed decrease in extraction on account of high hydration of anions [176]. Increasing concentration of lipophillic crown ethers in oil water emulsion droplets offered satisfactory separation for potassium [179]. The hydrophobicity of the membrane increased ion transport process for caesium when DC18C6 was used as the ionophore in the membrane transport separations.

However during separation of alkaline earths like calcium, strontium and barium, the presence of anion like picrate in dichloromethane promoted transport of interfering alkali ions in opposite direction for basic phase [181]. Within half an hour, calcium and strontium were separated with 18C6 picric acid with tetraethyl chloroethane and trichlorobenzene membrane [183]. Barium was successfully separated from industrial effluents on DB18C6 on octadecyl silica membrane disk but best results were achieved with the use of bis (crown ether carboxylate) derivative [185].

Amongst transition elements lanthanides, cobalt, silver, cadmium and from main group elements lead were separated by membrane transport. Neutral crown ethers proved to be ineffective for the separation of lanthanides. The proton ionisable crown ethers with varying lipophillicity and different cavity size could be used in bulk membrane for the efficient transport of cobalt amine cation. The aza crown ethers gave best separation of silver from divalent cations like cobalt, zinc, copper, cadmium and lead. While cadmium was separated from silver and magnesium with DC18C6 emulsion membrane in the presence of the surfactant Span-80., Aza 18C6 was used for silver and lead separations in liquid bulk membrane with palmitic acid, when copper was masked with thiosulphate. Lead was separated from gallium and iron with DC18C6 on propylene inclusion membrane with ONPOE as plasticizer. Such studies led to evaluation of the distribution coefficient and extraction constant for future prediction.

In comparison to macrocyclic compound the macrobicyclic compounds like cryptands were less commonly used specially for s-block metals. The control of pH facilitated the selective separation of potassium from other alkali metals with aza cryptand in liquid bulk membrane. Apart from these studies [193] no work was reported on cryptand in membrane transport studies.

The supramolecular compounds like calixarenes had become popular for membrane transport of selected cations. For instance calix(4)arene on black lipid membrane of synthetic amphiphiles was used to separate sodium on account of change in surface charge of bulk membrane. The stable complexes were formed with double amphiphiles. Double crowned calix(4)arene in 1, 3 alternate cone form on supported liquid membrane (SLM) gave excellent separation of caesium. For the first time calix(4)resorcinarene was used for separation of caesium from other alkali metals. Hexahomotrioxa calix(3)arene could separate scandium from yttrium at microgram concentration. The substituted derivative of calix(6)arene in bulk membrane was used to separate copper from iron, cobalt and nickel, by controlling pH during migration. The thio substituted calix(4) resorcinarene gave excellent separation of gold from interfering elements.

The details of various methods used for the separation of alkali alkaline earths, light transition elements and p-block metals like lead are best summarised in Table 10.12.1.

Table 10.12.1 Membrane Separations of metals with crown ethers

Element	Optimum Conditions of Separations
Alkali metal	The studies of ion transport through liquid membrane were carried out by using crown ether and double armed crown ethers. Cryptands and Calixarenes [171, 172] lipophillic symm. decyl DB16C5 oxyacetic acid on porous polypropylene supported liquid membrane was used for the separation [173] 2-nitrophenyl and octyl were best plasticizers. Azobis B15C5 which was photoresponsive on PVC based membrane gave unique separations [174].
Li(I)	Double armed diaza12C4 on liquid membrane gave excellent separation which were ionophorically selective [175].
Na(I)	2-3 benzo methyl 18C6 or 8, 15 dimethyl or 8, 11, 15 trimethyl derivative of 18C6 as liquid membrane when used showed faster transport rate with increasing concentration of crown ether while it decreased with the addition of methyl groups in 8, 15, or 8, 11, 15, position, the high hydration of anion also decreased transport rate [176] DB18C6 in liquid bulk membrane in chloroform or methylenechloride or nitrobenzene their mixture in binary separations gave good separation if equimolar chloroform and nitrobenzene was used [177] 4-nitro B15C5 in water or dichloroethane yielded best results [178]complex crossed the interface and not the crown ether.
K(I)	DC18C6 oil water microemulsion droplets when used gave better separation with increasing concentration of the lipophillic crown ether with [179] synergistic effect.
Cs(I)	DB18C6 on liquid membrane was used to study salt flux, the distribution coefficient and membrane potential during ion transport process. The hydrophobicity of the membrane had increased [180].
Ca(II)	18C6 derivatives on dichloromethane when used showed that picrate increased the transport of interfering potassium ion to the basic phase in opposite direction [181].
Sr(II)	18C6, DB18C6 or DC18C6 as carrier on emulsion membrane phase in toluene Span-80 hydrochloric picrate mixture accomplished its twenty fold separation from calcium [182] 18C6 picric acid with tetraethyl chloroethane and trichloro benzene membrane offered best separation of calcium and strontium [183] within half an hour.
Ba(II)	DB18C6 in picric acid on octadecylsilica membrane disk used for [184] the separation of barium from industrial effluents [185].
Ln(III)	Sym-DB16C5 oxyacetic acid in water chloroform liquid membrane was used for separation as the neutral crown ethers were unsuitable [186] for the purpose.
Co(II)	Six proton ionisable crown ether with differing lipophillicity and cavity size in chloroform or toluene when used in bulk membrane facilitated selective transport of cobalt cation with more lipophillic crown ethers [187].
Ag(I)	Aza 18C6 in bulk liquid membrane was separated in preference to divalent cations like cobalt, nickel, copper, zinc, cadmium, lead. Cadmium was complexed with diphosphate anion [188].
Cd(II)	DC18C6 in emulsion membrane system with surfactant Span-80 could separate it from magnesium, silver [189] analogue of 18C6 on supported

(Contd)

Element	Optimum Conditions of Separations
	liquid membrane was used to prepare a model of diffusion through membrane to show separation from lead and potassium [190].
Pb(II)	Aza 18C6 in bulk liquid membrane with palmitic acid was used to selectively separate it over iron and copper group elements, when copper was masked by thiosulphate [191] DC18C6 on propylene inclusion membrane with ONPOE as plasticizer was used to isolate it from gallium and iron. The transport mechanism was also studied in terms of different coefficient and extraction constant [192].

Table 10.12.2 Membrane separation with cryptands

Element	Optimum Conditions for Separations
K(I)	Monoazo cryptand liquid bulk membrane was used to selectivity transport it by controlling pH from mixture of s-block elements. The transport was compared with 18C6 or 20C6 ringed structures [193].

Table 10.12.3 Membrane separation with calixarenes

Element	Optimum Conditions for Separations
Na(I)	Calix(4)arene on black lipid membrane of synthetic amphiphiles when used showed change in adsorption-desorption properly due to change in surface charge bulk membrane. Double chain amphiphiles formed stable complexes [194].
Cs(I)	Doubly crowned calix(4)arene in 1, 3 alternate conformation supported liquid membrane gave good separations (195) calix(4) resorcinarene as liquid surfactant membrane was used to separate it from other alkali metals [196].
Sc(III)	p-chloro-oxacalix(3) arene, i.e. hexahomotrioxacalix(3) arene as liquid membrane afforded its separation from yttrium(III) lanthanum(III), lucetium(III) NMR titration gave binding constant [197].
Co(III)	Naturally occurring antibiotic lasaload A in chloroform membrane gave good isolation. Sarcophine cobalt complexes were transported [198].
Cu(II)	The 5, 11, 17, 23, 29, 35 hexa-t-butyl 37, 38, 39, 40, 41, 42 hexakis (n-2 hydroxyaminethyl methoxyl calix(6)arene derivative in chloroform bulk liquid membrane used to separate it from iron(III) cobalt(III) and nickel(III). The transport was also studied by controlling pH [199].
Au(III)	Tetrapodal calix(4) resorcinarene thiol on gold thiol monolayer indicated selectivity for absorption from dilute solutions with practically no absorption from bulk solution [200].

Conclusion

From the foregoing discussion it would appear that ion selective electrodes as well as membrane transport with crown compounds specially with bis (crown ethers) and calixarenes have great future in analytical chemistry. Although great strides have been made in the determination of cations by the use of ion selective electrodes, very insignificant contribution was visible in the area of analysis of anions with ion selective electrodes containing crown compounds. In future we except brisk activity in this field with different ionophores.

The fabrication of an electrode was an art, its potential measurement and interpretation of selectivity coefficient was easily correlated mathematically and great number of predictions were made. Crown ethers were used maximum as the ionophores in the ion selective electrodes. When large number of cation specific electrodes were fabricated greater attendation was focussed on the analysis of alkali and alkaline earths. This was justifiable as in analytical chemistry so far little attention was paid to the analysis of s-block metals and the analysis of anions from the view point of quantitative analysis. In comparison to crown ethers thia and aza crown ethers were used less frequently, only advantage being their were selective for light transition elements like silver, mercury, cobalt, iron over other s- or p-block metals. Lead was an exception to this rule. Calix(n) arene and bis (crown) ethers offered excellent methods for the quantitative analysis of several metal cations. The former compounds had an edge over conventional crown ether. One of the achievement was quantitative analysis of lithium with crown ethers as the ionophores, while no work on rubidium has been so far reported.

Amongst macrobicyclic compounds cryptand-222 found excellent applications for the analysis of zinc. In fact there was greater scope for the analysis of manganese, cadmium, lead, copper, thallium, and nickel by kryptofix-222. The great success depended obviously on the choice of counter anion although PVC as membrane support was expected to give rapid response.

The membrane transport phenomena was offshoot development in the field of ion selective electrodes. In fact this came in to great prominense because of its excellent end use in biological analysis. Solvent extraction as a matter of fact gave great boost to use of liquid membrane in metal separations. The thorough investigations on the source phase, utilising proton ionisable crown ethers in metal extraction, gave new dimension. The experimental setup for membrane transport could use either simple U-tube or concentric tube device, both gave good results. Apart from separations, membrane transport had many other uses. The crown ethers and podants played an impressive role in many such applications. The mechanism of transport of ions across the membrane was quite simple and it gave through insight to predict separation from the knowledge of permeation coefficient.

The major work was centered on biological fluids and other products. The lariat ethers, found end use for membrane transport studies. Calixarenes were used as the biomimics, and much more work was expected in future. However the major credit to gain popularity for membrane separation goes to analytical applications or the quantitative isolation of different metals. For instance apart from alkali and alkaline earths which were so extensively separated by membrane techniques, other transition metals like lanthanides, cobalt, silver, lead were studied with crown ethers or aza crown ethers. The potassium was sole exception where macrobicyclic compounds like monoaza cryptand was utilised for its separation.

However supramolecular compounds and specially calix(4)arene in its various forms as well as calix(4) resorcinarene was used for the first time for the separation of heavy alkali metal like caesium. Scandium, cobalt, gold were also separated using calix(4) arene or calix(6) arene in bulk liquid membrane by effectively controlling pH of the solution. In future we expect exciting investigations in the field of ion selective electrodes using macrobicyclic crown ethers and the supramolecular compounds as the ionophores.

References

1. E. Pretsch, D. Ammann, W. Simon, Design of ion carrier and application in ion selective electrodes. Research and Development **25**, 20 (1974).
2. R.B. Fischer, J. Chem. Ed. **51**, 387 (1974).
3. G.A. Rechnitz, E. Eydal, Anal. Chem. **44**, 370 (1972).
4. J. Petranek, O. Ryba, Anal. Chim. Acta **72**, 375 (1974).
5. M. Mascini, F. Pallozzi, Anal. Chim. Acta. **73**, 375 (1975).
6. J. Janata, R.J. Huber, Ion Selective Electrode Review **1**, 31 (1979).
7. G.J. Moody, R.B. Oke, J.D.R. Thomas, Analyst **95**, 910 (1970).
8. T. Teorell, Trans Faraday's Soc. **33**, 1051 (1973).
9. S. Ciani, G. Eisenman, G. Szabo, J. Memb. Biol. **1**, 1 (1969).
10. B.P. Nicolsky, Zhur Fiz khim **10**, 495 (1937).
11. M.R. Ganjail, A. Moghimi, M. Shamsipur, Anal. Chem. **70**, 5259 (1998).
12. G. Eisenman, S. Ciani, G. Szabo, J. Memb. Biology, **1**, 294 (1969).
13. W.E. Morf. D. Ammann, E. Pretsch, W. Simon, Pure and Applied Chem. **36**, 421 (1973).
14. J.H. Boles, R.P. Buck, Anal. Chem. **45**, 2057 (1973).
15. H.R. Wuhrmann, W.E. Morf, W. Simon, Hel. Chim. Acta **56**, 1011 (1923).
16. W.E. Morf, G. Kahr, W. Simon, Anal. Letter **7**, 9 (1974).
17. O. Ryba, J. Petranek, Coll Czech Chem. Comm. **49**, 2371 (1984).
18. O. Ryba, J. Petranek, J. Electroanal Chem. **44**, 425 (1973).
19. M. Yamauchi, A. Jyo, N. Ishibashi, Anal. Chim. Acta **136**, 339 (1982).
20. J. Jeng, J.S. Shih, Analyst **109**, 641 (1984).
21. U.S. Lal, M.C. Chattopadhya, A.K. Dey, J. Ind. Chem. Soc. **59**, 493 (1982).
22. K.M. Alamo, J. Krane, Acta. Chem. Scand. **A36**, 227 (1982).
23. U. Olsher, J. Am. Chem. Soc. **104**, 4006 (1982).
24. K. Tooda, H. Sakakura, K. Suzuki, T. Shirai, Procd 54th Annual Meet. Chem. Soc. Japan p. 473 (1987).
25. S.K. Norov, E.S. Gureev, A.K. Tashmukhamedov, O.G. Vartanova, N.Z. Saifullina, Electokhimia **15**, 943 (1979) CA **91**, 114413w
26. U.S. Lal, D. Phil. thesis University of Allahabad (1983).
27. H. Tamura, K. Kimura, T. Shono, Bull. Chem. Soc. Japan **53**, 547 (1980).
28. K. Kimura, H. Tamura, T. Shono, J. Electro analy Chem. **105**, 335 (1979).
29. K.W. Fung, K.H. Wong, J. Electroanal Chem. **111**, 359 (1980).
30. H. Tamura, K. Kimura, T. Shono, J. Electroanal. Chem. **115**, 115 (1980).
31. P. Bühlmann, E. Pretsh, E. Bakker, Chem. Review **98**, 1593 (1998).
32. J. Petranek, O. Ryba, Anal. Chim. Acta. **128**, 129 (1981).
33. M. Bochenska, J.F. Biernat, Anal. Chim. Acta **162**, 369 (1984).
34. Y. Yasaka, T. Yamamoto, K. Kimura, T. Shono, Chem. Letter 769 (1980).
35. S. Kamata, M. Higo, T. Kamibeppu, M. Tanaka, Chem. Letter 287 (1982).
36. M. Oue, K. Kimura, K. Akama, M. Tanaka, T. Shono, Chem. Letter. 409 (1988).

37. C.D. Gutsche, Acc. Chem. Res. 16, 161 (1983).
38. S.K. Chang, I. Cho, J. Chem. Soc. Perkin. Trans. 11, 211 (1986).
39. S.K. Chang, S.K. Kwon, Chem. Letter 947 (1987).
40. D. Diamond, S. Svehla, Trends Anal. Chem. 6, 46 (1987).
41. K. Kimura, M. Matsuo, T. Shono, Chem. Letter 615 (1988).
42. M. Bourgoin, K.H. Wong, J.Y. Hui, J. Smid, J. Am. Chem. Soc. 97, 3562 (1975).
43. K. Kimura, T. Maeda, T. Shono, Talanta. 26, 945 (1979).
44. K. Kimura, T. Tsuchida, T. Maeda, T. Shono, Talanta 27, 801 (1980).
45. H. Tamura, K. Kimura, T. Shono, Nippon Kagaku Kaishi 1648 (1980).
46. T. Ikeda, A. Abe, K. Kikukawa, T. Matsuda, Chem. Letter 369 (1983).
47. D. Huang, J. Zhang, C. Zhu, D. Wang, H. Hu, T. Fu, H. Ou, Z. Shen, Z. Yu, Huaxue Xuebao 42, 101 (1984).
48. S. Kapolow, T.E. Esch. Hogen, J. Smid, Macromolecules 6, 133 (1973).
49. T. Shono. M. Okahara, I. Ikeda, K. Kimura, H. Tamura, J. Electroanal. Chem. 132, 99 (1982).
50. O. Ryba, J. Petranek, Coll. Czech Chem. Comm. 49, 2371 (1984).
51. T. Shirai, K. Suzuki, H. Ariga, F. Naganori, H. Tanaka, T. Oshima, Japan Kokai Tokkyo Koho C.A. 106, 95226x (1987).
52. K. Toth, E. Linder, J. Jeney, E. Graf, M. Harrath, E. Pungor, I. Bitter, T. Meisel, Jr, B. Ayai Ion selec. Electrodes 5th Proc. Symp. 181 (1988) C.A. 113, 903702 (1990).
53. G.D. Blanco, A.P. Arias, A. Sanz-Medel, Quim Anal. 7 371 (1988) A.A. 53, 1A1 (1990).
54. E. Luboch, A. Cygan, J.F. Biernat, Tetrahedran 46, 2461 (1990).
55. I.B. Harberle, I. Spasojevic, A.L. Crumbliss, Inorg. Chem. 35, 2352 (1996).
56. S. Das, K.G. Thomas, K.J. Thomas, M.V. George, I. Bedja, P.V. Kamat, Anal. Proc, 32, 213 (1995).
57. T. Wickstroem, W. Lund, S. Buoeen, Anal. Chim, Acta. 219, 141 (1989).
58. K. Kimura, S. Kitazawa, T. Shono, Chem. Lett. 639 (1984).
59. T. Imato, M. Katahira, N. Ishibashi, Anal. Chim. Acta, 165, 285 (1984).
60. S. Kitazawa, K. Kimura, H. Yano, T. Shono, Analyst 110, 295 (1985).
61. A.S. Attiyat, G.D. Christian, R.Y. Xie, X. Wen, R.A. Bartsch, Anal. Chem. 60, 2561 (1988).
62. D.S. Newman, C. Lee, Proc. Electrochem. Soc. 743 (1990) C.A. 114, 85236b (1991).
63. H. Sakamoto, T, Miura, M. Tanaka, T. Shono, Buneski Kagaku 39, 779 (1990) A.A. 53, 10A55 (1991).
64. R. Kataky, P.E. Nicholoson, D. Paker, A.K. Cavington, Analyst 116, 135 (1991).
65. K. Suzuki, H. Yamada, K. Sato, K. Watanabe, H. Hisamoto, Y. Tobe, K. Kobivo, Anal. Chem. 65, 3404 (1993).
66. S. Faulkner, R., Kataky, D. Parker, A. Tcasdale, J. Chem. Soc. Perkin, Trans 2 1761 (1995).
67. T. Shono, K. Kimura, T. Maeda, Japan Kokai Tokkyo Koho JP C.A. 107, 167964g (1987).
68. J.B. Denton, K.F. Yip, Eur. Pat. C.A. 111, 153849u (1988).
69. K. Kimura, H. Oishi, H. Sakamoto, T. Shono, Nippon Kagaku Kaishi 277 (1987) C.A. 108, 48295a. (1988).
70. Q. Guoying, R. Wang, G. Wu, H. Shu T. Baozhi, Fenxi Huaxue 18, 424 (1990) C.A. 114, 1666be (1991).
71. A. Cygen, E. Luboch, J.F. Biernat, J. Coord. Chem. 27, 87 (1992).
72. N.G. Luk'yanenko, N.Yu. Titova, T.V. Golubenko, S.S. Basak, Zh. Anal. Khim 47, 331 (1992) A.A. 54, 12A88(1992).

73. A. Ohki, S. Maeda, J.P. Lu, R.A. Bartsch, Anal. Chem. **66**, 1743 (1994).
74. A. Ohki, J.P. Lu, J.L. Haluman, X. Huang, R.A. Bartsch, Anal. Chem. **67**, 2405 (1995).
75. K. Kimura, M. Yoshinga, K. Funaki, Y. Shibutani, K. Yakabe, T. Shono, M. Kasai, H. Mizufune, M. Tanaka, Anal. Sci. **12**, 67 (1996).
76. K. Suzuki, K. Sato, H. Hisamoto, D. Siswanta, K. Haysahi, N. Kasahara, K. Watanabe, N. Yamamoto, H. Sasakura, Anal. Chem. **68**, 208 (1996).
77. A. Ohki, K, Iwaki, K. Naka, S. Maeda, J.J. Collier, Y.C. Jang, H.S. Hwang, R.A. Bartsch Electroanalysis **8**, 615 (1996) A.A. **58**, 11D36 (1996).
78. Y. Shibutani, S. Mino, S.S. Long, T.M. Kawakami, K. Yakabe, T. Shono, Chem. Letters 49 (1997).
79. T. Nakamura, H. Mongi, Bull Chem. Soc. Japan **70**, 2449 (1997).
80. K. Kimura, A. Ishikawa, T. Shono, H. Tamura, Bull. Chem. Soc. Japan **56**, 1859 (1983).
81. H. Tamura, K. Kumami, K. Kimura, T. Shono, Mikrochim. Acta III 287 (1983).
82. E. Lindner, K. Toth, M. Horvath, E. Pungor, B. Agai, I. Bitter, L. Toke, Z. Hall, Z. Anal. Chem. **322**, 157 (1985).
83. C. Zhu, D. Wang, H. Hu, Wuji Huaxue **2**, 66 (1986) C.A. **107**, 16823j (1987).
84. Z. Xi, S. Huang, D. Zhang, Li. Hui, Faming Zhuanli Shenqing Gongkai, Shomingshu C.A. **109**, 85360j (1988).
85. D. Wang, X. Sun, J. Huang, H. Hu, Huaxue Xuebao **45**, 92 (1987) C.A. 106 187973 (1987).
86. A. Cygan, E. Luboch, J.F. Biernat J. Inclusion Phenom. **6**, 215 (1988).
87. K. Toth, E. Lindner, M. Horvath, J. Jeney, I. Bitter, B. Agai, T. Meisel, L. Toke, Anal. Lett. **22**, 1185 (1989).
88. T.L. Blair, S. Daunert, L.G. Bachas, Anal. Chim. Acta **222**, 253 (1989).
89. G. Wu, F. Wang, C. Shen, S. Haung, B. Tian Huaxue Xuebao **47**, 914 (1989) C.A. **112**, 171145m (1990).
90. E. Lindner, K. Toth, M. Horvath, J. Jeney, E. Pungar, I. Bitter, B. Agai, T. Meisel L. Toke, Magy. Kem. Foly **95**, 538 (1989) C.A. **112**, 232108a (1990).
91. E. Lindner, K. Toth, J. Jeney, M. Horvath, E. Pungor, I. Bitter, B. Agai, L. Toke, Mikrochim Acta I 157 (1990).
92. N. Yu, Nazarova, H. Holdt, J. Aurich, G. Kuntosch, N.G. Luk'yanenko, Zh. Anal. Khim **45**, 94 (1990) C.A. **113**, 33872r (1990).
93. E. Luboch, A. Cygan, J.F. Biernat, Tetrahedron **47**, 4101 (1991).
94. A.S. Attiyat, G.D. Christian, C.V. Cason, R.A. Bartsch, Electoanalysis **4**, 51 (1992) A.A. **55**, 6A73(1993).
95. Yu. E. Hi, M.G. Xie, A. Ottova, H.T. Tien, Anal. Lett. **28**, 443 (1995).
96. J.S. Shih, Chieh. Mien K'o Hsueh Hui Chih **10**, 11 (1987) C.A. **112**, 2992kw (1990).
97. M.G. Fallon, D. Mulcahy, W.S. Murphy, J.D. Glennon, Analyst, **121**, 127 (1996).
98. N.G. Lukyanenko; N. Yu. Nazarova; O.S. Karpinchik; O.T. Mel' nik, Anal. Chim. Acta **215**, 289 (1988).
99. K. Kimura, K. Kumami, S. Kitazawa, T. Shono, J. Chem. Soc. Chem. Commun. 442 (1984).
100. L. Cazaux, P. Tisnes, C. Picard, C. D'Silva, G. Williams, Analyst **119**, 2315 (1994).
101. N.N.L. Kirsch, R.J.J. Funck, W. Simon, Helv. Chim. Acta **61**, 2019 (1978).
102. J.R. Allen, T. Cynokowski, J. Desai, L.G. Bachas, Electroanalysis **4**, 533 (1992) A.A. **55**, 8A 106 (1993).
103. T.Y. Wang, J.S. Shih, J, Chin. Chem. Soc. **35**, 405 (1988) C.A. **111**, 69908e (1989).

104. A.K. Jain, V.K. Gupta, L.P. Singh, U. Khurana, Analyst. **122**, 583 (1997).
105. A.K. Singh, G. Bhattacharjee, M. Singh, S. Chandra, Bull. Chem. Soc. Japan, **70**, 2995 (1997).
106. A.K. Jain, V.K. Gupta, B.B. Sahoo, L.P. Singh, Anal. Proc. **32**, 99 (1995).
107. J. Cassbo, J.C. Perez, L. Escrichet, S. Alegret, F.E. Martinez, F. Teixidor, Chem. Letters 1107 (1990) A.A, **53**, 10A33 (1992).
108. C. Sarah, K. Wantae, B.P. Sung, Y. Ilyoon, S.L. Shim, D.S. Dae, J. Chem. Soc. Chem. Commun. 965 (1997).
109. L. Chen, D. Huang, Z. Huang, Yingyong Huaxue **8**, 97 (1991) C.A. **115**, 63511f (1991).
110. V.K. Gupta, S. Jain, U. Khurana, Electroanalysis **9**, 478 (1997).
111. Y. Masuda, E. Sekido, Bunseki Kagaku **39**, 683 (1990) A.A. **53**, 10 A45 (1991).
112. A.R. Fakhari, M.R. Ganjali, M. Shamispur, Anal. Chem. **69**, 3693 (1997).
113. S. Yang, H. Tong, P. Liu, Fenxi Huaxue **15**, 659 (1987) C.A. **108**, 86968e (1988).
114. Mikio. Ouchi, Tadao Hakushi, Coord. Chem. Reviews **148**, 171 (1996).
115. A. Hassan, H.A. Azab, S.A. El. Shatoury, Bull. Fac. Sci. **17**, 47 (1988) C.A. **112**, 209950k (1990).
116. T. Evalnora, S. Timofeera, A. Popov, Latr. Zinat Akad Vertis Kim. Ser 165 (1989) A.A. **52**, 1B48 (1990).
117. S.K. Srivastwa, V.K. Gupta, S. Jain, Analyst **120**, 495 (1993).
118. N. Tarakkoli, M. Shamsipur, Anal. Letter **29**, 2269 (1996).
119. A. Ohki, J.S. Kim, Y. Suzuki, T. Hayashita, S. Maeda, Talanta **44**, 1131 (1997).
120. S.K. Srivastava, V.K. Gupta, S. Jain, Anal. Chem. **68**, 1272 (1996).
121. K.M. O'Conner, D.W. Arrigan, G. Svehla, Electoanalysis **7**, 205 (1995) A.A. **57**, 9A89 (1995).
122. D. Diamond, M.A. McKervey, Chem. Soc. Reviews **25**, 15 (1996).
123. R.J. Forster, A. Cadogan, M.T. Diaz, D. Diamond, S.J. Harris, M.A. McKervey, Sens Actuators. B. 4 325 (1993) A.A. **55**, 2A81 (1993).
124. M. Careri, A. Casnati, A. Guarinoni, A. Mangia, G. Mori, A. Pochini, R. Ungaro, Anal. Chem. **65**, 3156 (1993).
125. K. Toth, B.T.T. Lan, J. Jeney, M. Horvath, I. Bitter, B. Agai, A. Grun, L. Toke, Talanta **41**, 1041 (1994).
126. H. Yamamoto, S. Shinkai, Chem. Lett. 1115 (1994).
127. E. Bakker, Anal. Chem. **69**, 1061 (1997).
128. K. Kimura, Y. Tsujimura, M. Yokoyama, T. Maeda, Bull. Chem. Soc. Japan **71**, 657 (1998).
129. R.J.W. Lugtenberg, Z. Brzoka, A. Casnati, R. Ungaro, J.F.J. Enbersen, D.N. Reinhoudt. Anal. Chim. Acta **310**, 263 (1995).
130. A. Casna, G. Mori, M. Careri, C. Bocchi, Anal. Chem. **67**, 4234 (1995).
131. T. McKittnak, D. Diamond, D.J. Marrs, P.O. Hayan, M.A. McKervey, Talanta **43**, 1145 (1996).
132. T. Grady, S. Maskula, D. Diamond, D.J. Marrs, M.A. Mckervey, P.O'Hayan, Anal. Proc. 32, 471 (1995) A.A. **58**, 1D32 (1996).
133. E. Malinowska, Z. Brzozka, K. Kasiura, R.J.M. Egberink, D.N. Reinhoudt. Anal. Chim. Acta **298**, 245 (1994).
134. K.M.O. Connor, W.O. Henderson, E. O'Neill, D.W.M. Arrigon, G. Svehla, M.A. McKervey, S.J. Harns, Electroanalysis **9**, 311 (1997) A.A. **59**, 11D27 (1997).
135. E. Malinowska Z. Brzozka, K. Kasiura, R.J.M. Egberink, D.N. Reinhoudt, Anal. Chim. Acta **298**, 253 (1994).
136. W. Wroblewski, E. Malinowska, Z. Brzozka, Electroanalysis **8**, 75 (1996) A.A. **58**, 6D109 (1996).

137. C. Moore, B.C. Pressman, Biochem. Biophy. Res. Comm. **15**, 562 (1964).
138. E. Racker, Acc. Chem. Res. Comm. **12**, 338 (1979).
139. W. Burgermeister, R. Winkler-Oswatisch, Top. Current. Chem. **69**, 91 (1977).
140. H. Diebler, M. Eigen, G. Ilgenfritz, G. Maass, R. Winkler, Pure and Applied Chem. **20**, 93 (1969).
141. Y.A. Orchinnikov, V.T. Ivanov, A.M. Shkrob, Membrane active complexation Vol. 12 Elsevier (1974).
142. W. Simon, W.E. Morf. P. Ch. Meier, Structure and bonding Vol. 16 p. 113 Springer Verlag (1973).
143. Y. Kabuke, Crown ether no Kagaku Ed. R. Oda. T. Shono. I. Tabush. Kagaku Zokan Kyto (1978).
144. M. Hiroaka, Seibistu, Butsui **17**, 156 (1977).
145. B.T. Kilborn, J.D. Dunitz, L.A.R. Pioda, W. Simon, J. Mol. Biol. **30**, 559 (1967).
146. M. Pinkerton, L.K. Steinrauf, P. Dawkins, Biochem. Biophys. Res. Comm. **35**, 512 (1969).
147. H. Lardy, Fed. Proc. **27**, 1278 (1968).
148. S. Estrada. O.A. Carabez, J. Bioenerg **3**, 429 (1972).
149. D.H. Haynes, H. Duncan, T. Weins, B.C. Pressman, J. Memb. Biol. **18**, 28 (1974).
150. G. Gokel, Crown ethers and cryptands Royal Society of Chemistry A research monograph p. 81 (1991).
151. B.G. Cox, H. Schneider, Coordination and transport properties of macrocyclic compounds in solution Research monograph p. 219 Elesvier (1992).
152. J.D. Lamb, J.J. Christensen, S.R. Izatt, Bedkel K., M.S. Astin, R.M. Izatt, J. Am. Chem. Soc. **102**, 3099 (1980).
153. H. Tuskube, K. Takagi, T. Higashiyama, T. Iwachido, N. Hayama, Bull. Chem. Soc. Japan. **59**, 2021 (1986).
154. T.B. Stolwijk, E.J.R. Sudholter, D.N. Reinhoudt, J. Am. Chem. Soc. **109**, 7042 (1987).
155. S. Shinkai, S. Nakamura, S. Tachiki, O. Manabe, T. Kajiyama, J. Am. Chem. **107**, 3363 (1985).
156. S. Shinkai, K. Torigoe, O. Manabe, T. Kajiyama, J. Am. Chem. Soc., **109**, 4458 (1987).
157. J.D. Lamb, R.M. Izatt, D.G. Garrick, J.S. Bradshaw, J.J. Christensen, J. Memb. Sci. **9**, 83 (1981).
158. J.D. Lamb, J.J. Christensen, J.L. Oscarson, B.L. Nielsen, B.W. Asay, R.M. Izatt, J. Am Chem. Soc. **102**, 6820 (1980).
159. J.J. Christensen, J.D. Lamb, S.R. Izatt, S.E. Starr, G.C. Weed, M.S. Astin, B.D. Stitt, R.M. Izatt, J. Am. Chem. Soc. **100**, 3219 (1978).
160. R.M. Izatt, R.L. Bruening, M.L. Brening, G.C. Lindh, J.J. Christensen, Anal. Chem. **61**, 1140 (1989).
161. M. Dicasa, L. Fabbrizzi, A. Perotti, A. Poggi, P. Tundo, Inorg. Chem. **24**, 1610 (1985).
162. E. Kimura, C.A. Dalimunte, A. Yumashita, R. Machida, J. Chem. Soc. Chem. Comm. 1014 (1985).
163. Y. Inoue, G.W. Gokel, Cation binding by macrocycles Marcel Dekker N.Y. (1960).
164. T.M. Fyles, Biorg Chem. Frontiers **1**, 71 (1990).
165. G.K. Gokel, L. Echegoyen, Biorg. Chem. Frontiers **1**, 115 (1990).
166. J.D. Lamb, R.M. Izatt, J.J. Christensen Proc Macrocycl. Chem. **2**, 41 (1981).
167. S. Lindenbaum, J.H. Rytting, L.A. Sterrison, Prog Macrocycl. Chem. **1**, 219 (1979).
168. G.R. Painter, B.C. Pressman, Top Current Chem. **101**, 83 (1982).

169. C.D. Gutsche, Calixarenes Royal Chemical Society London A research monograph p. 197 (1989).
170. M.N. Gandhi, S.M. Khopkar, J. Sci and Ind. Res. **55**, 139 (1996).
171. A.G. Gaikwad, H. Noguchi, M. Yoshio, Sepn. Sci. and Technology **26**, 853 (1991).
172. H.C. Tskube, Talanta **40**, 1313 (1993).
173. P.R. Brown, J.L. Hallman, L.W. Whaley, D.H. Desal, M.J. Pugia, R.A. Bartsch J. Memb. Sci. **56**, 195 (1991).
174. T. Koji, Y. Shinji, M. Kataka, O. Kazunori, U. Yoshio, Anal. Chem. **69**, 3360 (1997).
175. H. Tsukube, S. Shinoda, Y. Mizutan, M. Okano, K. Takagi, K. Hori, Tetrahedron **53**, 3487 (1997).
176. M. Shen, Z. Wang, Q. Luo, X. Gao, G. Lu, Huaxue Xuebao **49**, 718 (1991) C.A. **115**, 215541f (1991).
177. S. Dernini, S. Balmas, A.M. Poioaro, B. Maronglu, J. Chem. Eng. Data **37**, 281 (1992).
178. M.J. Crawford, J.G. Frey, T.J. Vandernoot, Y. Zhao, J. Chem. Soc. Faraday Trans. **92**, 1369 (1996).
179. X. Aristotelis, S. Clalude, T. Christian, Talanta **34**, 509 (1987).
180. Sakim Maria, T. Hayashita, T. Yamabe, M. Igawa, Bull. Chem. Soc. Japan **60**, 1289 (1987).
181. Y. Nakatsuj, T. Inoue, M. Wada, M. Okahara, J. Ind. Phen. Mol. Recog. **10**, 379 (1991) C.A. **115**, 79487p (1991).
182. V. Mikulaj. J. Hlatky, L. Vasekova, J. Radioana. and Nucl. Chem. **101**, 51 (1986).
183. V. Mikulaj, L. Vasekova, J. Radioanal. and Nucl. Chem. **150**, 281 (1991).
184. Y. Yamini, N. Alizadeh, M. Shamsipur, Sepn. Sci. and Techn. **32**, 2077 (1997).
185. T.M. Fyles, B. Zeng, J. Chem. Soc. Chem. Comm. 2295 (1996).
186. J. Tang, C. M. Wai, J. Mem. Sci. **35**, 339 (1988) C.A. **108**, 17449z (1988).
187. J. Strzelbicki, W.A. Charewicz, Y. Liu, R.A. Bartsch, J. Incl. Phen. and mol. Recog. **7**, 349 (1989).
188. M. Shamsipur, M. Akhond, Bull. Chem. Soc. Japan **70**, 339 (1997).
189. R.M. Izatt, J.J. Christensen, R.L. Bruening, M.H. Cho, W. Geng, J.D. Lamb, J. Memb. Sci. **33**, 169 (1987).
190. R.M. Izatt, R.L. Bruening, M.L. Bruening, G.C. Lind, J.J. Christensen Anal. Chem. **61**, 1140 (1989).
191. M. Akhond, M. Shamsipur, Sepn. Sci. and Techno. **32**, 1223 (1997).
192. J.D. Lamb, A.Y. Nazarenko, Sepn. Sci. and Technology **32**, 2749 (1997).
193. Y. Nakatsuji, T. Sunagawa, A. Masuyama, T. Kida, I. Ikeda, J. Incl. Phen. Mol. Recog. **29**, 289 (1997).
194. N. Kimizuka, T. Wakiyama, A. Yanagi, S. Shinkai, T. Kunitake, Bull. Chem. Soc. Japan **69**, 3681 (1996).
195. Z. Asfari, C. Bressot, J. Vicens, O. Hill, J.F. Dozol, H. Rouquette, S. Eymard V. Lamare, B. Tournois, Anal. Chem. Soc. Japan **67**, 3133 (1995).
196. Y. Koide, H. Sato, H. Shosenji, K. Yamada, Bull-Chem. Soc. Japan **69**, 315 (1996).
197. P.D. Hampton, C.E. Daitch, A.M. Shachter, Inorg Chem. **36**, 2956 (1997).
198. P.S.K. Chia, L.F. Linday, G.W. Walker, G.W. Evertt, Pure and applied Chem. **65**, 521 (1993).
199. U.S. Vural, Sepn Sci. and Techno. **31**, 787 (1996).
200. H. Adams, F. Davies, C.J.M. Stirling, J. Chem. Soc. Chem. Comm. 2527 (1994).

Subject Index

Acid catalysed synthesis, 36
Actinide exin, calixarene, 143
Adduct formation, 125
Advantages of ion chromatography, 173
Alkali metal extraction crown, 128
    cryptand, 135
Alkali metal extractive fluorimetry, 225
    extractive AAS analysis, 230
    extractive emission crown, 227
    ext photometry calixarene, 208
    ext photometry crown, 148
Alkaline earth ext, cryptands, 203
Alkaline earth extn, crown ethers, 230
Alternate, 1, 2, conformation
    for calixarene, 139
Ammonium ion ext photometry cryptand, 204
    extractive separation aza crown, 137
    extractive separation crown ether, 133
    calixarene, 142
Amphiphiles, 285
Anionic chromophores 188, 189
Anionic stripping voltammetry, 251, 252
Antenna effect, 221
Antigonastic effect, 132
Antimony extractive photometry crown, 202
Astantine extn crown, 134
Attovel level, 175
Autospace a tandem mass spectrometry, 74, 75
Aza crown electrodes, 266
Aza crown ether extraction, 136, 137
Aza crown ethers, 4
Aza crown ether synthesis, 28
Aza crown in photometry, 206
Aza, thia crown complex spect, 61
Azolinked crown, 194

Barium extractive photometry C
    cryptand, 203
    extractive separation crown, 30

Base catalysed synthesis, 34, 35
Bathochromic shift, 207, 208, 217, 218, 223
Benzoate for excitation, 220, 221
Bibranchiallanat, 72
Biomimics, 282
Bis crown effect, 128
Bis crown ether electrode, 268, 270
Bis crown ether extraction, 123
    calixarene, 209
    extractive photometry crown, 201
bpy cryptates, 222, 223
Branched macrocyclic ligand, 222
Bucki minister fullerene (C-60), 45
Bulk liquid membrane, 278
Butterfly crown ether, 234

Cadmium ext fluorimetry cryptand, 226
    extractive AAS crown ether, 230
    extractive photometry cryptand, 204
    aza crown ether, 206
Caesium effect, 29
Caesium extraction emission crown, 227
    extractive AAS crown, 230
    extractive sepn crown 129, calixarene, 142
Calcium ext fluorimetry cryptand, 226
    aza crown, 206
    cryptand, 203,
    extractive emission crown, 227
    extractive photo crown, 199
    extractive separation crown, 130
    cryptand, 135
Calixarene complex x-ray study, 95, 100
    spectral study, 59, 62
Calixarene electrophillic substitution, 39
Calixarene extractive photometry, 205
Calixarene metal complexes, 65
Calixarene (nitration), 40
Calixarene separations, 164
    synthesis, 33
Calix(4) arene tetra-acetamide, 223

Subject Index

Calixarenes carboxylate complexes, 91
Calixarenes in-AAS analysis, 231
   in fluorimetry, 226
   in ion selective electrodes, 277
Calixarenes in electrodes, 267, 277
Calix cone structure, 42, 139
Calix conformation, 42, 137
Calix crown ether, 43
Calix(n)arenes, 1, 2
Calixpartial cone structure, 139
Calix resorcinarene synthesis, 5, 36
Calix spherands, 68
Calorimetric titration, 102
Capillary electrophoresis, 175
   applications, 175-177
   wim calixarene, 176
Carrier relay mechanism, 273
Cation specific electrode, 266
Cation transport, 67, 280
Cavity in crown compounds, 54
Cavity size of crown ethers, 116
Cerium-extractive photometry, 19
   calixarene, 20
   crown, 199,
Chamel ions, 43
Characteristics of electrode, 262
Characteristic spectra, 58, 60
Chemfet-membrane, 273, 277-279
Chemical sensor, 91
Chiral calixarenes, 43
Chiral calixarene synthesis, 36
Chiral crownethers, 18
   synthesis, 31
Chiral diazo crown ether, 137
Chromium-ext photometry calixarene, 208
Chromogenic calixarenes, 68
Chromoionophores, classification, 186-188
Chrono amperiometry, 251
Chrono potentiometry, 251
Classian rearrangement, 19
Cobalt-extractive AAS crown, 230
   ext photometry crown, 200,
   ext separation crown, 133
Coloured counter anion, 196
Column chromatography, 156, 159, 160
Combined ion exchange solvent extraction (CIESE), 170
Comparison of extractants, 141
Conductivity applications, 246-248
Conformation changes, 142, 251

Copper-ext fluorimetry crown, 225
   aza crown, 206
   aza crown, 137, thia crown, 138
   calix, 208
   extractive AAS crown, 230
   extractive AAS cryp, 231
   ext photometry crown, 201
   extr sepn crown , 134
Coronands, 5
Coulometry applications, 253
Counter anion-in extraction, 115, 113
   structure, 120, 121
Crowned calixarenes, 285
Crown complex-organic, 55
Crown complex-x-ray study, 98
Crown compounds in ISE, 274
Crown compounds physical properties, 7, 8
Crown and podant as ionophores, 281
Crown ethers, 1, 2
Crown ethers extractive AAS, 228-230
   as mobile phase, 158
   as neutral carrier in ISE, 265
   as stationary phase, 158-159
   ext photometry, 194
   in membranes, 282
   in metal complexes, 52
   pH responsive, 231, 232
Cryptand-2, 2, 2 synthesis, 24, 25
Cryptands, 1-5,
Cryptates, 90
Cryptand-complexes x-ray, 99
   in atomic atsorphon spect, 229-231
   in electrodes, 271
   in emission spectroscopy, 228
   in extractive photometry, 197, 203
   in fluorimetry, 226
   spectral studies, 61, 94
Cryptand metal complexes, 51
Cryptand voltammetric uses, 251
Crypto cavitate clathrate complete, 88
Cryptohemiosphere, 203
Cryptos, 3
Cyclic voltammetry, charact, 249
Cyclocondensation, 26, 29
Cyclodextrins, 140
Cyclo-oligomerisation, 2, 4

Dibenzo 18crown6-synthesis, 19, 21
Diester crown ether, 31
Diluent in extraction, 121, 122
Diprotonic crown as ionophore, 192

Distribution of crown in solvents, 116
Distribution of 18C6 in solvents, 117
Distribution ratio of complex, 114
Distribution ratio of ligand in solvents, 118
Ditopic crown ethers, 84
Dragendorff reagent, 160
Dual phase potentiometry, 244

Electrical conductivity of complex, 246
Electroanalytical methods, 240
Electrochemical membrane, 264
Electrode (glass), 243
Emission spectroscopy of crown, 215, 227
Emulsion liquid membrane, 278
Energy transfer, 220
Entropy and stability, 55
Ether formation, 38
Europium-extractive fluorimetry
   calixarene, 226
   crown, 225
   extractive emission crown, 227
   ext separation calixarene, 143
Exo-endo configuration, 57
Exo-exo configuration, 57
Extraction and conformation, 68
Extraction chromatography, 167, 169
Extraction constant, 114, 117, 191
Extraction equilibria, 114
Extraction wim liquid membrane, 278
Extractive fluorimetry, 223, 225
Extractive photometry-thia, aza, 204
   separation with aza and thia, 135
   spectrophotometry, 186
   wim crown ethers, 191-201
   with cryptands, 135, calix, 138, 142

Fabrication of electrode, 262
Factors influencing extn, 17
Fast atom bombardment, 74, 148
Ferrocene calix complex, 72, 73
Fiber supported membrane, 278
Fick's law, 281
Field desorphon MS, 74, 75
Flow injection analysis, 198
Fluorogenic crown ethers, 215
Fluoroimmunoassay, 216
Fluoroionophores, 216, 217
Formation constant, 114
   in conductometry, 244
Friedel crafts reaction, 36

Gallium extraction with crown, 201
   calix, 149
   ext separation crown, 134
Gas chromatography, 162, 166
Germanium extractive photo crown, 301
Gold extn aza crown, 137
Grabb's free energy, 122

Hafnium-ext photometry crown, 200
Half wave potential, 247
Hexa-acetato calix(6) arene, 4
   photometry, 207, 208
High dilution reaction, 21
High performance liquid chromatography, 159, 161, 162
History of discovery of crown compounds, 2-4
Hole size of crown compounds, 54
Host guest chemistry, 5, 82
Hydration energy, 120
Hydration of complex, 119, 120
Hypsochromic effect, 188, 217, 219

In-in in-out structure, 3
Indium-ext photometry crown, 201
Intramolecular reaction, 20
Intermolecular reaction, 21
Internal standard in ISE, 274, 275
International symposium in macrocyclic comp, 10
Ion chromatography separations, 171, 175, 176
Ion exchange chromatography, 169
Ion exchange separation, 171, 172
Ion selective electrode, 261
Ion sensitive field effective electrode, 263
Ion transport phenomena, 279
Ionic diameter of metalion, 156
Ionic mobilities and transport of ions, 245
Ionic mobility, 243
Ionic size of metal ion, 116
Iron-ext photometry extn calix, 208
Iron-ext separation crown, 33, calix, 143
Isotope effect, 130, sepn, 168, 172
Isotophore analysis, 175, 178

Kolthoff reference cell, 242
Krypto fix, 221, polymer, 159
Lanthanide-complexes structure, 36
   calixarenes, 142
   ext photometry crown, 199, aza, 206
   ext sepn crown, 131, aza, 137

Lanthanum ext photometry crown, 199,
  calix, 208
  ext seperation crown, 132, aza, 137,
    calix, 142
Large crown ether synthesis, 24
Lariat ethers, 5, 83
  in cyclic voltammetry, 249
  in membrane, 282
  synthesis, 22
LD50 dose, 11, 12
Lead-extractive fluorimetry, calix, 226
  crown, 225, cryptand, 226
  extractive AAS cryptand, 231
  ext photo crown, 201, cryptand, 209
  ext sepn crown 134, calix, 144
Lipophillicity, 283
  of ISE, 270
Lithium-ext fluorimetry crown, 225
  ext photometry crown, 198, aza, 206
  extn separation crown, 129, cryptand, 135
Lower rim substitution, 37, 66
Luminescent lanthanide complexes, 219

Macrobicyclic compounds, 2, 4, 241
Macrocyclic compounds, 1
Macrocyclic effect, 54, 90
Magnesium-ext photometry cryptand, 203
  ext separation crown, 130
Magnetic resonance imaging, 69
Manganese-ext AAS cryptand, 231
  crown, 230
  ext photometry crown, 200, calix, 205
  ext sepn crown, 133
Mannich reaction, 40, 217, 193
Mass spectras of calixarene complex, 73-75
McReynold's constant, 155, 165-166
Mechanism of complexation, 53
Membrane permeability, 281
  membrane potential, 264
  proton ionisation of crown, 278
  separation of metals, 284, 285
  transport, 261, 273, 280
Mercury-ext fluorimetry cryptand, 226
  ext AAS crown, 230
  ext photometry crown, 201, aza, 206
  ext separation crown, 134, aza, 137
  thia crown, 138, calixarene, 143
Mesylate, 21
Metacyclophanes, 92, 140
Metal calixarene structure, 90

Metal complex anion extn, 120
Metal oxygen distance, 86
Metal pollutants, 127, 11
Misceller-electro kinetic CE, 175
Molar volume of species, 117
Molecular architecture, 9
  receptor, 10
Molecular recognisation, 7
Molecular tweezers, 234
Molecular weight of calixarenes, 73
Molybdenum-ext photometry crown, 200
  ext separation crown, 133
Monoprotonic crown ethers, 189
Monoprotonic ionophores, 187
Moody Thomas electrode, 261

Nafion-901, 271
Naked calixarene, 65, 66
Nature of crown during extn, 122
Neodynium-ext separation calixarene, 143
Nernst equation, 242, 247
Neutral chronoronopheres, 188
Nickel-extn AAS cryptand, 231
  ext photometry cryptand, 203
Niobium-ext photometry crown, 200
NMR-spectra of complexes, 59
  spectra of crown ethers, 63
  study of aza thia complexes, 65
  study of calixarene complex, 69-70, 73
  study of cryptand complex, 64
Nomenclature, 6
Nonprotonic ionophores, 193
NPOE-orthonitrophenyl octylether, 271
Nuclear over house effect, 64

Octa decyl silane, 161
Oligomers separation, 167
Osmium-exin photometry crown, 200
Overall distribution ratio, 114
Overall stability constant, 248

Palladium extractive AAS crown, 230
  ext photometry crown, 201, calix, 208
  ext separation crown, 133, aza, 137,
    calix, 143
Para-Claisen rearrangement, 40
Platinum-extractive separation calixarene, 143
Precursor, 34
Permeability coefficient, 281
Petrolite process, 34

Subject Index

Photo antenna, 233
Photoisomerisation, 61
Photoresponsive crown ethers, 123, 234
  switching crown, 233
Physical properties of crown ethers, 125
Physiology of calixarenes, 13, 14
Picrylamine proton, 192
Ping pong mechanism, 96
Plannar species, 120
Plasticizers, 271
Pleated loop conformation, 139
Podands, 5
Polarography of crown compounds, 247
Polyazo crown ethers syntheses, 30
Polymeric crown ethers, 156, 157
Potassium crown complex, 56
Potassium ext fluorimetry crown, 225
  extn AAS crown, 238
  ext emission crown, 227
  ext photometry crown, 198, crypt, 203
  ext sepn crown, 129
Potentiometry applications, 243, 244
  of crown compound, 241
  selective coefficient, 264, 265
Pressman cell, 279
Programmed temp chromatography, 160, 162
Proton dissociation constant, 191
Proton ionisable crown ether, 31
Pseudo cavity, 83

Radiochemical study of complexes, 103, 104
Radium ext photometry crown, 131
Ramon spectroscopy characterisation, 60
Receiving phase, 279
Redox switching crowns, 232
Reversed Friedel Craft reaction, 39
Reversed phase extraction chromatography, 167, 169
Reversed phase HPLC, 161
Rhenium-ext photometry crown, 200
  ext separation crown, 133
Rhodium extractive separation aza, 137
Ring compounds, 2
Rotaxane, 19
  synthesis, 41
Rubidium ext emission crown, 227
  ext photometry crown, 200
  ext separation crown, 129

Sandwich complex, 269, 87

Scandium extractive photometry crown, 199
Schiff's base, 200
Selectivity and potential in electrode, 263
Selectivity in extraction, 123
Separation of zirconium and hafnium, 200
Separation of alkaline earths, 227
Separations with crown ethers, 128-134
Sequential separation, 127, 202
Side armed aza crown ether, 190
Silver ext fluorimetry crown, 225
  ext AAS calix, 231
  ext photometry crown, 201, aza, 206, calix, 208
  ext separation crown, 134, thia
Sodium ext fluorimetry crown
  ext photometry crown, 198
  ext separation crown, 129, cryptand, 135
  aza crown, 139, calix, 142
Solid complex, 85
Solubility of crown compounds, 8, 57
  of complex, 118
Source phase, 279
  pairing of ions, 278
Spherands (in electrodes), 4, 140, 267
Spirobenzopyrane, 198
Stabilisation factor synersium, 124
Stability of complexes, 53, 56, 245
  constant, 242, alkali, 18, 66-116
Steric hindrance, 216
Strontium ext fluo cryptand, 226
  ext AAS crown, 230
  ext emission crown, 227
Strontium-ext photo crown, 199 cryp, 203
  ext separation crown, 130
Structure of B15C5 and potassium complex, 85
  cydic polyether, 88
  elucidation classical, 30
  in solutions, 87
  of DB18C6, 85
  of RbSCN-DB18C6 complex, 85
  of 18C6 comp & UO_2^{2+}, 87
  study by x-ray, 94
Sulphonation of calixarene, 66
Super critic fluid chromatography, 134
Super cryptands, 27
Suppressor column, 171

Supramolecular aggregation, 255
Supramolecular compounds, 2, 5
Supramolecular luminescence compound, 221
Supramolecule and electrochem, 254
Synergic extraction, 124
   and fluorimetry, 225
Synthesis hazards for 18C6, 23
Synthesis of crown compounds, 18
Switch on crownethers, 216, 231

Tandem mass spectrometry, 52
Tellurium-ext separation crown, 133
Template effect, 20, 22, 24, 28, 319
Temperature and extraction, 123
Terbium ext fluorimetry crown, 225
Thallium ext fluorimetry crown, 225
   ext AAS cryptand, 231
   ext photo crown, 201, calix, 209
   ext separation crown, 134
Thermal constant, 57
Thermal studies, 101, 102
Thermal switching crowns, 232
Thermodynamic factor, 123, 24
   stability constant, 242
Thermometric titration, 102
Thia crown ethers, 4
   and electrodes, 266
   synthesis, 30
Thin sheet supported membrane, 278
Thorium ext fluorimetry crown, 225
   extractive emission calix, 228
   ext photometry crown, 200, calix, 200
   ext separation calixarene, 143

Tin ext photometry crown, 201
Titanium calixarene complex, 88, 89
Tosylate synthesis, 21
Toxicity of crown compounds
   15C5 and 18C6, 13
   12C4 and cryptands, 13, 14
   DB18C6, DC18C6, 13
Toxicology, 11, 12
Transportation in membrane, 281
Trögers base, 44, 128, 19

Upper rim substitution, 37, 39
Uvanium ext photo crown, 200, cry, 203
   calixarene, 208
   ext separation crown, 132, calix, 143

Vanadium ext photo crown, 200
   ext emission crown, 227
Vander waal's interaction, 71
Voltammetry and uses, 250, 252
   of calix complex, 254

Williamson reaction, 19, 21

Xenon calixarene complexes, 72, 73

Yttrium ext separation crown, 131

Zetapotential, 175
Zinc extractive fluorimetry cryptand, 226
   ext separation crown, 134
Zinke procedure, 34
Zirconium ext photometry
   crown, 200, cryptand, 203